WITHDRAWN

Date

SPECTROSCOPY SOURCE BOOK

THE McGRAW-HILL SCIENCE REFERENCE SERIES

Acoustics Source Book
Communications Source Book
Computer Science Source Book
Fluid Mechanics Source Book
Meteorology Source Book
Nuclear and Particle Physics Source Book
Optics Source Book
Physical Chemistry Source Book
Solid-State Physics Source Book

SPECTROSCOPY SOURCE BOOK

Sybil P. Parker, *Editor in Chief*

McGRAW-HILL BOOK COMPANY

New York   St. Louis   San Francisco
Auckland   Bogotá   Caracas   Colorado Springs   Hamburg
Lisbon   London   Madrid   Mexico   Milan   Montreal
New Delhi   Oklahoma City   Panama   Paris   San Juan
São Paulo   Singapore   Sydney   Tokyo   Toronto

Cover: White light reflected by a diffraction grating and separated into its visible components. (*Bausch & Lomb, Inc.*)

This material has appeared previously in the McGRAW-HILL ENCYCLOPEDIA OF SCIENCE AND TECHNOLOGY, 6th Edition, copyright © 1987 by McGraw-Hill, Inc. All rights reserved.

SPECTROSCOPY SOURCE BOOK, copyright © 1988 by McGraw-Hill, Inc. All rights reserved. Printed in the United States of America. Except as permitted under the United States Copyright Act of 1976, no part of this publication may be reproduced or distributed in any form or by any means, or stored in a data base or retrieval system, without prior written permission of the publisher.

1 2 3 4 5 6 7 8 9 0   DOC/DOC   8 9 5 4 3 2 1 0 9 8

ISBN 0-07-045505-8

Library of Congress Cataloging in Publication Data:

Spectroscopy source book / Sybil P. Parker, editor in chief.
    p. cm. — (McGraw-Hill science reference series)
   "This material has appeared previously in the McGraw-Hill encyclopedia of science and tecnology, 6th edition"—CIP t.p. verso.
  Bibliography: p.
  Includes index.
  ISBN 0-07-045505-8
   1. Spectrum analysis—Dictionaries. I. Parker, Sybil P. II. McGraw-Hill Book Company. III. Title: McGraw-Hill encyclopedia of science and technology (6th ed.)
IV. Series.
QD95.S6368 1988
543'.0858—dc19   87-35254

TABLE OF CONTENTS

Introduction	1
Origin of Spectra	11
Instrumentation and Techniques	79
Atomic and Molecular Spectroscopy	113
Nuclear Spectroscopy	177
Microwave and Radio-Frequency Spectroscopy	201
Mass Spectroscopy	229
Analytic Techniques	245
Contributors	275
Index	281

SPECTROSCOPY SOURCE BOOK

INTRODUCTION

SPECTROSCOPY IS an analytic technique concerned with the measurement of the interaction (usually the absorption or the emission) of radiant energy with matter, with the instruments necessary to make such measurements, and with the interpretation of the interaction both at the fundamental level and for practical analysis. Mass spectroscopy is not concerned with the interaction of light with matter, but was so named because the appearance of the data resembles that of the spectroscopic data as just defined.

A display of such data is called a spectrum, that is, a plot of the intensity of emitted or transmitted radiant energy (or some function of the intensity) versus the energy of that light. Spectra due to the emission of radiant energy are produced as energy is emitted from matter, after some form of excitation, then collimated by passage through a slit, then separated into components of different energy by transmission through a prism (refraction) or by reflection from a ruled grating or a crystalline solid (diffraction), and finally detected. Spectra due to the absorption of radiant energy are produced when radiant energy from a stable source, collimated and separated into its components in a monochromator, passes through the sample whose absorption spectrum is to be measured, and is detected. Instruments which produce spectra are variously called spectroscopes, spectrometers, spectrographs, and spectrophotometers.

Interpretation of spectra provides fundamental information on atomic and molecular energy levels, the distribution of species within those levels, the nature of processes involving change from one level to another, molecular geometries, chemical bonding, and interaction of molecules in solution. At the practical level, comparisons of spectra provide a basis for the determination of qualitative chemical composition and chemical structure, and for quantitative chemical analysis.

Early history. In the historical development of spectroscopy, following the fundamental studies of crude spectra of sunlight by Isaac Newton in 1672, certain contributions and achievements are especially noteworthy. The significance of using a narrow slit instead of a pinhole or round aperture so as to produce spectral lines, each one an image of the slit and representing a different color or wavelength, was demonstrated independently by W. H. Wollaston in 1802 and by Joseph Fraunhofer in 1814. Fraunhofer made many subsequent contributions to optics and spectroscopy, including first observation of stellar spectra, discovery and construction of transmission diffraction gratings, first accurate measurements of wavelengths of the dark lines in the solar spectrum, and invention of the achromatic telescope. The origin of the dark Fraunhofer lines in the solar spectrum was accounted for by G. R. Kirchhoff in 1859 on the basis of absorption by the elements in the cooler Sun's atmosphere of the continuous spectrum emitted by the hotter interior of the Sun. Further studies by Kirchhoff with R. Bunsen demonstrated the great utility of spectroscopy in chemical analysis. By systematically comparing the Sun's spectrum with flame or spark spectra of salts and metals, they made the first chemical analysis of the Sun's atmosphere. In 1861, while investigating alkali metal spectra, they discovered two new alkali metals, cesium and rubidium. These achievements by Kirchhoff and Bunsen provided tremendous stimulus to spectroscopic researches. The adoption in 1910 of the first international standards of wavelength gave further impetus. These and later standards made possible the measurement of wavelengths of any electromagnetic radiation with unprecedented accuracy. Since World War II, remarkable developments in spectroscopy have occurred in instrumentation, achieved largely through advances in electronics and in manufacturing technology. Direct reading, automatic recording, improved sensitivity with good stability, simplicity of operation, and extended capabilities are features provided by many commercial instruments, many of which now are microprocessor-controlled. Many newer instruments have dedicated data systems. Predictably, these developments, by facilitating widespread use of spectroscopic techniques, have had an enormous influence in promoting developments in both applied and theoretical spectroscopy.

The ultimate standard of wavelength is that of the meter, defined since 1983 as the length of the path traveled by light in vacuum during a time interval of 1/299,792,458 of a second.

Spectroscopic units. The change in energy of an ion, atom, or molecule associated with absorption and emission of radiant energy may be measured by the frequency of the radiant energy according to Max Planck, who described an equality $E = h\nu$, where E is energy, ν is the frequency of the radiant energy, and h is Planck's constant. The frequency is related to the wavelength λ by the relation $\nu\lambda = c/n$, where c is the velocity of radiant energy in a vacuum, and n is the refractive index of the medium through which it passes; n is a measure of the retardation of radiant energy passing through matter. The units most commonly employed to describe these characteristics of light are:

Wavelength: 1 micrometer (μm) = 10^{-6} m
1 nanometer (nm) = 10^{-9} m (= 10 angstroms)
Frequency: 1 hertz (Hz) = 1 s^{-1}

For convenience the wave number $\bar{\nu}$ (read nu bar), the reciprocal of the wavelength, may be used; for this, the common units are cm^{-1}, read as reciprocal centimeters or occasionally kaysers. This number equals the number of oscillations per centimeter.

Spectral regions. Visible light constitutes only a small part of the spectrum of radiant energy, or electromagnetic spectrum; the human eye responds to electromagnetic waves with wavelengths from about 380 to 780 nm, though there is individual variation in these limits. The eye cannot measure color and intensity quantitatively; even for the visible portion of the electromagnetic spectrum, therefore, instruments are used for measurements and recording.

Origin of spectra. Atoms, ions, and molecules emit or absorb characteristically; only certain energies of these species are possible; the energy of the photon (quantum of radiant energy) emitted or absorbed corresponds to the difference between two permitted values of the energy of the species, or energy levels. (If the flux of photons incident upon the species is great enough, simultaneous absorption of two or more photons may occur.) Thus the energy levels may be studied by observing the differences between them. The absorption of radiant energy is accompanied by the promotion of the species from a lower to a higher energy level; the emission of radiant energy is accompanied by falling from a higher to a lower state; and if both processes occur together, the condition is called resonance.

Transitions. Transitions between energy levels associated with electrons or electronic levels range from the near infrared, the visible, and ultraviolet for outermost, or highest-energy, electrons, that is, those which can be involved in bonding, to the x-ray region for the electrons nearest the nucleus. At low pressures such transitions in gaseous atoms produce sharply defined lines because the energy levels are sharply defined. Transitions between energy levels of the nucleus are observed in the gamma-ray region. In the absence of an applied electric or magnetic field, these electronic and nuclear transitions are the only ones which atoms can undergo.

Electronic transitions in molecules are also observed. In addition, transitions can occur between levels associated with the vibrations and rotations of molecules. Spacings between electronic levels are greater than between vibrational levels, and those between vibrational levels are greater than between rotational levels; each vibrational level has a set of rotational levels associated with it, and each electronic level a set of vibrational levels. Transitions between vibrational levels of the same electronic level correspond to photons in the infrared region; transitions between rotational levels, to photons in the far infrared and microwave region. Rotational spectra of gaseous molecules consist of sharp lines; vibrational spectra consist of bands, each of which arises from a given transition between vibrational levels altered in energy by increments due to changes in rotation occurring when the vibrational level changes. Likewise, molecular electronic spectra consist of bands due to transitions between electronic energy levels, altered by increments resulting from changes in vibration and rotation of the molecule on changing electronic state.

External fields. The application of a magnetic or electric field to the sample often separates normally indistinguishable, or degenerate, states in energy from each other. Thus, the orientation of the spin of an unpaired electron in an atom, ion, or molecule with respect to an applied magnetic field may have different values, and in the magnetic field the atom or molecule may have different energies. For typical field strengths, the difference in energy between these newly separated energy levels occurs in the microwave region. Similarly, for an atomic nucleus in an ion or molecule, differences in the orientation of the spin of the nucleus with respect to a magnetic field give rise to different energy levels of the nucleus in that field, so that energy differences will be found in the microwave region. The former phenomenon produces electron spin resonance or electron paramagnetic resonance spectra, and the latter produces nuclear magnetic resonance spectra.

Nuclear magnetic resonance has been advanced by pulsed irradiation techniques which permit the identification of solid samples and complex biological molecules, and which are useful for the production of three-dimensional plots to permit analysis across the interior of objects without destruction. In medicine the imaging of internal organs is known as magnetic resonance imaging. Electron paramagnetic spectroscopy is used to establish structures of species containing unpaired electrons in solution or in a crystalline solid; some qualitative and quantitative analysis is also performed.

Electronic spectra are also altered by external fields; removal of degeneracy by an externally applied electric field is termed the Stark effect, and removal of degeneracy by an externally applied magnetic field is termed the Zeeman effect.

Spontaneous emission. The spontaneous emission of light from a sample by the decay of an electron from a higher level to the lowest available level is called either fluorescence or phosphorescence. The fundamental difference between these two terms is associated with the orientation of the spin of this electron. If the spin of the electron is such that it falls to the lowest state without a change of spin (the atom or molecule having the same spin before and after the transition), fluorescence occurs, and the process is characterized as allowed and is relatively fast. If the spin is such that the electron can fall to the lower state only with a change in spin dictated by the Pauli exclusion principle, the process is phosphorescence, characterized as a forbidden process, which is relatively slow. In practice, other factors also govern the time the electron spends in the higher level, and the time intervals associated with each process overlap; that for fluorescence is 10^{-4} to 10^{-8} s, and that for phosphorescence is 10^{-4} to 10 s. The spontaneous emission process thus cannot be completely distinguished merely on the basis of the time interval associated with the emission. Time-resolved emission studies may allow separation of the processes. Fluorescence and phosphorescence are measured at a right angle to the incident beam exciting the sample, in order to avoid background problems. The emission of light beyond blackbody radiation when a sample is heated (thermoluminescence) or when it is strained or fractured (triboluminescence) has limited applicability. Similar emission during chemical reactions (chemiluminescence; in biochemical systems, bioluminescence) has broader analytical use if the reactions in which products emit light are, for example, catalyzed by a small amount of the substance to be analyzed.

Light rotation. Compounds whose molecules have a structure which cannot be superimposed on its reflection in a mirror (asymmetric molecules) rotate the plane of polarized light. If a device can be included in an instrument to plane-polarize the light from the source, that is, to produce light oscillating in only one plane, the amount of rotation of that plane by the sample can be studied as a function of the wavelength of light used in the experiment. A plot of this rotation versus wavelength is called an optical rotatory dispersion curve; such curves are useful in studying fundamentals of light rotation and in establishing the absolute structure of asymmetric molecules empirically. Asymmetric samples whose absorption of the components of plane-polarized light differs are said to exhibit circular dichroism; analysis of a sample in a magnetic field, which is known as magnetic circular dichroism, has a broader applicability in structural analysis.

Instrumentation. Spectrometers require a source of radiation, a dispersion device, and a detector. For emission spectra, the source may be taken as the sample to be measured itself, although another source of radiation may be needed to excite the sample to an excited state. Absorption spectra require a source to generate appropriate radiation across the range of energies needed for the experiment. These radiant sources produce continuous spectra and can be distinguished from those which produce discontinuous spectra.

Continuous spectra. Of the sources listed in the table, most of those from the vacuum ultraviolet region to the far-infrared region produce continuous spectra. Within this type, sources for the near ultraviolet, visible, and infrared regions consist of hot, incandescent solids; the maximum intensity of their emission occurs at a wavelength dependent on their temperature and emissivity. Electrical discharges in gases at higher pressures provide sources of continuous radiation for the vacuum ultraviolet and ultraviolet regions; hydrogen or deuterium gas discharges are especially used for the latter. X-ray continuous emission can be produced by collision of electrons or other particles accelerated through tens of kilovolts with a target.

Discontinuous spectra. The sources of high-energy radiation which yield discontinuous spectra emit because of discrete processes in individual atoms, ions, or molecules. Such sources may include flames and furnaces to vaporize and atomize samples, and electrical discharges in gases at lower pressures. Other atomizing sources may be produced by high-frequency radio discharges and microwave discharges through a flowing gas. For sources of radiation connected with nuclear processes, solid samples may be used containing the element to be studied, either a radioactive isotope or one which can be made radioactive by particle bombardment.

Among these so-called line sources are hollow-cathode lamps, in which a few layers of a cylindrical cathode containing the same element as is to be analyzed are vaporized by collisions with ions generated from rare gases and accelerated electrons. The gaseous atoms, also excited to higher energy states by this process, emit characteristic line spectra useful for analysis of that element. Lasers have become an especially useful source, and have opened different areas of spectroscopy, not only because of the intensity of the radiation supplied to the sample, which permits absorption of several photons in a single process, but also because of the coherent properties of the radiation, which permit the study of events occurring on a picosecond time scale.

The sources for radio-frequency studies are tunable radio-frequency oscillators, designed with special regard for stability. The microwave sources may be spark discharges, magnetrons, or most commonly klystrons. An electron beam traveling through a closed cavity of defined geometry sets up electromagnetic oscillations, of which a certain value will be reinforced as a function of the dimension of the containing unit; the value can be varied somewhat by adjusting the size of the cavity.

Dispersive elements. There are two kinds of dispersive elements, prisms and diffraction elements. The earlier and more commonly available prism has been supplanted in many cases by diffraction elements.

After collimation of the source radiation through a slit, diffraction is achieved in x-ray spectroscopy by the use of a pure crystal. The distances between atomic nuclei in the crystal are on the same order as the wavelength of the radiant energy; under these conditions an efficient dispersion of the radiation over an arc of many degrees is possible. For a distance d between repeating planes of atoms, the wavelength λ will be diffracted through an angle θ related by the Bragg law $n\lambda = 2d \sin \theta$, where n is an integer called the diffraction order. Materials used include gypsum ($CaSO_4 \cdot 2H_2O$), ammonium dihydrogen phosphate ($NH_4H_2PO_4$), and alkali halides (lighter elements). Each of these gives a wide angular dispersion in a portion of the 0.03–1.4-nm range of the spectrum commonly used for x-ray studies; the appropriate one must be chosen according to the experiment to be performed.

For spectroscopic techniques using visible light and the regions adjacent to it, ultraviolet and infrared, ruled diffraction gratings are often employed. Because their production has become economical, they have to a large extent replaced prisms, which may require the use of several materials to get adequate angular dispersion and wavelength resolution over the entire range of interest, for example, 2–15 μm for an infrared spectrum. The dispersion of a prism increases near the limit of its transparency, and there is always a compromise in efficiency between transmission and dispersion. The resolution R, defined as the quotient of the wavelength of a line and the wavelength difference between it and another line just separated from it, is given by $T/(dn/d\lambda)$, where T is the thickness of the prism base and $dn/d\lambda$ is the variation of the refractive index with respect to wavelength. Efficient prisms are made of quartz for the ultraviolet, glass for the visible, and various salts for the infrared.

Very short- or very long-wavelength spectra are usually produced without a dispersive element; gamma-ray detection uses a wavelength-sensitive detector (a pulse height discriminator, for example), and microwave and radio-frequency detection may use a variable tuning radio receiver; in the latter case the source is tuned to emit a highly resolved frequency as well.

Detectors. Detectors commonly used for the various spectral regions include Geiger and scintillation counters; photomultipliers, tubes, and cells; thermocouples; bolometers; silicon-tungsten crystals; and radio receivers. The devices for the infrared and microwave regions are basically heat sensors. Except for the eye and photographic methods used sometimes in the regions for gamma ray to visible, each detector converts the signal into an electrical response which is amplified before transmittal to a recording device. Direct imaging of the response by vidicon detection has also been used. Otherwise an oscilloscope or oscillograph is used to record very rapidly obtained signals.

Instruments. Spectroscopic methods involve a number of instruments designed for specialized applications.

Spectroscope. An optical instrument consisting of a slit, collimator lens, prism or grating, and a telescope or objective lens which produces a spectrum for visual observation is called a spectroscope. The first complete spectroscope was assembled by Bunsen in 1859.

Spectrograph. If a spectroscope is provided with a photographic camera or other device for recording the spectrum, the instrument is called a spectrograph. For recording ultraviolet and visible spectra, two types of quartz spectrographs are in common use. One type utilizes a Cornu quartz prism constructed of two 30° prisms, left- and right-handed, as well as left- and right-handed lenses, so that the rotation occurring in one-half the optical path is exactly compensated by the reverse rotation in the other. The other type of quartz spectrograph employs a Littrow 30° quartz prism with a rear reflecting surface that reverses the path of the light through prism and lens, thus compensating for rotation of polarization in one direction by equal rotation in the opposite direction. Thus, in either type the effect of optical activity and birefringence in crystal quartz which produces double images of the slit is eliminated.

Grating spectrographs cover a much broader range of wavelengths (vacuum ultraviolet to far infrared) than do prism instruments. Various-type mountings are employed. The most common mounting for a plane-reflection grating is in a Littrow mount, entirely analogous to that of the Littrow quartz spectrograph. Mountings for concave reflection gratings require that the grating, slit, and camera plate all lie on the Rowland circle (imaginary circle of radius equal to one-half the radius of curvature of the grating) in order to achieve proper focus of spectral lines (slit images) on the plate. Paschen, Rowland, Eagle, and Wadsworth mountings are common in grating spectrographs.

Spectrometers. A spectroscope that is provided with a calibrated scale either for measurement of wavelength or for measurement of refractive indices of transparent prism materials is called a spectrometer. Also, the term frequently is used to refer to spectrographs which incorporate photoelectric photometers instead of photographic means to measure radiant intensities.

Spectrophotometer. A spectrophotometer consists basically of a radiant-energy source, monochromator, sample holder, and detector. It is used for measurement of radiant flux as a function of wavelength and for measurement of absorption spectra.

Interferometer. This optical device divides a beam of radiant energy into two or more parts which travel different paths and then recombine to form interference fringes. Since an optical path is the product of the geometric path and the refractive index, an interferometer measures differences of geometric path when two beams travel in the same medium, or the difference of refractive index when the geometric paths are equal. Interferometers are employed for high-resolution measurements and for precise determination of relative wavelengths: they are capable of distinguishing between two spectral lines that differ by less than 10^{-6} of a wave.

Quantitative relationships. For practical analysis, the absorption of radiant energy is related to concentration of sample by the relationship $-\log P/P_0 = abc$, where P and P_0 are respectively the power of the light attenuated after passing through the sample and its unattenuated power before reaching the sample, c is the concentration of the sample, b is the path length of the light through the sample, and a is a proportionality constant called the absorptivity of the sample. If c is in moles per liter, the absorptivity is called the molar absorptivity and symbolized ϵ. This relationship is known as the Beer-Lambert-Bouguer law, or simply Beer's law. For x-ray spectroscopy, the right-hand variables are grouped differently.

The practical relation between emission of light and concentration is given by the equation $F = k\phi I_0 abc$, where F is the fluorescence intensity, k is an instrumental parameter, ϕ is the fluorescence efficiency (quanta emitted per quantum absorbed), I_0 is the intensity of the incident light upon the sample, and the other units are as defined previously. The choice between absorption and fluorescence techniques for determining concentration is sometimes made on the basis of the accuracy of the linear method (fluorescence) versus the versatility of the logarithmic method (absorption).

Other methods and applications. Since the early methods of spectroscopy there has been a proliferation of techniques, often incorporating sophisticated technology.

Acoustic spectroscopy. When modulated radiant energy is absorbed by a sample, its internal energy increases. The loss of that excess produces a temperature increase that can be monitored as a periodic pressure change in the gas around the sample by using a microphone transducer. This is the optoacoustic effect. Its application provides a rapid method for study of some difficult samples.

Astronomical spectroscopy. The radiant energy emitted by celestial objects can be studied by combined spectroscopic and telescopic techniques to obtain information about their chemical composition, temperature, pressure, density, magnetic fields, electric forces, and radial velocity. Radiation of wavelengths much shorter than 300 nm is absorbed by the Earth's atmosphere, and can only be studied by spectrographs transported into space by rockets.

Atomic absorption and fluorescence spectroscopy. This branch of electronic spectroscopy uses line spectra from atomized samples to give quantitative analysis for selected elements at levels down to parts per million, on the average. However, detection limits vary greatly from element to element. Plasmas, especially inductively coupled plasmas, provide a stable, efficient, novel method for atomization and ionization.

Attenuated total reflectance spectroscopy. Spectra of substances in thin films or on surfaces can be obtained by the technique of attenuated total reflectance or by a closely related technique called frustrated multiple internal reflection. In either method the sample is brought into contact with a total-reflecting trapezoid prism and the radiant-energy beam is directed in such a manner that it penetrates only a few micrometers of the sample one or more times.

The technique is employed primarily in infrared spectroscopy for qualitative analysis of coatings and of opaque liquids.

Electron spectroscopy. This area includes a number of subdivisions, all of which are associated with electronic energy levels. The outermost or valence levels are studied in photoelectron spectroscopy, which uses photons of the far-ultraviolet region to remove electrons from molecules and to infer the energy levels of the remaining ion from the kinetic energy of the expelled electron. This technique is used mostly for fundamental studies of bonding. Electron impact spectroscopy uses low-energy electrons (0–100 eV) to yield information similar to that observed by visible and ultraviolet spectroscopy, but governed by different selection rules. X-ray photoelectron spectroscopy (electron spectroscopy for chemical analysis or ESCA) uses x-ray photons to remove inner-shell electrons in a similar way. In Auger spectroscopy, an x-ray photon removes an inner electron, and when another electron falls from a higher level to take its place and a third is ejected to conserve the energy gained by the second, the kinetic energy of the third is measured. Both of these techniques are used to study surfaces, since x-rays penetrate many objects for only a few layers. Ion neutralization spectroscopy uses protons or other charged particles instead of photons.

Fourier transform spectroscopy. This family of techniques consists of a fundamentally different method of irradiation of the sample, in which all pertinent wavelengths simultaneously irradiate it for a short period of time and the absorption spectrum is obtained by mathematical manipulation of the cyclical power pattern so obtained. It has been applied particularly to infrared spectrometry and nuclear magnetic resonance spectrometry, and allows the acquisition of spectra from smaller samples in less time, with high resolution and wavelength accuracy. The infrared technique is carried out with an interferometer of conventional instrumentation.

Gamma-ray spectroscopy. Of special note in this category are the techniques of activation analysis, which are performed on a sample without destroying it by subjecting it to a beam of particles, often neutrons, which react with stable nuclei present to form new unstable nuclei. The emission of gamma rays for each element at different wavelengths with different half-lives as these radioactive nuclei decay is characteristic.

Mössbauer spectroscopy results from resonant absorption of photons of extremely high resolution emitted by radioactive nuclei in a crystalline lattice. The line width is extremely narrow because there is no energy loss for the nucleus recoiling in the lattice, and differences in the chemical environment of the emitting and absorbing nuclei of the same isotope may be detected by the slight differences in wavelength of the energy levels in the two samples. The difference is made up by use of the Doppler effect as the sample to be studied is moved at different velocities relative to the source, and the spectrum consists of a plot of gamma-ray intensity as a function of the relative velocity of the carriage of the sample. It yields information on the chemical environment of the sample nuclei and has found application in inorganic and physical chemistry, solid-state physics, and metallurgy.

Laser spectroscopy. Laser radiation is nearly monochromatic, of high intensity, and coherent. Some or all of these properties may be used to acquire new information about molecular structure. In multiphoton absorption the absorption of more than one photon whose energy is an exact fraction of the distance between two energy levels allows study of transitions ordinarily forbidden when only a single photon is involved. Multiphoton ionization refers to the removal of an election after the absorption of several photons. Raman spectroscopy has also become more common because of the use of lasers as sources, especially with the development of resonance Raman spectroscopy.

Information on processes which occur on a picosecond time scale can be obtained by making use of the coherent properties of laser radiation, as in coherent anti-Stokes-Raman spectroscopy. Lasers may also be used to evaporate even refractory samples, permitting analysis of the atoms so obtained, and since the radiation may be focused, such a technique may be applied to one small area or a sample surface at a time, permitting profiling of the analysis across the sample surface.

Mass spectrometry. The source of a mass spectrometer produces ions, often from a gas, but also in some instruments from a liquid, a solid, or a material adsorbed on a surface. Ionization of a gas is most commonly performed by a current of electrons accelerated through about 70–80 V, but many other methods are used. The dispersive unit may consist of an electric field, or sequential electric and magnetic fields, or a number of parallel rods upon which a combination of dc and ac voltages are applied;

these provide either temporal or spatial dispersion of ions according to their mass-to-charge ratio.

Multiplex or frequency-modulated spectroscopy. The basis of this family of techniques is to encode or modulate each optical wavelength exiting the spectrometer output with an audio frequency that contains the optical wavelength information. Use of a wavelength analyzer then allows recovery of the original optical spectrum. The primary advantage of such a system is that it enables the detector to sense all optical wavelengths simultaneously, resulting in greatly decreased scan times and the possibility for signal averaging. Frequency modulation can be achieved, for example, by means of a special encoding mask as in Hadamard transform spectroscopy or by use of a Michelson interferometer as in Fourier transform spectroscopy.

Raman spectroscopy. Consider the passage of a beam of light, of a wavelength not corresponding to an absorption, through a sample. A small fraction of the light is scattered by the molecules, and so exits the sample at a different angle. This is called Rayleigh scattering if the wavelength of the scattered light is the same as the original wavelength; but if the wavelength is different, it is called Raman scattering. The fraction of photons thus scattered by the Raman effect is even smaller than by Rayleigh scattering. Differences in wavelength correspond to the wavelengths of certain vibrational and rotational processes. If the scattered light has a longer wavelength, the new line in the spectrum is called a Stokes line; if shorter, an anti-Stokes line. The information obtained is of use in structural chemistry, because the rules allowing and forbidding transitions are different from those in absorption spectroscopy. Use of lasers for sources has revived this technique. A related process, resonance Raman spectroscopy, makes use of the fact that Raman probabilities are greatly increased when the exciting radiation has an energy which approaches the energy of an allowed electronic absorption. Raman scattering is also enhanced by surfaces, providing a valuable tool for surface analysis.

X-ray spectroscopy. The excitation of inner electrons in atoms is manifested as x-ray absorption; emission of a photon as an electron falls from a higher level into the vacancy thus created is x-ray fluorescence. The techniques are used for chemical analysis. Extended x-ray absorption fine structure, obtained under conditions of high resolution, shows changes in the energy of inner electrons due to the presence of neighboring atoms, and can be used in structural chemistry for the determination of the nature and location of neighbor atoms nearest to that being studied. Generation of x-rays by bombarding a sample with a proton beam gives the microanalytical technique of proton-induced x-ray emission.

MAURICE M. BURSEY

1

ORIGIN OF SPECTRA

Spectrum	12
Energy level	13
Ritz's combination principle	15
Linewidth	16
Atomic structure and spectra	17
Line spectrum	32
Isoelectronic sequence	33
Auger effect	33
Fine structure	34
Hyperfine structure	35
Stark effect	36
Zeeman effect	37
Isotope shift	41
Nuclear moments	42
Molecular structure and spectra	45
Band spectrum	55
Crystal absorption spectra	56
Luminescence	59
Fluorescence	69
Phosphorescence	69
Optical activity	71
Rotatory dispersion	74
Cotton effect	75

SPECTRUM
W. F. Meggers and W. W. Watson

The term spectrum is applied to any class of similar entities or properties strictly arrayed in order of increasing or decreasing magnitude. In general, a spectrum is a display or plot of intensity of radiation (particles, photons, or acoustic radiation) as a function of mass, momentum, wavelength, frequency, or some other related quantity. For example, a beta-ray spectrum represents the distribution in energy or momentum of negative electrons emitted spontaneously by certain radioactive nuclides, and when radionuclides emit alpha particles, they produce an alpha-particle spectrum of one or more characteristic energies. A mass spectrum is produced when charged particles (ionized atoms or molecules) are passed through a mass spectrograph in which electric and magnetic fields deflect the particles according to their charge-to-mass ratios. The distribution of sound-wave energy over a given range of frequencies is also called a spectrum. SEE MASS SPECTROSCOPE.

In the domain of electromagnetic radiation, a spectrum is a series of radiant energies arranged in order of wavelength or of frequency. The entire range of frequencies is subdivided into wide intervals in which the waves have some common characteristic of generation or detection, such as the radio-frequency spectrum, infrared spectrum, visible spectrum, ultraviolet spectrum, and x-ray spectrum. Spectra are also classified according to their origin or mechanism of excitation, as emission, absorption, continuous, line, and band spectra.

An emission spectrum is produced whenever the radiations from an excited light source are dispersed. Excitation of emission spectra may be by thermal energy, by impacting electrons and ions, or by absorption of photons. Depending upon the nature of the light source, an emission

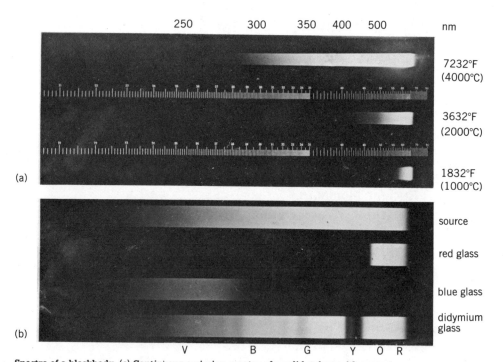

Spectra of a blackbody. (a) Continuous emission spectra of a solid, taken with a quartz spectrograph. (b) Continuous absorption spectra of the source alone and three kinds of colored glass interposed on it. (From F. A. Jenkins and H. E. White, Fundamentals of Optics, 4th ed., McGraw-Hill, 1976)

spectrum may be a continuous or a discontinuous spectrum, and in the latter case, it may show a line spectrum, a band spectrum, or both.

An absorption spectrum is produced against a background of continuous radiation by interposing matter that reduces the intensity of radiation at certain wavelengths or spectral regions. The energies removed from the continuous spectrum by the interposed absorbing medium are precisely those that would be emitted by the medium if properly excited. This reciprocity of absorption and emission is known as Kirchhoff's principle; it explains, for example, the absorption spectrum of the Sun, in which thousands of lines of gaseous elements appear dark against the continuous-spectrum background.

A continuous spectrum contains an unbroken sequence of waves or frequencies over a long range (see **illus**.).

In illus. *a* the spectra for 1000 and 2000°C (1832 and 3632°F) were obtained from tungsten filament, that for 4000°C (7232°F) from a positive pole of carbon arc. The upper spectrum in illus. *b* is that of the source alone, extending roughly from 400 to 650 nanometers. The others show the effect on this spectrum of interposing three kinds of colored glass.

All incandescent solids, liquids, and compressed gases emit continuous spectra, for example, an incandescent lamp filament or a hot furnace. In general, continuous spectra are produced by high temperatures, and under specified conditions the distribution of energy as a function of temperature and wavelength is expressed by Planck's law.

Line spectra are discontinuous spectra characteristic of excited atoms and ions, whereas band spectra are characteristic of molecular gases or chemical compounds. SEE ATOMIC STRUCTURE AND SPECTRA; BAND SPECTRUM; LINE SPECTRUM; MOLECULAR STRUCTURE AND SPECTRA.

ENERGY LEVEL
DAVID PARK

Self-contained physical systems of molecular size or smaller are found to have states of internal motion or excitation in which their energies take on definite stationary values. The energies of these states are called energy levels.

The existence of stationary states was first clearly seen in the study of atomic spectra. In 1900 Johannes Rydberg remarked that the frequencies of many observed spectral lines could be expressed in terms of the differences between pairs of numbers in a certain set called spectral terms; in 1908 Walter Ritz proposed this as a fundamental physical law, which later became known as the combination principle. Five years later Niels Bohr explained the combination principle in dynamical terms: Atoms exist in stationary states; when an atom passes from one stationary state to another of lower energy, it delivers the extra energy to a single quantum of light. Since according to Max Planck the energy of a quantum is proportional to its frequency, it follows that the spectral terms, written in appropriate units, are the energy values of the stationary states and that Ritz's combination principle merely expresses the conservation of energy.

Experimentally it is found that most terms consist of several energy levels spaced close together. This multiplet structure is now explained as largely an effect of magnetic interactions associated with spinning electrons. It has further been discovered that certain transitions between levels are far more likely to occur than others. This is an effect of the conservation of angular momentum. Like electrons, photons (light quanta) carry the intrinsic angular momentum known as spin, but photons carry a full unit of $\hbar$ ($\hbar$ is Planck's constant divided by 2π), whereas electrons have only half as much. Thus, ordinarily, the angular momentum of an atom changes by only one unit in a radiative transition. Occasionally a photon may be emitted from a point off-center in the atom; if so, it carries orbital angular momentum as well as its spin, and the atom's angular momentum may change by more than one unit. Rules which state that certain transitions are excluded, or at least are much less probable than others, are called selection rules. Normally, as here, they have their origin in the requirement that some dynamical quantity be conserved, but certain of the selection rules governing transitions among elementary particles have not yet been understood in this way.

The simplest of all atoms is hydrogen. Its energy levels, ignoring multiplet structure, are shown in **Fig. 1**. The obvious regularity of the terms can be summarized in a formula first guessed

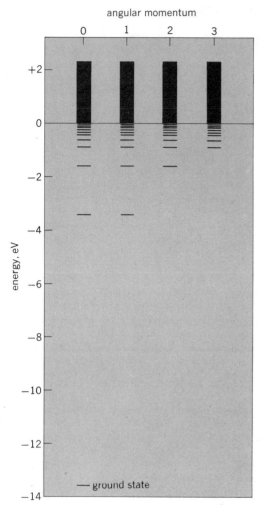

Fig. 1. Energy levels of the hydrogen atom, classified by the orbital angular momentum of the electron, expressed in units of $\hbar$. Lines show energy levels of bound states; shading shows continuous spectrum corresponding to positive energies. Allowed radiative transitions correspond to changes of one unit in the angular momentum of the electron.

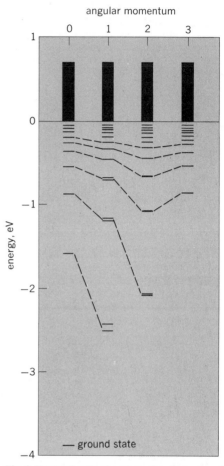

Fig. 2. Energy levels of cesium, showing some of the multiplet structure. All levels with orbital angular momentum greater than zero are double, with the states in which the electron's spin is in the same direction as its orbital motion having higher energy. Broken lines connect corresponding levels, which, in absence of the inner core of electrons, would lie at the same energy.

in 1885 by Johann Jakob Balmer on the basis of the four spectral lines then known. As derived by Bohr from the first primitive version of quantum mechanics in 1913, the formula is stated below,

$$E_n = -\frac{me^4}{2n^2\hbar^3} \quad n = 1, 2, 3, \ldots$$

where m is the electron's mass (or, more accurately, the reduced mass of an electron and a proton), e is the electronic charge, and n is an integer called the principal quantum number. The same formula is given by modern quantum mechanics, and in hydrogen the multiplet structure is explained quantitatively as the effect of two perturbations—the moving electron's intrinsic magnetic moment interacting with the Coulomb field of the proton, and the relativistic variation of the electron's mass with velocity. Only the states with negative energy have discrete levels. Positive energies correspond to a proton and an electron with too much energy to be bound together. For them any energy is allowed, so that the levels above zero form a continuum. SEE FINE STRUCTURE (SPECTRAL LINES).

Figure 2 shows for contrast the energy levels of cesium. This is an atom in which one electron, orbiting outside a cloud of more tightly bound electrons, usually does all the radiating. The differences between this and the hydrogen spectrum are explainable by the Pauli exclusion principle and the perturbations due to the central cloud.

As might be expected, the energy levels of molecules are far more complex in their structure than those of atoms, since in addition to electronic transitions like those in atoms, the molecule as a whole can rotate, stretch, and twist in various ways, and the stationary stages of these motions give rise to sets of energy levels related by selection rules similar to those governing atomic transitions. SEE ATOMIC STRUCTURE AND SPECTRA; MOLECULAR STRUCTURE AND SPECTRA.

Bibliography. A. Beiser, *Concepts of Modern Physics*, 3d ed., 1981; R. T. Weidner and R. L. Sells, *Elementary Modern Physics*, 3d ed., 1985; H. D. Young, *Fundamentals of Waves, Optics and Modern Physics*, 1975.

RITZ'S COMBINATION PRINCIPLE
EDWARD GERJUOY

The empirical rule, formulated by W. Ritz in 1905, that sums and differences of the frequencies of spectral lines often equal other observed frequencies. The rule is an immediate consequence of the quantum-mechanical formula $hf = E_i - E_f$ relating the energy hf of an emitted photon to the initial energy E_i and final energy E_f of the radiating system; h is Planck's constant and f is the frequency of the emitted light. For example, the **illustration** shows the photon energies hf_{32}, hf_{31},

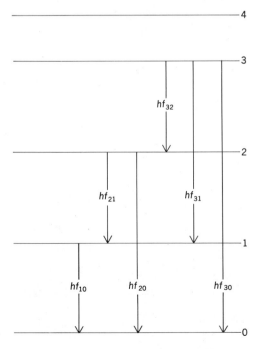

Energy levels and emitted frequencies.

hf_{30} associated with transitions from level 3 to lower-lying levels, and so on. Level 3 may radiate directly to the ground state 0, emitting f_{30}, or it may first make a transition to level 2, which subsequently radiates to the ground state, and so on. Since the total energy emitted in these two alternative means of making transitions from 3 to 0 is exactly the same, namely $E_3 - E_0$, it follows that $hf_{30} = hf_{32} + hf_{20}$. Similarly, $hf_{30} = hf_{32} + hf_{21} + hf_{10}$, and so forth. SEE ATOMIC STRUCTURE AND SPECTRA; ENERGY LEVEL.

LINEWIDTH
FRANCIS M. PIPKIN

A measure of the width of the band of frequencies of radiation emitted or absorbed in an atomic or molecular transition. One of the dominant sources of electromagnetic radiation of all frequencies is transitions between two energy levels of an atomic or molecular system. The frequency of the radiation is related to the difference in the energy of the two levels by the Bohr relation, Eq. (1),

$$\nu_0 = (E_1 - E_2)/h \qquad (1)$$

where ν_0 is the frequency of the radiation, h is Planck's constant, and E_1 and E_2 are the energies of the levels. This radiation is not monochromatic, but consists of a band of frequencies centered about ν_0 whose intensity $I(\nu)$ can be characterized by the linewidth. The linewidth is the full width at half height of the distribution function $I(\nu)$. The simplest case is for a transition from an excited state to the ground state for an atom or molecule at rest. For this case, the normalized distribution function is the lorentzian line profile given by Eq. (2). Here $\Delta\nu$ is the full width at half maximum (FWHM). The FWHM is related to the lifetime τ of the excited level through Eq. (3). This is a

$$I(\nu) = \frac{1}{\pi} \frac{\Delta\nu/2}{(\nu - \nu_0)^2 + (\Delta\nu/2)^2} \qquad (2)$$

manifestation of the quantum-mechanical uncertainty principle, and the linewidth $\Delta\nu$ is referred to as the natural linewidth. For a transition between two unstable levels with lifetimes τ_1 and τ_2, the FWHM is given by Eq. (4). SEE ENERGY LEVEL.

$$(\Delta\nu)(\tau) = \frac{1}{2\pi} \qquad (3) \qquad\qquad \Delta\nu = \frac{1}{2\pi}\left(\frac{1}{\tau_1} + \frac{1}{\tau_2}\right) \qquad (4)$$

Another major source of line broadening for atomic and molecular transitions is the Doppler shift due to thermal motion. To first-order in v/c, the frequency of the radiation depends upon whether the radiating object is moving away from or toward the observer according to Eq. (5).

$$\nu = \nu_0\left(1 - \frac{v}{c}\right) \qquad (5)$$

Here v is the component of the velocity along the line joining the observer and the radiating object taken as positive when the separation is increasing, and c is the speed of light. For a gaseous radiator whose velocity distribution is a maxwellian one characterized by a most probable velocity given by Eq. (6), where k is the Boltzmann constant, T the absolute temperature, and M the mass of the radiating molecule, the distribution of radiation is given by Eq. (7). This is a gaussian line profile with FWHM given by Eq. (8). For most situations the Doppler width is greater than

$$v_p = (2kT/M)^{1/2} \qquad (6) \qquad\qquad I(\nu) = I_0 \exp\{-[c(\nu - \nu_0)/\nu_0 v_p]^2\} \qquad (7)$$

$$(\Delta\nu)_D = (2\sqrt{\ln 2})\,\nu_0 v_p/c = \left(\frac{\nu_0}{c}\right)\left(\frac{8kT \ln 2}{M}\right)^{1/2} \qquad (8)$$

the natural linewidth. A more accurate line profile is the Voigt line profile, a convolution of the lorentzian and gaussian profiles given by Eq. (9). SEE DOPPLER EFFECT.

$$I(\nu) = (\text{const}) \int_0^\infty \frac{\exp\{-[c(\nu' - \nu_0)/\nu_0 v_p]^2\}}{(\nu - \nu')^2 + (\Delta\nu/2)^2} d\nu' \qquad (9)$$

A third major source of line broadening is collisions of the radiating molecule with other molecules. This broadens the line, shifts the center of the line, and shortens the lifetime of the radiating state. The net result is a lorentzian line profile of the form of Eq. (10), where Γ is given by Eq. (11). Here $(\Delta\nu)_N$ is the natural linewidth, $(\Delta\nu)_{IC}$ the width due to inelastic collisions, $(\Delta\nu)_{EC}$

$$I(\nu) = I_0 \frac{\Gamma^2}{[\nu - \nu_0 - (\delta\nu)_{EC}]^2 + \Gamma^2} \qquad (10) \qquad \Gamma = [(\Delta\nu)_N + (\Delta\nu)_{IC}]/2 + (\Delta\nu)_{EC} \qquad (11)$$

the width due to elastic collisions, and $(\delta\nu)_{EC}$ the shift of the center due to elastic collisions. The combination of the collision width, the natural width, and the Doppler width yields a generalized Voigt profile. This form is used to analyze data from plasmas and stellar atmospheres to determine the temperature of the radiator and the density of the perturbers.

In many measurements the linewidth limits the precision with which the center of the line can be determined. A number of techniques have been invented to circumvent the limitation due to the Doppler width. One technique, called Dicke narrowing, is to confine the radiating atoms to a volume whose characteristic dimensions are much less than the wavelength of the radiation. SEE LASER SPECTROSCOPY.

For radiating atoms in a liquid or solid the width is usually dominated by the strong interaction of the radiator with the surrounding molecules. This usually results in a spectrum with a wide distribution which cannot be characterized by a simple lorentzian or gaussian line profile. The inelastic interactions of the excited molecule with its surroundings produce radiationless transitions and shorten the lifetime of the excited state. There are often a variety of radiation sites with different characteristic frequencies. The net result is a broad line profile with a complex structure. SEE CRYSTAL ABSORPTION SPECTRA.

The above discussion deals principally with the situations where radiation is emitted by an excited molecule. The considerations are similar for the inverse reaction where a ground-state molecule absorbs radiation and makes a transition to an excited state. SEE ATOMIC STRUCTURE AND SPECTRA; MOLECULAR STRUCTURE AND SPECTRA.

Bibliography. R. C. Breen, Jr., Line width, in S. Flugge (ed.), *Handbuch der Physik*, vol. 27, p. 1, 1964; R. C. Breen, Jr., *Theories of Spectral Line Shape*, 1981; W. Demtröder, *Laser Spectroscopy*, 1982; D. S. McClure, Electronic spectra of molecules and ions in crystals, in F. Seitz and D. Turnbull (eds.), *Solid State Physics*, vols. 8 and 9, 1959.

ATOMIC STRUCTURE AND SPECTRA
IVAN A. SELLIN

The idea that matter is subdivided into discrete and further indivisible building blocks called atoms dates back to the Greek philosopher Democritus, whose teachings of the fifth century B.C. are commonly accepted as the earliest authenticated ones concerning what has come to be called atomism by students of Greek philosophy. The weaving of the philosophical thread of atomism into the analytical fabric of physics began in the late eighteenth and the nineteenth centuries. Robert Boyle is generally credited with introducing the concept of chemical elements, the irreducible units which are now recognized as individual atoms of a given element. In the early nineteenth century John Dalton developed his atomic theory, which postulated that matter consists of indivisible atoms as the irreducible units of Boyle's elements, that each atom of a given element has identical attributes, that differences among elements are due to fundamental differences among their constituent atoms, that chemical reactions proceed by simple rearrangement of indestructible atoms, and that chemical compounds consist of molecules which are reasonably stable aggregates of such indestructible atoms.

Electromagnetic nature of atoms. The work of J. J. Thomson in 1897 clearly demonstrated that atoms are electromagnetically constituted and that from them can be extracted fundamental material units bearing electric charge that are now called electrons. These pointlike charges have a mass of 9.110×10^{-31} kg and a negative electric charge of magnitude 1.602×10^{-19} coulomb. The electrons of an atom account for a negligible fraction of its mass. By virtue of overall electrical neutrality of every atom, the mass must therefore reside in a compensating, positively charged atomic component of equal charge magnitude but vastly greater mass.

Thomson's work was followed by the demonstration by Ernest Rutherford in 1911 that nearly all the mass and all of the positive electric charge of an atom are concentrated in a small nuclear core approximately 10,000 times smaller in extent than an atomic diameter. Niels Bohr in 1913 and others carried out some remarkably successful attempts to build solar system models of atoms containing planetary pointlike electrons orbiting around a positive core through mutual electrical attraction (though only certain "quantized" orbits were "permitted"). These models were ultimately superseded by nonparticulate, matter-wave quantum theories of both electrons and atomic nuclei.

The modern picture of condensed matter (such as solid crystals) consists of an aggregate of atoms or molecules which respond to each other's proximity through attractive electrical interactions at separation distances of the order of 1 atomic diameter (approximately 10^{-10} m) and repulsive electrical interactions at much smaller distances. These interactions are mediated by the electrons, which are in some sense shared and exchanged by all atoms of a particular sample, and serve as a kind of interatomic glue which binds the mutually repulsive, heavy, positively charged atomic cores together.

Planetary atomic models. Fundamental to any planetary atomic model is a description of the forces which give rise to the attraction and repulsion of the constituents of an atom. Coulomb's law describes the fundamental interaction which, to a good approximation, is still the basis of modern theories of atomic structure and spectra: the force exerted by one charge on another is repulsive for charges of like sign and attractive for charges of unlike sign, is proportional to the magnitude of each electric charge, and diminishes as the square of the distance separating the charges.

Also fundamental to any planetary model is Thomson's discovery that electrically neutral atoms in some sense contain individual electrons whose charge is restricted to a unique quantized value. Moreover, Thomson's work suggested that nuclear charges are precisely equally but oppositely quantized to offset the sum of the constituent atomic electron charges. Boyle's elements, of which over a hundred have been discovered, may then be individually labeled by the number of quantized positive-charge units Z (the atomic number) residing within the atomic nucleus (unit charge $= +1.602 \times 10^{-19}$ C). Each nucleus is surrounded by a complement of Z electrons (with charges of -1.602×10^{-19} C each) to produce overall charge neutrality. Molecules containing two, three, ..., atoms can then be thought of as binary, ternary, ..., planetary systems consisting of heavy central bodies of atomic numbers $Z_1, Z_2, \ldots$, sharing a supply of $Z_1 + Z_2 + \ldots$ electrons having the freedom to circulate throughout the aggregate and bond it together as would an electronic glue.

Atomic sizes. All atoms, whatever their atomic number Z, have roughly the same diameter (about 10^{-10} m), and molecular sizes tend to be the sum of the aggregate atomic sizes. It is easy to qualitatively, though only partially, account for this circumstance by using Coulomb's law. The innermost electrons of atoms orbit at small radii because of intense electrical attractions prevailing at small distances. Because electrons are more than 2000 times lighter than most nuclei and therefore move less sluggishly, the rapid orbital motion of inner electrons tends to clothe the slow-moving positive nuclear charge in a negative-charge cloud, which viewed from outside the cloud masks this positive nuclear charge to an ever-increasing extent. Thus, as intermediate and outer electrons are added, each experiences a diminishing attraction. The Zth, and last, electron sees $+Z$ electronic charge units within the nucleus compensated by about $(Z-1)$ electronic charges in a surrounding cloud, so that in first approximation the Zth electron orbits as it would about a bare proton having $Z = 1$, at about the same radius. Crudely speaking, an atom of any atomic number Z thus has a size similar to that of the hydrogen atom. When the molecular aggregates of atoms of concern to chemistry are formed, they generally do so through extensive sharing or exchanging of a small number (often just one) of the outermost electrons of each atom. Hence, the interatomic spacing in both molecules and solids tends to be of the order of one to a very few atomic diameters.

Chemical reactions. In a microscopic planetary atomic model, chemical reactions in which rearrangements result in new molecular aggregates of atoms can be viewed as electrical phenomena, mediated by the electrical interactions of outer electrons. The reactions proceed by virtue of changing positions and shapes of the orbits of the binding electrons, either through their internal electrical interactions or through their being electrically "bumped" from outside (as by collisions with nearby molecules).

Difficulties with the models. Before continuing with a detailed account of the successes of detailed planetary atomic models, it is advisable to anticipate some severe difficulties associated with them. Some of these difficulties played a prominent role in the development of quantum theory, and others present as yet unsolved and profound challenges to classical as well as modern quantum theories of physics.

A classical planetary theory fails to account for several atomic enigmas. First, it is known that electrons can be made to execute fast circular orbits in large accelerating machines, of which modern synchrotrons are excellent examples. As they are then in accelerated motion, they radiate light energy, as predicted by Maxwell's theory, at a frequency equal to that of their orbital circular motions. Thus, a classical planetary theory fails to explain how atoms can be stable and why atomic electrons do not continuously radiate light, falling into the nucleus as they do so. Such a theory also does not account for the observation that light emitted by atomic electrons appears to contain only quantized frequencies or wavelengths. Furthermore, a classical planetary theory would lead one to expect that all atomic electrons of an atom would orbit very close to their parent nuclei at very small radii, instead of distributing themselves in shells of increasing orbital radius, which seem able to accommodate only small, fixed numbers of electrons per shell.

Any theory of atomic structure must deal with the question of whether the atom is mostly empty space, as a planetary system picture suggests, or whether the entire atomic volume is filled with electronic charges in smeared out, cloudlike fashion. Such a theory must also be concerned with whether electrons and nuclei are pointlike, structureless, and indivisible, or whether they can be subdivided further into smaller constituents, in analogy to the way in which Thomson was able to extract electrons as constituents of atoms. Questions which still have not received definitive answers concern how much energy is needed to construct electrons and nuclei, what their radii are, and why they do not fly apart under the explosive repulsion of half their charge distribution for the other half. A more advanced theory than the classical planetary model is also required to determine whether electrons really interact with each other and with nuclei instantaneously, as Coulomb's law would have it, or whether there is a finite interaction delay time. Finally, a fundamental question, which can be answered only by quantum theory, is concerned with whether electrons and nuclei behave as pointlike objects, as light does when it gives rise to sharp shadows, or whether they behave as extended waves which can exhibit diffraction (bending) phenomena, as when light or water waves bend past sharp edges and partially penetrate regions of shadow in undulatory intensity patterns.

Scattering experiments. A key experimental technique in answering many important questions, such as those just posed, is that of scattering, in which small, high-speed probe projectiles are used to interact with localized regions of more extended objects from which they rebound, or scatter. A study of the number of scattered projectiles, and their distributions in speed and angle, gives much information about the structure of target systems, and the internal distributions, speeds, and concentrations of constituent bodies or substructures.

The scattering of x-rays by solids, for example, was used by C. Barkla to indicate that the number of electrons in an atom is approximately half the atomic weight. The mechanism of the production of the secondary (scattered) x-rays is indicated in **Fig. 1**a. X-rays are electromagnetic waves of wavelength considerably smaller than the size of the atom. If an x-ray sets an atomic electron into vibration, there follows the emission of a wave of lesser amplitude which can be observed at directions outside the incident beam.

Rutherford's experiments on the scattering of alpha particles represented an important step in understanding atomic structure. Alpha particles are helium ($Z = 2$) nuclei emitted at high velocities by some radioactive materials. A beam of these particles was directed at thin foils of different metals, and the relative numbers scattered at various angles were observed. While most of the particles passed through the foil with small deflections, such as the lower particle in Fig. 1b, a considerable number of very large deflections occurred. The upper particle in Fig. 1b has

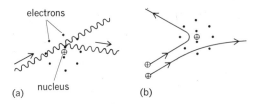

Fig. 1. Scattering by an atom (a) of x-rays, (b) of alpha particles.

undergone such a deflection. The precise results could be explained only if the positive electric charge of the atom is concentrated in the very small volume of the nucleus, which also contains almost all of the atom's mass. The diameter of the nucleus, found to depend on the atomic mass, was about 10^{-14} m for heavy nuclei. The nuclear charge was found to be the atomic number Z times the magnitude of the electron's charge.

The results of the scattering experiments therefore established the model of the atom as consisting of a small, massive, positively charged nucleus surrounded by a cloud of electrons to provide electrical neutrality. As noted above, such an atom should quickly collapse. The first step toward an explanation as to why it did not, came from the Bohr picture of the atom.

BOHR ATOM

The hydrogen atom is the simplest atom, and its spectrum (or pattern of light frequencies emitted) is also the simplest. The regularity of its spectrum had defied explanation until Bohr solved it with three postulates, these representing a model which is useful, but quite insufficient, for understanding the atom.

Postulate 1: The force that holds the electron to the nucleus is the Coulomb force between electrically charged bodies.

Postulate 2: Only certain stable, nonradiating orbits for the electron's motion are possible, those for which the angular momentum is an integral multiple of $h/2\pi$ (Bohr's quantum condition on the orbital angular momentum). Each stable orbit represents a discrete energy state.

Postulate 3: Emission or absorption of light occurs when the electron makes a transition from one stable orbit to another, and the frequency ν of the light is such that the difference in the orbital energies equals $h\nu$ (A. Einstein's frequency condition for the photon, the quantum of light).

Here the concept of angular momentum, a continuous measure of rotational motion in classical physics, has been asserted to have a discrete quantum behavior, so that its quantized size is related to Planck's constant h, a universal constant of nature. The orbital angular momentum of a point object of mass m and velocity v, in rotational motion about a central body, is defined as the product of the component of linear momentum mv (expressing the inertial motion of the body) tangential to the orbit times the distance to the center of rotation.

Modern quantum mechanics has provided justification of Bohr's quantum condition on the orbital angular momentum. It has also shown that the concept of definite orbits cannot be retained except in the limiting case of very large orbits. In this limit, the frequency, intensity, and polarization can be accurately calculated by applying the classical laws of electrodynamics to the radiation from the orbiting electron. This fact illustrates Bohr's correspondence principle, according to which the quantum results must agree with the classical ones for large dimensions. The deviation from classical theory that occurs when the orbits are smaller than the limiting case is such that one may no longer picture an accurately defined orbit. Bohr's other hypotheses are still valid.

Quantization of hydrogen atom. According to Bohr's theory, the energies of the hydrogen atom are quantized (that is, can take on only certain discrete values). These energies can be calculated from the electron orbits permitted by the quantized orbital angular momentum. The orbit may be circular or elliptical, so only the circular orbit is considered here for simplicity. Let the electron, of mass m and electric charge $-e$, describe a circular orbit of radius r around a nucleus of charge $+e$ and of infinite mass. With the electron velocity v, the angular momentum is mvr, and the second postulate becomes Eq. (1). The integer n is called the principal quantum

$$mvr = n(h/2\pi) \quad (n = 1, 2, 3, \ldots) \tag{1}$$

number. The centripetal force required to hold the electron in its orbit is the electrostatic force described by Coulomb's law, as shown in Eq. (2). Here ϵ_0 is the permittivity of free space, a

$$mv^2/r = e^2/4\pi\epsilon_0 r^2 \tag{2}$$

constant included in order to give the correct units to the mks statement of Coulomb's law.

The energy of an electron in an orbit consists of both kinetic and potential energies. For these circular orbits, the potential energy is twice as large as the kinetic energy and has a negative sign, where the potential energy is taken as zero when the electron and nucleus are at rest and separated by a very large distance. The total energy, which is the sum of kinetic and potential energies, is given in Eq. (3). The negative sign means that the electron is bound to the nucleus

$$E = (mv^2/2) - mv^2 = -mv^2/2 \tag{3}$$

and energy must be provided to separate them.

It is possible to eliminate v and r from these three equations. The result is that the possible energies of the nonradiating states of the atom are given by Eq. (4).

$$E = -\frac{me^4}{8\epsilon_0^2 h^2} \cdot \frac{1}{n^2} \tag{4}$$

The same equation for the hydrogen atom's energy levels, except for some small but significant corrections, is obtained from the solution of the Schrödinger equation for the hydrogen atom.

The frequencies of electromagnetic radiation or light emitted or absorbed in transitions are given by Eq. (5) where E' and E'' are the energies of the initial and final states of the atom.

$$\nu = (E' - E'')/h \tag{5}$$

Spectroscopists usually express their measurements in wavelength λ or in wave number σ in order to obtain numbers of a convenient size. The frequency of a light wave can be thought of as the number of complete waves radiated per second of elapsed time. If each wave travels at a fixed velocity c (approximately 3×10^8 m/s in vacuum), then after t seconds, the distance traveled by the first wave in a train of waves is ct, the number of waves in the train is νt, and hence the length of each must be $ct/\nu t = c/\nu = \lambda$. The wave number, defined as the reciprocal of the wavelength, therefore equals ν/c. The quantization of energy $h\nu$ and of angular momentum in units of $h/2\pi$ does not materially alter this picture. The wave number of a transition is shown in Eq. (6). If $T = -E/hc$, then Eq. (7) results. Here T is called the spectral term.

$$\sigma = \frac{\nu}{c} = \frac{E'}{hc} - \frac{E''}{hc} \tag{6} \qquad \sigma = T'' - T' \tag{7}$$

The allowed terms for hydrogen, from Eq. (4), are given by Eq. (8). The quantity R is the

$$T = \frac{me^4}{8\epsilon_0^2 ch^3} \cdot \frac{1}{n^2} = \frac{R}{n^2} \tag{8}$$

important Rydberg constant. Its value, which has been accurately measured by laser spectroscopy, is related to the values of other well-known atomic constants, as in Eq. (8).

The effect of finite nuclear mass must be considered, since the nucleus does not actually remain at rest at the center of the atom. Instead, the electron and nucleus revolve about their common center of mass. This effect can be accurately accounted for and requires a small change in the value of the effective mass m in Eq. (8). The mass effect was first detected by comparing the spectrum of hydrogen with that of singly ionized helium, which is like hydrogen in having a single electron orbiting the nucleus. For this isoelectronic case, the factor Z^2 must be included in the numerator of Eqs. (2) and (8) to account for the greater nuclear charge. The mass effect was used by H. Urey to discover deuterium, one of three hydrogen isotopes, by the shift of lines in its spectrum because of the very small change in its Rydberg constant. SEE ISOELECTRONIC SEQUENCE; ISOTOPE SHIFT.

Elliptical orbits. In addition to the circular orbits already described, elliptical ones are also consistent with the requirement that the angular momentum be quantized. A. Sommerfeld showed that for each value of n there is a family of n permitted elliptical orbits, all having the same major axis but with different eccentricities. **Figure 2a** shows, for example, the Bohr-Sommerfeld orbits for $n=3$. The orbits are labeled s, p, and d, indicating values of the azimuthal quantum number $l = 0$, 1, and 2. This number determines the shape of the orbit, since the ratio of the major to the minor axis is found to be $n/(l + 1)$. To a first approximation, the energies of all orbits of the same n are equal. In the case of the highly eccentric orbits, however, there is a slight lowering of the energy due to precession of the orbit (Fig. 2b). According to Einstein's theory of relativity, the mass increases somewhat in the inner part of the orbit, because of greater velocity. The velocity increase is greater as the eccentricity is greater, so the orbits of higher eccentricity have their energies lowered more. The quantity l is called the orbital angular momentum quantum number or the azimuthal quantum number.

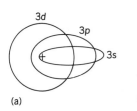

Fig. 2. Possible elliptical orbits, according to the Bohr-Sommerfeld theory. (a) The three permitted orbits for $n = 3$. (b) Precession of the 3s orbit caused by the relativistic variation of mass. (After A. P. Arya, *Fundamentals of Atomic Physics*, Allyn and Bacon, 1971)

A selection rule limits the possible changes of l that give rise to spectrum lines (transitions of fixed frequency or wavelength) of appreciable intensity. The rule is that l may increase or decrease only by one unit. This is usually written as $\Delta l = \pm 1$ for an allowed transition. Transitions for which selection rules are not satisfied are called forbidden; these tend to have quite low intensities. The quantum number n may change by any amount. Selection rules for hydrogen may be derived from Bohr's correspondence principle. However, the selection rules, as well as the relation between n and l, arise much more naturally in quantum mechanics.

MULTIELECTRON ATOMS

In attempting to extend Bohr's model to atoms with more than one electron, it is logical to compare the experimentally observed terms of the alkali atoms, which contain only a single electron outside closed shells, with those of hydrogen. A definite similarity is found but with the striking difference that all terms with $l > 0$ are double. This fact was interpreted by S. A. Goudsmit and G. E. Uhlenbeck as due to the presence of an additional angular momentum of $\frac{1}{2}$ $(h/2\pi)$ attributed to the electron spinning about its axis. The spin quantum number of the electron is $s = \frac{1}{2}$.

The relativistic quantum mechanics developed by P. A. M. Dirac provided the theoretical basis for this experimental observation.

Exclusion principle. Implicit in much of the following discussion is W. Pauli's exclusion principle, first enunciated in 1925, which when applied to atoms may be stated as follows: no more than one electron in a multielectron atom can possess precisely the same quantum numbers. In an independent, hydrogenic electron approximation to multielectron atoms, there are $2n^2$ possible independent choices of the principal (n), orbital (l), and magnetic (m_l, m_s) quantum numbers available for electrons belonging to a given n, and no more. Here m_l and m_s refer to the quantized projections of l and s along some chosen direction. The organization of atomic electrons into shells of increasing radius (the Bohr radius scales as n^2) follows from this principle, answering the question raised above as to why all the electrons of a heavy atom do not collapse into the most tightly bound orbits.

Examples are: helium ($Z = 2$), two $n = 1$ electrons: neon ($Z = 10$), two $n = 1$ electrons, eight $n = 2$ electrons; argon ($Z = 18$), two $n = 1$ electrons, eight $n = 2$ electrons, eight $n = 3$ electrons. Actually, in elements of Z greater than 18, the $n = 3$ shell could in principle accommodate 10 more electrons but, for detailed reasons of binding energy economy rather than fundamental symmetry, contains full $3s$ and $3p$ shells for a total of eight $n = 3$ electrons, but often a partially empty $3d$ shell. The most chemically active elements, the alkalies, are those with just one outer orbital electron in an n state, one unit above that of a completely full shell or subshell.

Spin-orbit coupling. This is the name given to the energy of interaction of the electron's spin with its orbital angular momentum. The origin of this energy is magnetic.

A charge in motion through either "pure" electric or "pure" magnetic fields, that is, through fields perceived as "pure" in a static laboratory, actually experiences a combination of electric and magnetic fields, if viewed in the frame of reference of a moving observer with respect to whom the charge is momentarily at rest. For example, moving charges are well known to be deflected by magnetic fields. But in the rest frame of such a charge, there is no motion, and any acceleration of a charge must be due to the presence of a pure electric field from the point of view of an observer analyzing the motion in that reference frame.

A spinning electron can crudely be pictured as a spinning ball of charge, imitating a circulating electric current (though Dirac electron theory assumes no finite electron radius—classical pictures fail). This circulating current gives rise to a magnetic field distribution very similar to that of a small bar magnet, with north and south magnetic poles symmetrically distributed along the spin axis above and below the spin equator. This representative bar magnet can interact with external magnetic fields, one source of which is the magnetic field experienced by an electron in its rest frame, owing to its orbital motion through the electric field established by the central nucleus of an atom. In multielectron atoms, there can be additional, though generally weaker, interactions arising from the magnetic interactions of each electron with its neighbors, as all are moving with respect to each other, and all have spin. The strength of the bar magnet equivalent to each electron spin, and its direction in space are characterized by a quantity called the magnetic moment, which also is quantized essentially because the spin itself is quantized. Studies of the effect of an external magnetic field on the states of atoms show that the magnetic moment associated with the electron spin is equal in magnitude to a unit called the Bohr magneton.

The energy of the interaction between the electron's magnetic moment and the magnetic field generated by its orbital motion is usually a small correction to the spectral term, and depends on the angle between the magnetic moment and the magnetic field or, equivalently, between the spin angular momentum vector and the orbital angular momentum vector (a vector perpendicular to the orbital plane whose magnitude is the size of the orbital angular momentum). Since quantum theory requires that the quantum number j of the electron's total angular momentum shall take values differing by integers, while l is always an integer, there are only two possible orientations for s relative to l: s must be either parallel or antiparallel to l. (This statement is convenient but not quite accurate. Actually, orbital angular momentum is a vector quantity represented by the quantum number l and of magnitude $\sqrt{l(l+1)} \cdot (h/2\pi)$. There are similar relations for spin and total angular momentum. These statements all being true simultaneously, the spin vector cannot ever be exactly parallel to the orbital angular momentum vector. Only the quantum numbers themselves can be described as stated.) **Figure 3a** shows the relative orientations of these two vectors and of their resultant j for a p electron (one for which $l = 1$). The corresponding spectral term designations are shown adjacent to the vector diagrams, labeled with the customary spectroscopic notation, to be explained later.

For the case of a single electron outside the nucleus, the Dirac theory gives Eq. (9) for the

$$\Delta T = \frac{R\alpha^2 Z^4}{n^3} \cdot \frac{j(j+1) - l(l+1) - s(s+1)}{l(2l+1)(l+1)} \tag{9}$$

spin-orbit correction to the spectral terms. Here $\alpha = e^2/2\epsilon_0 hc \cong 1/137$ is called the fine structure constant. The fine structure splitting predicted by Eq. (9) is present in hydrogen, although its observation requires instruments of very high precision. A relativistic correction must be added.

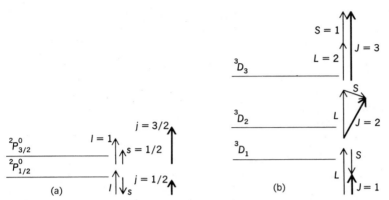

Fig. 3. Vector model for spectral terms arising from (a) a single p electron, and (b) two electrons, either two p electrons, or an s and a d electron.

In atoms having more than one electron, this fine structure becomes what is called the multiplet structure. The doublets in the alkali spectra, for example, are due to spin-orbit coupling; Eq. (9), with suitable modifications, can still be applied. These modifications may be attributed to penetration of the outer orbital electron within the closed shells of other electrons.

The various states of the atom are described by giving the quantum numbers n and l for each electron (the configuration) as well as the set of quantum numbers which depend on the manner in which the electrons interact with each other.

Coupling schemes. When more than one electron is present in the atom, there are various ways in which the spins and orbital angular momenta can interact. Each spin may couple to its own orbit, as in the one-electron case; other possibilities are orbit–other orbit, spin-spin, and so on. The most common interaction in the light atoms, called LS coupling or Russell-Saunders coupling, is described schematically in Eq. (10). This notation indicates that the l_i are coupled

$$\{(l_1, l_2, l_3, \ldots)(s_1, s_2, s_3, \ldots)\} = \{L, S\} = J \tag{10}$$

strongly together to form a resultant L, representing the total orbital angular momentum. The s_i are coupled strongly together to form a resultant S, the total spin angular momentum. The weakest coupling is that between L and S to form J, the total angular momentum of the electron system of the atom in this state. Suppose, for example, it is necessary to calculate the terms arising from a p ($l = 1$) and a d ($l = 2$) electron. The quantum numbers L, S, and J are never negative, so the only possible values of L are 1, 2, and 3. States with these values of L are designated P, D, and F. This notation, according to which $L = 0, 1, 2, 3, 4, 5$, etc. correspond to S, P, D, F, G, H, etc. terms, survives from the early empirical designation of series of lines as sharp, principal, diffuse, and so on.

The spin $s = 1/2$ for each electron, always, so the total spin $S = 0$ or $S = 1$ in this case. Consider only the term with $S = 1$ and $L = 2$. The coupling of L and S gives $J = 1, 2$, and 3, as shown in Fig. 3b. These three values of J correspond to a triplet of different energy levels which lie rather close together, since the LS interaction is relatively weak. It is convenient to use the notation ^3D for this multiplet of levels, the individual levels being indicated with J values as a subscript, as $^3D_1$, $^3D_2$, and $^3D_3$. The superscript has the value $2S + 1$, a quantity called the multiplicity. There will also be a term with $S = 0$ and $L = 2$ and the single value of $J = 2$. This is a singlet level, written as ^1D or $^1D_2$. The entire group of terms arising from a p and d electron includes ^1P, ^3P, ^1D, ^3D, ^1F, ^3F, with all the values of J possible for each term.

The ^2P state shown in Fig. 3a is derived from a single electron which, since $L = l = 1$ and $S = js = 1/2$, has J values of 3/2 and 1/2, forming a doublet. If there are three or more electrons, the number of possible terms becomes very large, but they are easily derived by a similar procedure. The resulting multiplicities are shown in the **table**. If two or more electrons are equivalent,

Possible multiplicities with different numbers of electrons				
Number of electrons:	1	2	3	4
Values of S:	½	1, 0	3/2, ½	2, 1, 0
Multiplicities	Doublets	Singlets Triplets	Doublets Quartets	Singlets Triplets Quintets

that is, have the same n and l, the number of resulting terms is greatly reduced, because of the requirement of the Pauli exclusion principle that no two electrons in an atom may have all their quantum numbers alike. Two equivalent p electrons, for example, give only ^1S, ^3P, and ^1D terms, instead of the six terms ^1S, ^3S, ^1P, ^3P, ^1D, and ^3D possible if the electrons are nonequivalent. The exclusion principle applied to equivalent electrons explains the observation of filled shells in atoms, where $L = S = J = 0$.

Coupling of the LS type is generally applicable to the low-energy states of the lighter atoms. The next commonest type is called jj coupling, represented in Eq. (11). Each electron has its spin

$$\{(l_1, s_1)(l_2, s_2)(l_3, s_3) \ldots\} = \{j_1, j_2, j_3 \ldots\} = J \qquad (11)$$

coupled to its own orbital angular momentum to form a j_i for that electron. The various j_i are then more weakly coupled together to give J. This type of coupling is seldom strictly observed. In the heavier atoms it is common to find a condition intermediate between LS and jj coupling; then either the LS or jj notation may be used to describe the levels, because the number of levels for a given electron configuration is independent of the coupling scheme.

SPECTRUM OF HYDROGEN

Figure 4a shows the terms R/n^2 in the spectrum of hydrogen resulting from the simple Bohr theory. These terms are obtained by dividing the numerical value of the Rydberg constant for hydrogen (109,678 cm^{-1}) by n^2, that is, by 1, 4, 9, 16, etc. The equivalent energies in electronvolts may then be found by using the conversion factor 1 eV = 8066 cm^{-1}. These energies, in contrast to the term values, are usually measured from zero at the lowest state, and increase for successive levels. They draw closer together until they converge at 13.598 eV. This corresponds to the orbit with $n = \infty$ and complete removal of the electron, that is, ionization of the atom. Above this ionization potential is a continuum of states representing the nucleus plus a free electron possessing a variable amount of kinetic energy.

The names of the hydrogen series are indicated in **Fig. 4**a. The spectral lines result from transitions between these various possible energy states. Each vertical arrow on the diagram connects two states and represents the emission of a particular line in the spectrum. The wave number of this line, according to Bohr's frequency condition, is equal to the difference in the term values for the lower and upper states and is therefore proportional to the length of the arrow. The wave numbers of all possible lines are given by Eq. (12), known as the Balmer formula, where the

$$\sigma = T'' - T' = R\left(\frac{1}{n''^2} - \frac{1}{n'^2}\right) \qquad (12)$$

double primes refer to the lower energy state (larger term value) and the single primes to the upper state. Any particular series is characterized by a constant value of n'' and variable n'.

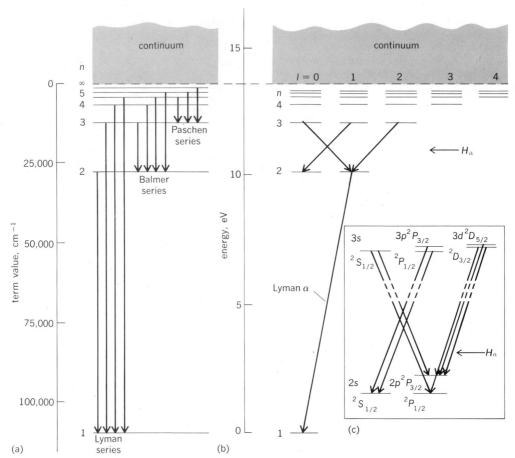

Fig. 4. Terms and transitions for the hydrogen atom. (a) Bohr theory. (b) Bohr-Sommerfeld theory. (c) Dirac theory.

Figure 5 shows the spectrum of a hydrogen discharge tube taken with a quartz spectrograph of resolution insufficient to resolve the fine structure. The Balmer series, represented by Eq. (12) with $n'' = 2$, is the only one shown. Its first line, that for $n' = 3$, is the bright red line at the wavelength 656.3 nanometers and is called H_α. Succeeding lines H_β, H_γ, and so on, proceed toward the ultraviolet with decreasing spacing and intensity, eventually converging at 364.6 nm. Beyond this series limit there is a region of continuous spectrum. The other series, given by $n'' = 1, 3, 4$, etc., lie well outside the visible region. The Lyman series covers the wavelength range 121.5–91.2 nm

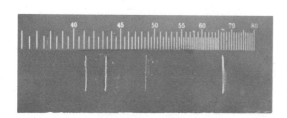

Fig. 5. Line emission spectrum of hydrogen. The scale gives the wavelengths in units of 10^{-8} m.

in the vacuum ultraviolet, and the Paschen series 1875.1–820.6 nm in the infrared. Still other series lie at even longer wavelengths.

Since hydrogen is by far the most abundant element in the cosmos, its spectrum is extremely important from the astronomical standpoint. The Balmer series has been observed as far as H_{31} in the spectra of hot stars. The Lyman series appears as the strongest feature of the Sun's spectrum, as photographed by rockets and orbiting satellites such as Skylab above the Earth's atmosphere.

NUCLEAR MAGNETISM AND HYPERFINE STRUCTURE

Most atomic nuclei also possess spin, but rotate about 2000 times slower than electrons because their mass is on the order of 2000 or more times greater than that of electrons. Because of this, very weak nuclear magnetic fields, analogous to the electronic ones that produce fine structure in spectral lines, further split atomic energy levels. Consequently, spectral lines arising from them are split according to the relative orientations, and hence energies of interaction, of the nuclear magnetic moments with the electronic ones. The resulting pattern of energy levels and corresponding spectral-line components is referred to as hyperfine structure. SEE NUCLEAR MOMENTS.

The fine structure and hyperfine structure of the hydrogen terms are particularly important astronomically. In particular, the observation of a line of 21-cm wavelength, arising from a transition between the two hyperfine components of the $n = 1$ term, gave birth to the science of radio astronomy. SEE ASTRONOMICAL SPECTROSCOPY.

Investigations with tunable lasers. The enormous capabilities of tunable lasers have allowed observations which were impossible previously. For example, high-resolution saturation spectroscopy, utilizing a saturating beam and a probe beam from the same laser, has been used to measure the hyperfine structure of the sodium resonance lines (called the D_1 and D_2 lines). Each line was found to have components extending over about 0.017 cm^{-1}, while the separation of the D_1 and D_2 lines is 17 cm^{-1}. The smallest separation resolved was less than 0.001 cm^{-1}, which was far less than the Doppler width of the lines. SEE DOPPLER EFFECT; HYPERFINE STRUCTURE; LASER SPECTROSCOPY.

Isotope shift. Nuclear properties also affect atomic spectra through the isotope shift. This is the result of the difference in nuclear masses of two isotopes, which results in a slight change in the Rydberg constant. There is also sometimes a distortion of the nucleus. Isotope shifts are frequently measured by using an atomic beam as a light source. The element studied evaporates out from a heated oven into a cooled evacuated region which condenses all the material except those atoms passing out through a cooled slit. The atoms in this beam may then be excited by a beam of electrons at right angles; the emitted radiation is then viewed at right angles to the beam. The Doppler effect is reduced because of the low velocities of the atoms at right angles to the motion of the beam. Since it generally desired to measure the separation of closely spaced wavelengths, the Doppler shift produced by the atomic beam, shifting all wavelengths the same amount, does not affect the measurement. SEE MOLECULAR BEAMS.

LAMB SHIFT AND QUANTUM ELECTRODYNAMICS

The Bohr-Sommerfeld theory, which permitted elliptical orbits with small relativistic shifts in energy, yielded a fine structure for H_α that did not agree with experiment. The selection rule $\Delta l = \pm 1$ permits three closely spaced transitions. Actually, seven components have been observed within an interval of 0.5 cm^{-1}. According to the Dirac theory, the spin-orbit and the relativistic correction to the energy can be combined in the single formula shown as Eq. (13), so that

$$\Delta T = \frac{R\alpha^2 Z^4}{n^3} \cdot \left(\frac{2}{2j + 1} - \frac{3}{4n} \right) \tag{13}$$

levels with the same n and j coincide exactly, as shown in Fig. 4c. The splittings of the levels are greatly exaggerated in the figure. Selection rules, described later, limit the transitions to $\Delta j = 0$ and ± 1, so there are just seven permitted transitions for the H_α line. Unfortunately for the theory, two pairs of these transitions coincide, if states of the same j coincide, so only five lines of the seven observed would be accounted for.

The final solution of this discrepancy came with the experimental discovery by W. Lamb, Jr., that the $2s^2S$ level is shifted upward by 0.033 cm^{-1}. The discovery was made by extremely sensitive microwave measurements, and it has subsequently been shown to be in accord with the general principles of quantum electrodynamics. The Lamb shift is present to a greater or lesser degree in all the hydrogen levels, and also in those of helium, but is largest for the levels of smallest l and n. Accurate optical measurements on the wavelength of the Lyman α line have given conclusive evidence of a shift of the $1s^2S$ level.

Though an extended discussion of the highly abstract field theory of quantum electrodynamics is inappropriate here, a simple picture of the lowest-order contribution to the Lamb shift in hydrogen can be given. No physical system is thought ever to be completely at rest; even at temperatures near absolute zero (about −273°C or −523°F), there is always a zero-point oscillatory fluctuation in any observable physical quantity. Thus, electromagnetic fields have very low-level, zero-point fluctuations, even in vacuum, at absolute zero. Electrons respond to these fluctuations with a zero-point motion: in effect, even a point charge is smeared out over a small volume whose mean square radius, though tiny, is not exactly zero. The smearing out of the charge distribution changes the interaction energy of a charged particle with any additional external field in which it may reside, including that of an atomic nucleus. As s electrons spend more time near the nucleus, where the nuclear-field Coulomb electric field is most intense, than do, say, electrons having more extended orbits, the change in interaction energy is greater for s electrons than, say, p electrons, and is such as to raise the energy of s electrons above the exact equality with that for p electrons of the same n arising in Dirac's theory of the electron.

RADIATIONLESS TRANSITIONS

It would be misleading to think that the most probable fate of excited atomic electrons consists of transitions to lower orbits, accompanied by photon emission. In fact, for at least the first third of the periodic table, the preferred decay mode of most excited atomic systems in most states of excitation and ionization is the electron emission process first observed by P. Auger in 1925 and named after him. For example, a singly charged neon ion lacking a 1s electron is more than 50 times as likely to decay by electron emission as by photon emission. In the process, an outer atomic electron descends to fill an inner vacancy, while another is ejected from the atom to conserve both total energy and momentum in the atom. The ejection usually arises because of the interelectron Coulomb repulsion. *See Auger effect.*

The vast preponderance of data concerning levels of excited atomic systems concerns optically allowed, single-electron, outermost-shell excitations in low states of ionization. Since most of the mass in nature is found in stars, and most of the elements therein rarely occupy such ionization-excitation states, it can be argued that presently available data provide a very unrepre-

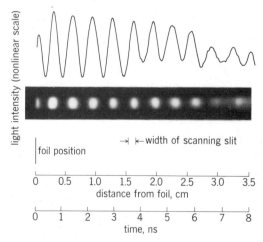

Fig. 6. Example of atomic process following excitation of a beam of ions in a thin solid target less than 1 μm in thickness. Quantum fluctuations in the light emitted by a 2-mm-diameter beam of such particles, traveling at a few percent of the speed of light, are shown as the atoms travel downstream from the target. (After I. A. Sellin et al., *Periodic intensity fluctuations of Balmer lines from single-foil excited fast hydrogen atoms, Phys. Rev., 184:56–63, 1969*)

sentative description of the commonly occurring excited atomic systems in nature. When the mean lives of excited atomic systems are considered, the relative rarity of lifetime measurements on Auger electron-emitting states is even more striking. The experimentally inconvenient typical lifetime range (10^{-12} to 10^{-16} s) accounts for this lack.

An effective means of creating such high ionization-excitation states is provided by beam-foil spectroscopy, in which ions accelerated by a Van de Graaff accelerator pass through a thin carbon foil. The resulting beam of ions can be analyzed so the charge states and velocities are accurately known. Highly excited states are produced by the interaction with the foil. The subsequent emission, detected with a spectrometer, allows the measurement of lifetimes of states which cannot be produced in any other source. When, for example, electromagnetic atomic decays are observed, optical or x-ray spectrometers may be useful (**Fig. 6**). When, as is most frequently the case, Auger processes are dominant, less familiar but comparably effective electron spectrometers are preferred. SEE BEAM-FOIL SPECTROSCOPY.

DOPPLER SPREAD

In both cases, a common problem called Doppler broadening of the spectral lines arises, which can cause overlapping of spectral lines and make analysis difficult. The broadening arises from motion of the emitted atom with respect to a spectrometer. Just as in the familiar case of an automobile horn sounding higher pitched when the auto is approaching an observer than when receding, so also is light blue shifted (to higher frequency) when approaching and redshifted when receding from some detection apparatus. The percentage shift in frequency or wavelength is approximately $(v/c) \cos \theta$, where v is the emitter speed, c the speed of light, and θ the angle between $\vec{v}$ and the line of sight of the observer. Because emitting atoms normally are formed with some spread in v and θ values, there will be a corresponding spread in observed frequencies or wavelengths. Several ingenious ways of isolating only those atoms nearly at rest with respect to spectrometric apparatus have been devised. The most powerful employ lasers in the saturation spectroscopy mode mentioned above.

RECOIL ION SPECTROSCOPY

Because violent, high-velocity atomic collisions are needed to reach the highest ionization-excitation states possible, as for example in the beam-foil method, Doppler spread problems have distinctly hindered both the optical and electron spectroscopy of highly ionized matter. These problems have also hindered corresponding lifetime measurements obtained by monitoring the rate of decay in flight of atoms in necessarily well-identified upper states. Because highly ionized atoms are very important in stellar atmospheres, and in terrestrially produced highly ionized gases (plasmas) on which many potentially important schemes for the thermonuclear fusion energy production schemes are based, it is important to overcome the Doppler spread problem. I. A. Sellin and collaborators noted that for sufficiently highly ionized, fast, heavy projectiles, lighter target atoms in a gaseous target can be ionized to very high states of ionization and excitation under single-collision conditions, while incurring relatively small recoil velocities. Sample spectra indicate that a highly charged projectile can remove as many as $Z - 1$ of the Z target electrons while exciting the last one remaining, all under single-collision-event conditions. Subsequent studies showed that the struck-atom recoil velocities are characterized by a velocity distribution no worse than a few times thermal, possibly permitting a new field of precision spectroscopy on highly ionized atoms to be developed.

RELATIVISTIC DIRAC THEORY AND SUPERHEAVY ELEMENTS

Dirac's theory of a single electron bound to a point nucleus implies a catastrophe for hydrogen atoms of high Z, if Z could be made on the order of $137 \cong 1/\alpha$ or larger (in which $\alpha = e^2/2\epsilon_0 hc$ is the fine-structure constant). Dirac's expression for the electronic binding energy becomes unphysical at that point. Nuclear theorists have, however, contemplated the possible stable existence of superheavy elements, appreciably heavier than any observed heretofore, say, in the range $Z = 114$ to 126 or even larger. Though searches for such superheavy elements of either primordial or artificial origin have thus far proved unsuccessful, the advent of heavy-ion accelerators capable of bringing heavy particles to a distance of closest approach much smaller than the radius of the innermost shell in any atom, even a heavy one, raises the possibility of transient

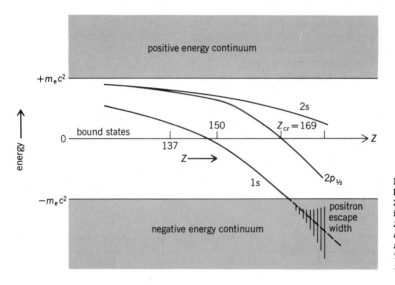

Fig. 7. Behavior of the binding energies of the $1s$, $2p_{1/2}$, and $2s$ electrons with increasing atomic number Z. (After J. H. Hamilton and I. A. Sellin, Heavy ions: Looking ahead, Phys. Today, 26(4):42–49, April 1973)

creation of a superheavy atom with a combined atomic number, $Z_{comb} = Z_1 + Z_2$, of about 180 or even greater (in which Z_1 and Z_2 are the individual atomic numbers of the colliding system).

The binding energy of a $1s$ electron in an atom increases rapidly with increasing atomic number. As already noted, in the usual linear version of the Dirac equation, a catastrophe occurs when the nuclear charge reaches $(1/\alpha) \cong 137$. When $Z_1 + Z_2$ is sufficiently large, the $1s$ binding energy can reach and exceed twice the rest energy of the electron $m_e c^2$, in which m_e is the electron mass and c is the speed of light. By varying the target or projectile, one can trace the path of the electron energies as they dive toward the negative-energy sea, or, in the case of the $1s$ electron, into the negative-energy sea, to give rise to bound states degenerate with the negative-energy continuum (**Fig. 7**). The negative-energy sea at $-m_e c^2$ was introduced by Dirac and used to predict the existence of a new particle, the positron, represented by a hole in the sea. According to W. Greiner and colleagues, and independently, V. Popov, the $1s$ level acquires a width because of the admixture of negative-energy continuum states, and the $1s$-state width corresponds to a decaying state, provided there is nonzero final-state density. When a hole is present in the K shell, spontaneous positron production is predicted to take place. Production of an electron-positron pair is possible, since a final $1s$ ground state is available for the produced electron; the positron escapes with kinetic energy corresponding to overall energy balance.

Because of the effects of finite nuclear size in modifying the potential, $Z_1 + Z_2$ must reach some initial value Z_{CR} greater than 137 at which the "splash" into the continuum is made. According to nuclear-model-dependent estimates, the splash would occur at about 170. Some types of nonlinear additions to the Dirac equation may remove the diving into the negative-energy continuum, and these could be tested by observation of positron production as a function of $Z_1 + Z_2$. Other "limiting field" forms of nonlinear electrodynamics lead to larger values of Z_{CR}, and since the positron escape width turns out to be approximately proportional to the square of the difference between the effective quasi-nuclear charge and Z_{CR}, these nonlinearities could be tested too. Intensive searches for spontaneous positron production have been undertaken.

UNCERTAINTY PRINCIPLE AND NATURAL WIDTH

A principle of universal applicability in quantum physics, and of special importance for atomic physics, is W. Heisenberg's 1927 uncertainty principle. In its time-energy formulation, it may be regarded as holding that the maximum precision with which the energy of any physical system (or the corresponding frequency ν from $E = h\nu$) may be determined is limited by the time interval Δt available for a measurement of the energy or frequency in question, and in no case can ΔE be less than approximately $h/2\pi\Delta t$. The immediate consequence is that the energy of an excited state of an atom cannot be determined with arbitrary precision, as such states have a finite lifetime available for measurement before decay to a lower state by photon or electron emission. It follows that only the energies of particularly long-lived excited states can be determined with great precision. The energy uncertainty ΔE, or corresponding frequency interval $\Delta \nu$, is referred to as the natural width of the level. For excited atoms radiating visible light, typical values of $\Delta \nu$ are on the order of 10^9 Hz. For inner electrons, which give rise to x-ray emission, and also for Auger electron-emitting atoms, 10^{14} Hz is more typical. An energy interval $\Delta E = 1$ eV corresponds to about 2.4×10^{14} Hz.

The time-energy uncertainty principle is perhaps best regarded as a manifestation of the basic wave behavior of electrons and other quantum objects. For example, piano tuners have been familiar with the time-frequency uncertainty principle for hundreds of years. Typically, they sound a vibrating tuning fork of accepted standard frequency in unison with a piano note of the same nominal frequency, and listen for a beat note between the tuning fork and the struck tone. Beats, which are intensity maxima in the sound, occur at a frequency equal to the difference in frequency between the two sound sources. For example, for a fork frequency of 440 Hz and a string frequency of 443 Hz, three beat notes per second will be heard. The piano tuner strives to reduce the number of beats per unit time to zero, or nearly zero. To guarantee a frequency good to 440 ± .01 Hz would require waiting for about 100 s to be sure no beat note had occurred ($\Delta \nu \Delta t \gtrsim 1$).

SUCCESSFUL EXPLANATIONS AND UNRESOLVED PROBLEMS

Many of the atomic enigmas referred to above have been explained through experimental discoveries and theoretical inspiration. Examples include Rutherford scattering, the Bohr model, the invention of the quantum theory, the Einstein quantum relation between energy and frequency $E = h\nu$, the Pauli exclusion and Heisenberg uncertainty principles, and many more. A few more are discussed below, but only in part, as the subject of atomic structure and spectra is by no means immune from the internal contradictions and defects in understanding that remain to be resolved.

In 1927 C. Davisson and L. Germer demonstrated that electrons have wave properties similar to that of light, which allows them to be diffracted, to exhibit interference phenomena, and to exhibit essentially all the properties with which waves are endowed. It turns out that the wavelengths of electrons in atoms are frequently comparable to the radius of the Bohr orbit in which they travel. Hence, a picture of a localized, point electron executing a circular or elliptical orbit is a poor and misleading one. Rather, the electrons are diffuse, cloudlike objects which surround nuclei, often in the form of standing waves, like those on a string. Crudely speaking, the average acceleration of an electron in such a pure standing wave, more commonly called a stationary state in the quantum theory, has no time-dependent value or definite initial phase. Since, as in Maxwell's theory, only accelerating charges radiate, atomic electrons in stationary states do not immediately collapse into the nucleus. Their freedom from this catastrophe can be interpreted as a proof that electrons may not be viewed as pointlike planetary objects orbiting atomic nuclei.

Atoms are by no means indivisible, as Thomson showed. Neither are their nuclei, whose constituents can be ejected in violent nucleus-nucleus collision events carried out with large accelerators. Nuclear radii and shapes can, for example, be measured by high-energy electron scattering experiments, in which the electrons serve as probes of small size.

However, a good theory of electron structure still is lacking, although small lower limits have been established on its radius. Zero-point fluctuations in the vacuum smear their physical locations in any event. There is still no generally accepted explanation for why electrons do not explode under the tremendous Coulomb repulsion forces in an object of small size. Estimates of the amount of energy required to "assemble" an electron are very large indeed. Electron structure is an unsolved mystery, but so is the structure of most of the other elementary objects in nature, such as protons, neutrons, neutrinos, and mesons. There are hundreds of such objects, many of which have electromagnetic properties, but some of which are endowed with other force fields as well. Beyond electromagnetism, there are the strong and weak forces in the realm of elementary particle physics, and there are gravitational forces. Electrons, and the atoms and molecules whose structure they determine, are only the most common and important objects in terrestrial life.

Though the action-at-a-distance concept built into a Coulomb's-law description of the interactions of electrons with themselves and with nuclei is convenient and surprisingly accurate, even that model of internal atomic interactions has its limits. It turns out that electrons and nuclei do not interact instantaneously, but only after a delay comparable to the time needed for electromagnetic waves to travel between the two. Electrons and nuclei do not respond to where they are with respect to each other, but only to where they were at some small but measurable time in the past.

Bibliography. E. U. Condon and H. Odabasi, *Atomic Structure*, 1980; R. D. Cowan, *The Theory of Atomic Structure and Spectra*, 1981; R. Eisberg and R. Resnick, *Quantum Physics*, 2d ed., 1985; H. Haken and H. C. Wolf, *Atomic and Quantum Physics: An Introduction to the Fundamentals of Experiment and Theory*, 1984; H. Kuhn, *Atomic Spectra*, 2d ed., 1969; J. Slater, *Quantum Theory of Atomic Structure*, 1960; G. K. Woodgate, *Elementary Atomic Structure*, 1980.

LINE SPECTRUM
George R. Harrison

A discontinuous spectrum characteristic of excited atoms, ions, and certain molecules in the gaseous phase at low pressures, to be distinguished from band spectra, emitted by most free molecules, and continuous spectra, emitted by matter in the solid, liquid, and sometimes gaseous

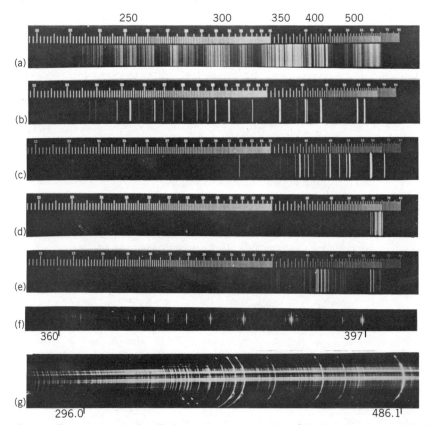

Common line spectra, wavelengths in angstroms (1 A = 0.1 nm). Emission spectra a–e all taken with the same quartz spectrograph. (a) Spectrum of iron arc. (b) Mercury spectrum from an arc enclosed in quartz. (c) Helium, (d) neon, (e) argon in a glass discharge tube. (f) Balmer series of hydrogen in the ultraviolet, photographed with a grating spectrograph. (g) Emission spectrum from gaseous chromosphere of the Sun, a grating spectrum taken without a slit at the instant preceding a total eclipse. Two strongest images, H and K lines of calcium, show marked prominences, or clouds, of calcium vapor. Other strong lines are caused by hydrogen and helium. (*From F. A. Jenkins and H. E. White, Fundamentals of Optics, 4th ed., McGraw-Hill, 1976*)

phase. If an electric arc or spark between metallic electrodes, or an electric discharge through a low-pressure gas, is viewed through a spectroscope, images of the spectroscope slit are seen in the characteristic colors emitted by the atoms or ions present.

To avoid the overlapping of close spectral images, the slit illuminated by the light source is made very narrow. The spectrum then appears as an array of bright line slit images on a dark background (see **illus**.). Under certain conditions spectra show dark absorption lines against a bright background. SEE ATOMIC STRUCTURE AND SPECTRA.

ISOELECTRONIC SEQUENCE
F. A. JENKINS AND W. W. WATSON

A term used in spectroscopy to designate the set of spectra produced by different chemical elements ionized in such a way that their atoms or ions contain the same number of electrons. The sequence in the **table** is an example. Since the neutral atoms of these elements each contain Z

Example of isoelectronic sequence

Designation of spectrum	Emitting atom or ion	Atomic number, Z
CaI	Ca	20
ScII	Sc^+	21
TiIII	Ti^{2+}	22
VIV	V^{3+}	23
CrV	Cr^{4+}	24
MnVI	Mn^{5+}	25

electrons, removal of one electron from scandium, two from titanium, and so forth, yields a series of ions all of which have 20 electrons. Their spectra are therefore qualitatively similar, but the spectral terms (energy levels) increase approximately in proportion to the square of the core charge, just as they depend on Z^2 in the one-electron sequence H, He^+, Li^{2+}, and so forth. As a result, the successive spectra shift progressively toward shorter wavelengths, soon reaching the vacuum ultraviolet region. Isoelectronic sequences are useful in predicting unknown spectra of ions belonging to a sequence in which other spectra are known. SEE ATOMIC STRUCTURE AND SPECTRA.

AUGER EFFECT
LEONARD C. FELDMAN

One of the two principal processes for the filling of an inner-shell electron vacancy in an excited or ionized atom. The Auger effect is a two-electron process in which an electron makes a discrete transition from a less bound shell to the vacant but more tightly bound electron shell. The energy gained in this process is transferred, via the electrostatic interaction, to another bound electron which then escapes from the atom. This outgoing electron is referred to as an Auger electron and is labeled by letters corresponding to the atomic shells involved in the process. For example, a KL_IL_{III} Auger electron corresponds to a process in which an L_I electron makes a transition to the

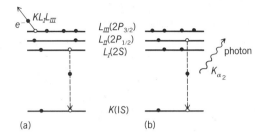

Two principal processes for the filling of an inner-shell electron vacancy. (a) Auger emission; a KL_IL_{III} Auger process in which an L_I electron fills the K-shell vacancy with the emission of a KL_IL_{III} Auger electron from the L_{III} shell. (b) Photon emission; a radiative process in which an L_{II} electron fills the K-shell vacancy with the emission of a K_{α_2} photon.

K shell and the energy is transferred to an L_{III} electron (**illus**. a). By the conservation of energy, the Auger electron kinetic energy E is given by $E = E(K) - E(L_I) - E(L_{III})$, where $E(K,L)$ is the binding energy of the various electron shells. Since the energy levels of atoms are discrete and well understood, the Auger energy is a signature of the emitting atom. SEE ENERGY LEVEL.

The other principal process for the filling of an inner shell hole is a radiative one in which the transition energy is carried off by a photon (illus. b). Inner shell vacancies in elements with large atomic number correspond to large transition energies and usually decay by such radiative processes; vacancies in elements with low atomic number or outer shell vacancies with low transition energies decay primarily by Auger processes. SEE ATOMIC STRUCTURE AND SPECTRA.

Auger electron spectroscopy is an important tool in the analysis of the near-surface region of solids. Inner-shell vacancies in the atoms of solids are created by incident radiation, usually an energetic electron beam, and the energy spectrum of the outgoing Auger electrons is then used to determine the types of atoms present, and hence the elemental composition of the material. In solids, the outgoing Auger electron has a high probability of undergoing an inelastic or energy loss event which changes the Auger electron energy. The probability of an inelastic process is characterized by a mean free path, a thickness in which approximately 63% of the Auger electrons undergo an energy-changing event. The mean free path can vary from approximately 0.5 nanometer to 100 nm, depending on the electron energy and the solid; thus Auger electron spectroscopy is useful for measurements of the elemental composition of a material in its first 0.5–100 nm. Elemental composition as a function of depth is determined by using Auger electron spectroscopy combined with some erosion process which slowly removes the surface layers; the Auger spectrum measured at various stages of the erosion process yields a depth profile of the elemental composition of the material. SEE ELECTRON SPECTROSCOPY; SURFACE PHYSICS.

Bibliography. P. Auger, The compound photoelectric effect, *J. Phys. Radium*, 6:205–208, 1925; W. Bambynek et al., X-ray fluorescence yields, Auger and Coster-Kronig transition probabilities, *Rev. Mod. Phys.*, 44:716–814, 1972; D. Briggs and M. P. Seah, *Practical Surface Analysis by Auger and Photo-Electron Spectroscopy*, 1983; L. A. Casper and C. J. Powell (eds.), *Industrial Applications of Surface Analysis*, 1982; M. P. Seah and W. A. Dench, Quantitative electron spectroscopy of surfaces: A standard data base for electron inelastic mean free paths in solids, *Surf. Interface Anal.* 1:2–11, 1979; A. Temkin (ed.), *Autoionization: Recent Developments and Applications*, 1985; M. Thompson et al., *Auger Electron Spectroscopy*, 1985.

FINE STRUCTURE
F. A. JENKINS AND W. W. WATSON

A term referring to the closely spaced groups of lines observed in the spectra of the lightest elements, notably hydrogen and helium. The components of any one such group are characterized by identical values of the principal quantum number n, but different values of the azimuthal quantum number l and the angular momentum quantum number j.

According to P. A. M. Dirac's relativistic quantum mechanics, those energy levels of a one-electron atom that have the same n and j coincide exactly, but are displaced from the values predicted by the simple Bohr theory by an amount proprotional to the square of the fine-structure constant α. The constant α is dimensionless, and nearly equal to 1/137. Its value is actually 0.007297351 ± 0.000000006. In 1947 deviations from Dirac's theory were found, indicating that the level having $l = 0$ does not coincide with that having $l = 1$, but is shifted appreciably upward. This is the celebrated Lamb shift named for its discoverer, Willis Lamb, Jr. Modern quantum electrodynamics accounts for this shift as being due to the interaction of the electron with the zero-point fluctuations of the electromagnetic field. SEE ATOMIC STRUCTURE AND SPECTRA.

In atoms having several electrons, this fine structure becomes the multiplet structure resulting from spin-orbit coupling. This gives splittings of the terms and the spectral lines that are "fine" for the lightest elements but that are very large, of the order of an electronvolt, for the heavy elements.

HYPERFINE STRUCTURE
Louis D. Roberts

A closely spaced structure of the spectrum lines forming a multiplet component in the spectrum of an atom or molecule, or of a liquid or solid. In the emission spectrum for an atom, when a multiplet component is examined at the highest resolution, this component may be seen to be resolved, or split, into a group of spectrum lines which are extremely close together. This hyperfine structure may be due to a nuclear isotope effect, to effects related to nuclear spin, or to both.

Isotope effect. The element zinc, for example, has three relatively abundant naturally occurring nuclear isotopes, ^{64}Zn, ^{66}Zn, and ^{68}Zn. The radius of a nucleus increases with the nuclear mass and, for a given element, the Coulomb interaction of the nucleus with the atomic s-electrons will be slightly weaker when the nuclear size is larger. This nuclear size effect causes a slight shift of certain of the spectrum lines, and this shift will be different for each isotope. For a mixture of ^{64}Zn, ^{66}Zn, and ^{68}Zn, certain of the multiplet components will thus consist of three closely spaced lines, one line for each isotope. A study of the isotope effect for an element leads, for example, to information about the dependence of the nuclear size on isotope mass, that is, on the number of neutrons in the isotope. SEE ISOTOPE SHIFT.

Structure due to nuclear spin. For the zinc isotopes discussed above, the nuclear spin $I = 0$, and these nuclei will be nonmagnetic and, in effect, have a spherical shape. If $I \neq 0$, however, two new nuclear properties may be observed. The nucleus may have a magnetic moment, and the shape of the nucleus may not be spherical but rather may be that of a prolate or oblate spheroid; that is, it may have a quadrupole moment.

Atoms and molecules. If the electrons in an atom or a molecule have an angular momentum, the electron system may likewise have a magnetic moment. An electron quadrupole moment may also exist. The magnetic moment of the nucleus may interact with the magnetic moment of the electrons to produce a magnetic hyperfine structure. The quadrupole moments of the nucleus and of the electrons may couple to give an electric quadrupole hyperfine structure. In a simple example, the magnetic and the electric quadrupole hyperfine structure may be described by an energy operator H_{hfs} which has the form below, where $\bar{S}$ is an operator describing electron spin, $\bar{I}$

$$H_{\text{hfs}} = A\bar{I} \cdot \bar{S} + P(\bar{I}_z^2 - 3I(I + 1))$$

and $\bar{I}_z$ are operators describing the nuclear spin and its z component, and I gives the magnitude of the nuclear spin. A and P are coupling constants which may take positive or negative values, and may range in magnitude from zero to a few hundred meters^{-1}. The term in A describes a magnetic, and the term in P a quadrupole, hyperfine structure.

The measurement of a hyperfine structure spectrum for a gaseous atomic or molecular system can lead to information about the values for A and P. These values may be interpreted to obtain information about the nuclear magnetic and quadrupole moments, and about the atomic or molecular electron configuration.

Important methods for the measurement of hyperfine structure for gaseous systems may employ an interferometer, or use atomic beams, electron spin resonance, or nuclear spin resonance. SEE ELECTRON PARAMAGNETIC RESONANCE (EPR) SPECTROSCOPY; MAGNETIC RESONANCE; MOLECULAR BEAMS; NUCLEAR MAGNETIC RESONANCE (NMR).

Liquid and solid systems. Hyperfine structure coupling may also occur and may be measured for liquid and solid systems. For liquids and solids, measurements are often made by electron spin or nuclear spin resonance methods. For solids, and for radioactive nuclei, one may, for example, also employ the Mössbauer effect or the angular correlation of nuclear gamma rays. SEE GAMMA RAYS; MÖSSBAUER EFFECT.

For a diamagnetic solid, $A = 0$ in the equation above, and if the crystalline environment of an atom is cubic, $P = 0$ also. If this environment is not cubic, P may have a finite measurable value.

If the solid is paramagnetic, ferromagnetic, or antiferromagnetic, A may be finite and measurable, and again P may or may not be zero depending on whether the atomic environment is cubic or not.

One may gain information about the nuclear moments and about electron bonding and magnetic structure from measurements of hyperfine structure for liquids and solids. Such measurements are extensively used, for example, in atomic and condensed matter physics, chemistry, and biology. SEE ATOMIC STRUCTURE AND SPECTRA; NUCLEAR MOMENTS.

Bibliography. A. Abragam, *The Principles of Nuclear Magnetism*, 1961, reprint 1983; L. C. Biedenharn and J. D. Louck, *Angular Momentum in Quantum Physics*, 1984; R. S. Raghavan and D. E. Murnick (eds.), *Hyperfine Interactions IV*, 1978; M. E. Rose, *Elementary Theory of Angular Momentum*, 1957.

STARK EFFECT
F. A. JENKINS AND W. W. WATSON

The effect of an electric field on spectrum lines. The electric field may be externally applied, but in many cases it is an internal field caused by the presence of neighboring ions or atoms in a gas, liquid, or solid. Discovered in 1913 by J. Stark, the effect is most easily studied in the spectra of hydrogen and helium, by observing the light from the cathode dark space of an electric discharge. Because of the large potential drop across the region, the lines are split into several components. For observation perpendicular to the field, the light of these components is linearly polarized. The splitting can be easily resolved with a spectrograph of moderate dispersion, amounting in the case of the H_γ line at 434 nanometers to a total spread of 3.2 nanometers for a field of 10,400 kV/m.

Linear Stark effect. This effect exhibits large, nearly symmetrical patterns. Examples are shown in the **illustration**, where the symbols π and σ refer to the two states of polarization. The interpretation of the linear Stark effect was one of the first successes of the quantum theory. According to this theory, the effect of the electric field on the electron orbit is to split each energy level of the principal quantum number n into $2n - 1$ equidistant levels, of separation proportional to the field strength. Thus the higher members of a series show a larger number of components and greater overall splittings. The criterion for the occurrence of the linear Stark effect is that the splitting of the levels shall be large compared to the natural separation of levels of the same n but different L, where L is the quantum number of orbital angular momentum. SEE ZEEMAN EFFECT.

Quadratic Stark effect. This occurs in lines resulting from the lower energy states of many-electron atoms. Here the large separation of states of different L results from the penetration of the valence electrons into the core of other electrons, with the result that the permanent dipole moment associated with hydrogenlike orbits no longer exists. There is, however, a small induced dipole moment due to polarization of the atom. This moment is proportional to the electric field strength, and since the energy change is proportional to the product of the dipole moment and the field strength, the energy levels shift by an amount depending on the square of the field. Thus all levels have shifts of the same sign and therefore are displaced to lower energies. Each field-free level is also split, as a result of the space quantization of the angular momentum vector J, so that there are $J + 1$ components if J is integral, or $J + \frac{1}{2}$ components if it is half-integral. Because the lower levels are usually less displaced than the higher ones, the quadratic Stark effect ordinarily shows itself as a shift of the lines toward the red end of the spectrum, with an accompa-

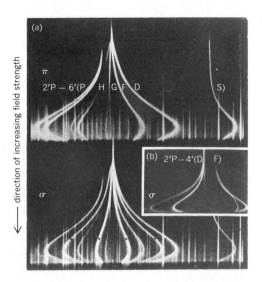

Stark effect for helium lines. (*a*) The 414.4- and 416.9-nm lines at field strengths 0–85,000 V/cm. Note the disappearance of ^1P–^1H line at a certain field strength, and symmetry of the σ components (those with polarization perpendicular to the field), indicated by dots. (*b*) The 492.2-nm line at 0–40,000 V/cm. Note the crossing over of two ^1P–^1F components. (*J.S. Foster and C. Foster, McGill University*)

nying separation into several components. The quadratic Stark effect is basic to the explanation of the formation of molecules from atoms, of dielectric constants, and of the broadening of spectral lines.

Intermolecular Stark effect. Produced by the action of the electric field from surrounding atoms or ions on the emitting atom, the intermolecular effect causes a shifting and broadening of spectrum lines. The molecules being in motion, these fields are inhomogeneous in space and also in time. Hence the line is not split into resolved components but is merely widened. Particularly in the electric discharge through gases with high currents, the large ion density may cause very wide lines. The amount of the broadening is found to run parallel to the sensitivity of the line to the Stark effect and thus is greatest for those lines susceptible to the linear effect.

Inverse Stark effect. This is the effect as observed with absorption lines. It has been detected, for example, by applying an electric field to potassium vapor, and measuring a small displacement of the absorption lines toward the red. The displacements are found to be proportional to the square of the field strength, as in the quadratic Stark effect. In addition, certain lines of large n, which are normally forbidden by the selection rule for L, are observed to appear in the presence of the field. This type of transition is said to be produced by "forced dipole radiation." Such forbidden lines also may be obtained in emission. In the illustration, the lines $2^1P - 6^1S$ and $2^1P - 4^1F$ are examples of forced dipole transitions.

Bibliography. R. D. Cowan, *The Theory of Atomic Structure and Spectra*, 1981; G. Herzberg, *Atomic Spectra and Atomic Structure*, 2d ed., 1944; J. D. McGervey, *Introduction to Modern Physics*, 2d ed., 1984; M. R. Wehr, J. A. Richards, and T. W. Adair, *Physics of the Atom*, 4th ed., 1984.

ZEEMAN EFFECT
F. A. Jenkins, G. H. Dieke, and W. W. Watson

A splitting of spectral lines when the light source being studied is placed in a magnetic field. Discovered by P. Zeeman in 1896, the Zeeman effect furnishes information of prime importance in the analysis of spectra. Each kind of spectral term has its characteristic mode of splitting, and the types of terms are most definitely identified by this property. Furthermore, the effect allows an evaluation of the ratio of charge to mass of the electron and an evaluation of its precise magnetic moment.

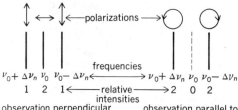

Fig. 1. Triplet observed in normal Zeeman effect.

Normal Zeeman effect. This is a splitting into two or three lines, depending on the direction of observation, as shown in **Fig. 1**. The light of these components is polarized in ways indicated in the figure. The normal effect is observed for all lines belonging to singlet systems, those for which the spin quantum number $S = 0$. The change of frequency $\Delta \nu_n$ of the shifted components can be evaluated on classical electromagnetic principles as follows. Assume that the electron of charge e revolves in a circular orbit of radius r and circular frequency ω radians per second (**Fig. 2**). If a magnetic field H is applied perpendicular to the plane of the orbit, the electron will be speeded up or slowed down because of the changing flux through its orbit, just as in the electron accelerator known as the betatron.

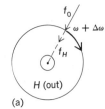
(a)

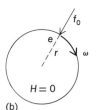
(b)

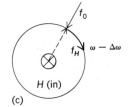

Fig. 2. Effect of a magnetic field H applied perpendicular to a circular electron orbit. Diagrams a–c are explained in the text.

Denoting the centripetal force holding the electron on its orbit before application of the field by f_0 (Fig. 2b), and the additional force due to the motion of the electron across the field by f_H (Fig. 2a and c), one has Eqs. (1). In Fig. 2a, where the two forces are in the same direction, one

$$f_0 = m\omega^2 r \qquad\qquad f_H = He(\omega \pm \Delta\omega)r \qquad (1)$$

may equate their sum to the centripetal force on an electron of frequency $\omega + \Delta\omega$, obtaining Eq. (2). Solution of this equation for $\Delta\omega$, under the assumption that it is small compared to ω itself, yields Eqs. (3), the latter expression following from the fact that $\nu = \omega/2\pi$. This relation, although

$$f_0 + f_H = m\omega^2 r + He(\omega + \Delta\omega)r = m(\omega + \Delta\omega)^2 r \qquad (2)$$

$$\Delta\omega = eH/2m \qquad\qquad \Delta\nu_n = eH/4\pi m \qquad (3)$$

derived for a special case, is generally valid for any system of particles having a particular value of e/m and moving under the action of a central force.

On substitution of the radio of charge to mass of an electron, one obtains Eq. (4). Con-

$$\Delta\nu_n = 1.3996 \times 10^6 \, H \, s^{-1} \qquad (4)$$

ORIGIN OF SPECTRA **39**

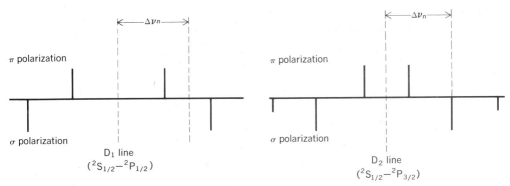

Fig. 3. Anomalous Zeeman effect of the sodium lines. $\Delta\nu_n$ denotes the normal Zeeman splitting, while π and σ refer to polarizations like those of the central and outer components in the normal effect illustrated in Fig. 1. The heights of the lines indicate their relative intensities.

versely, from the observed spectroscopic splitting $\Delta\nu_n$ and measurement of the field strength, the value of e/m for the electron has been evaluated as 1.7572 ± 0.0007 emu/g. This is in good agreement with the figure determined by other methods.

Anomalous Zeeman effect. This effect is a more complicated type of line splitting, so named because it did not agree with the predictions of classical theory. It occurs for any spectral line arising from a combination of terms of multiplicity greater than one. As examples, **Fig. 3** gives diagrams of the theoretical patterns for the yellow lines of sodium, belonging to a doublet system, while **Fig. 4** shows some actual patterns observed for doublets and quartets in rhodium.

Since multiplicity in spectral lines is caused by the presence of a resultant spin vector S of the electrons, the anomalous effect must be attributed to a nonclassical magnetic behavior of the electron spin. While classical theory associates with the vector L of the orbital angular momentum

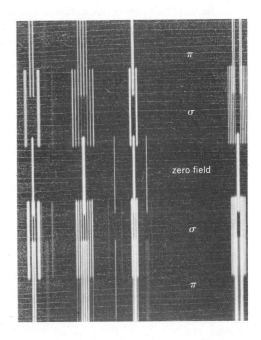

Fig. 4. Zeeman effect of the rhodium spectrum in the wavelength range 347.9–346.2 nanometers. Field strengths, (lower exposure) 70,000 oersteds and (upper exposure) 90,500. (*From G. R. Harrison and F. Bitter, Massachusetts Institute of Technology*)

a magnetic moment as in Eq. (5), it is necessary, in explaining the anomalous Zeeman effect, that the magnetic moment corresponding to S be as in Eq. (6). Thus the spin generates twice as much

$$\mu_L = (eh/4\pi mc)L \qquad (5) \qquad\qquad \mu_S = (eh/2\pi mc)S \qquad (6)$$

magnetic moment, relative to its angular momentum, as does the orbital motion. In an atom for which both L and S are finite, the effective magnetic moment may be written as Eq. (7), where J

$$\mu_J = g(eh/4\pi mc)J = g\mu_0 J \qquad (7)$$

measures the total angular momentum (in LS coupling, the resultant of L and S), and μ_0 is the Bohr magneton, $eh/4\pi mc$.

Theory gives values of g, the Landé g factor, which are characteristic of the type of spectral term. In LS coupling, the value is given by Eq. (8).

$$g = 1 + \frac{J(J+1) + S(S+1) - L(L+1)}{2J(J+1)} \qquad (8)$$

For a classical electron orbit $g = 1$, which yields the normal Zeeman effect. When the spin S is present, however, the changes in energy produced by the magnetic field, which are proportional to μ_J, are just g times as great, and this fact is responsible for the anomalous Zeeman effect. It also should be mentioned that both theory and experiment now show that the g factor for the electron is not exactly 2, but 2.00229.

The component of μ_J in the field direction is $g\mu_0 M$, where M is the quantized component of J in this direction. The energy terms become $T = T_0 + g\mu_0 M$, where T_0 is the term value with no field. This magnetic quantum number M has only the $2J + 1$ values, $J, J - 1, J - 2, \ldots, -J$, and the allowed transitions between energy terms must obey the selection rule $\Delta M = 0$, ± 1. To explain the statement that the normal Zeeman effect is observed for all lines in singlet systems, recall that S is then zero, $J = L$, and hence $g = 1$.

Quadratic Zeeman effect. The quadratic effect, which depends on the square of the field strength, is of two kinds. The first results from the second-order terms that were neglected in the preceding derivation, and the second, from the diamagnetic reaction of the electron when revolving in large orbits.

Inverse Zeeman effect. This is the Zeeman effect of absorption lines. It is closely related to the Faraday effect, the rotation of plane-polarized light by matter situated in a magnetic field. SEE ATOMIC STRUCTURE AND SPECTRA.

Zeeman effect in molecules. This effect is, in general, so small as to be unobservable, even for molecules which have a permanent magnetic moment. Each level with a total angular momentum J splits into $2J + 1$ components, as in the case of atoms.

The component of the magnetic moment along the direction of the external field is small, however, because the rotation of the molecule, which carries the magnetic moment along with it, causes the principal part of the magnetic moment to average out to zero. The consequence is that the magnetic levels have an extremely narrow spacing except for cases where the molecule has either very little rotation or none at all. An exception occurs for some light molecules where the magnetic moment is coupled so lightly to the frame of the molecule that it can orient itself freely in the magnetic field just as for atoms (**Fig. 5**).

Zeeman effect in crystals. A clear Zeeman effect also can be observed in many crystals with sharp spectrum lines in absorption or fluorescence. Such crystals are found particularly among the salts of the rare earths. In these cases the internal electric field in the crystal splits and shifts the level of the free ion. When the number of electrons is even and the crystal symmetry low, this electric splitting is complete. No degeneracy remains, and there can be no further splitting by a magnetic field. If the number of electrons is odd, or if for an even number there is high crystal symmetry, the levels occur in degenerate pairs which are split by a magnetic field. Each line is then split into four components (**Fig. 6**). For cubic crystal symmetry and when angular momentum is due only to electron spin, splitting into more than four components may occur.

Nuclear Zeeman effect. The magnetic moment of the nucleus causes a Zeeman splitting in atomic spectra which is of an order of magnitude a thousand times smaller than the ordinary Zeeman effect. This Zeeman effect of the hyperfine structure usually is modified by a

ORIGIN OF SPECTRA 41

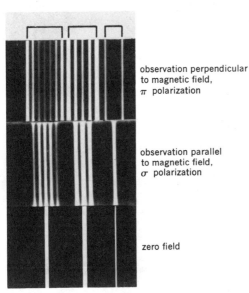

Fig. 5. Zeeman effect of three adjacent rotational lines of the hydrogen molecule. (*Johns Hopkins University*)

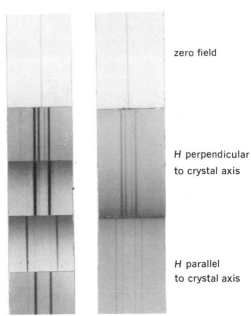

Fig. 6. Zeeman effect of an absorption line of neodymium chloride, (left) with polarization and (right) without polarization. The splitting is quite different, depending on whether the trigonal crystal axis is parallel or perpendicular to the magnetic field. $H = 35{,}000$ oersteds. (*Johns Hopkins University*)

nuclear Paschen-Back effect, first studied by E. Back and S. A. Goudsmit for the spectral lines of bismuth.

A strong magnetic field actually may modify the intensity and selection rules so that usually absent lines may appear. For example, the J selection rule is no longer valid in a magnetic field.

Bibliography. R. D. Cowan, *The Theory of Atomic Structure and Spectra*, 1981; G. Herzberg, *Atomic Spectra and Atomic Structure*, 1944; F. A. Jenkins and H. E. White, *Fundamentals of Optics*, 4th ed., 1976; J. H. Van Vleck, *The Theory of Electric and Magnetic Susceptibilities*, 1932; M. R. Wehr, J. A. Richards, and T. W. Adair, *Physics of the Atom*, 4th ed., 1984.

ISOTOPE SHIFT
Peter M. Koch

A small difference between the different isotopes of an element in the transition energies corresponding to a given spectral line transition. For a spectral line transition between two energy levels a and b in an atom or ion with atomic number Z, the small difference $\Delta E_{ab} = E_{ab}(A') - E_{ab}(A)$ in the transition energy between isotopes with mass numbers A' and A is the isotope shift. It consists largely of the sum of two contributions, the mass shift (MS) and the field shift (FS), also called the volume shift. The mass shift is customarily divided into a normal mass shift (NMS) and a specific mass shift (SMS); each is proportional to the factor $(A' - A)/A'A$. The normal mass shift is a reduced mass correction that is easily calculated for all transitions. The specific mass shift is produced by the correlated motion of different pairs of atomic electrons and is, therefore, absent in one-electron systems.

It is generally difficult to calculate precisely the specific mass shift, which may be 30 times larger than the normal mass shift for some transitions. The field shift is produced by the change in the finite size and shape of the nuclear charge distribution when neutrons are added to the nucleus. Since electrons whose orbits penetrate the nucleus are influenced most, S-P and P-S transitions generally have the largest field shift.

For very light elements, $Z \leq 37$, the mass shift dominates the field shift. For $Z = 1$, the 0.13 nanometer shift in the red Balmer line led to the discovery of deuterium, the $A = 2$ isotope of hydrogen. For medium-heavy elements, $38 \leq Z \leq 57$, the mass shift and field shift contributions to the isotope shift are comparable. For heavier elements, $Z \geq 58$, the field shift dominates the mass shift. A representative case is shown in the **illustration**. See ATOMIC STRUCTURE AND SPECTRA.

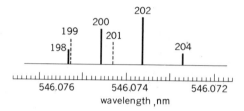

Some isotope shifts in the green line of mercury, $Z = 80$. In this heavy element the contribution of the field shift is much larger than that of the mass shift.

When isotope shift data have been obtained for at least two pairs of isotopes of a given element, a graphical method introduced by W. H. King in 1963 can be used to evaluate quantitatively the separate contributions of the mass shift and the field shift. Experimentally determined field shifts can be used to test theoretical models of nuclear structure, shape, and multipole moments. Experimentally determined specific mass shifts can be used to test detailed theories of atomic structure and relativistic effects. See NUCLEAR MOMENTS.

Experimental techniques that have greatly increased both the amount and the precision of isotope shift data that can be obtained include on-line isotope separators for the study of isotopes with half-lives as short as a few seconds and spectroscopic methods employing high-resolution tunable lasers. Active development of isotope separation schemes based on these laser techniques has been undertaken. Isotope shift data have also been obtained for x-ray transitions of electrons in inner atomic shells and of muons in muonic atoms. See LASER SPECTROSCOPY.

NUCLEAR MOMENTS
NOÉMIE KOLLER

Intrinsic properties of atomic nuclei; electric moments result from deviations of the nuclear charge distribution from spherical symmetry; magnetic moments are a consequence of the intrinsic spin and the rotational motion of nucleons within the nucleus. The classical definitions of the magnetic and electric multipole moments are written in general in terms of multipole expansions.

Parity conservation allows only even-rank electric moments and odd-rank magnetic moments to be nonzero. The most important terms are the magnetic dipole, given by Eq. (1), and the electric monopole, quadrupole, and hexadecapole, given by Eq. (2), for $l = 0, 2, 4$. Here m is

$$\vec{\mu} = \int \vec{M}(\vec{r}) \, dv \qquad (1) \qquad Q = \frac{1}{e} \int r^l Y_{lm}(\theta, \phi) \rho(\vec{r}) \, dv \qquad (2)$$

the projection of the orbital angular momentum l on a z axis appropriately chosen in space, $\vec{M}\vec{r}$ is the magnetization density of the nucleus and depends on the space coordinates $\vec{r}$, e is the electronic charge, $\rho\vec{r}$ is the charge density in the nucleus, and Y_{lm} are normalized spherical harmonics that depend on the angular coordinates θ and ϕ.

Quantum-mechanically only the z components of the effective operators have nonvanishing values. Magnetic dipole and electric quadrupole moments are usually expressed in terms of the nuclear spin I through Eqs. (3) and (4), where g, the nuclear gyromagnetic factor, is a measure of

$$\mu/\mu_N = gI \quad (3) \qquad Q(m_I) = \frac{[3m_I^2 - I(I+1)]Q}{I(2I-1)} \quad (4)$$

the coupling of nuclear spins and orbital angular momenta, $\mu_N = eh/4\pi M_p c = 5.0508 \times 10^{-24}$ erg/gauss = 5.0508×10^{-27} joule/tesla, is the nuclear magneton, M_p is the proton mass, h is Planck's constant, e is the electron charge, and c is the speed of light. $Q(m_I)$ is the effective quadrupole moment in the state m_I, and Q is the quadrupole moment for the state $m_I = I$. All angular momenta are expressed in units of $h/2\pi$. The magnitude of g varies between 0 and 1.8, and Q is of the order of 10^{-25} cm^2.

In special cases nuclear moments can be measured by direct methods involving the interaction of the nucleus with an external magnetic field or with an electric field gradient produced by the scattering of high-energy charged particles. In general, however, nuclear moments manifest themselves through the hyperfine interaction between the nuclear moments and the fields or field gradients produced by either the atomic electrons' currents and spins, or the molecular or crystalline electronic and lattice structures. *See* HYPERFINE STRUCTURE.

Effects of nuclear moments. In a free atom the magnetic hyperfine interaction between the nuclear spin $\vec{I}$ and the effective magnetic field $\vec{H}_e$ associated with electronic angular momentum $\vec{J}$ results in an energy $W = -\vec{\mu} \cdot \vec{H}_e = ha\vec{I} \cdot \vec{J}$, which appears as a splitting of the energy levels of the atom. The magnetic field at the nucleus due to atomic electrons can be as large as 10–100 teslas for a neutral atom. The constant a is of the order of 1000 MHz. *See* ATOMIC STRUCTURE AND SPECTRA.

The electric monopole moment is a measure of the nuclear charge and does not give rise to hyperfine interactions. The quadrupole moment Q reflects the deviation of the nuclear charge distribution from a spherical charge distribution. It is responsible for a quadrupole hyperfine interaction energy W_Q, which is proportional to the quadrupole moment Q and to the spatial derivative of the electric field at the nucleus due to the electronic charges, and is given by Eq. (5), where q is the average of expression (6). Here, r_i is the radius vector from the nucleus to the ith electron,

$$W_Q = \frac{e^2 Qq}{2I(2I-1)J(2J-1)} \cdot [3(\vec{I} \cdot \vec{J}) + 3/2\,(\vec{I} \cdot \vec{J}) - I(I+1)] \quad (5)$$

$$\sum_i (3\cos^2\theta - 1)/r_i^{-3} \quad (6)$$

and θ_i is the angle between r_i and the z axis. *See* MÖSSBAUER EFFECT.

In free molecules the hyperfine couplings are similar to those encountered in free atoms, but as the charge distributions and the spin coupling of valence electrons vary widely, depending on the nature of the molecular bonding, a greater diversity of magnetic dipole and quadrupole interactions is met. *See* MOLECULAR STRUCTURE AND SPECTRA.

In crystals the hyperfine interaction patterns become extremely complex, because the crystalline electric field is usually strong enough to compete with the spin orbit coupling of the electrons in the ion. Nevertheless, the energy-level structure can often be resolved by selective experiments at low temperatures on dilute concentrations of the ion of interest. *See* ELECTRON PARAMAGNETIC RESONANCE (EPR) SPECTROSCOPY.

Measurement. The hyperfine interactions affect the energy levels of either the nuclei or the atoms, molecules, or ions, and therefore can be observed either in nuclear parameters or in the atomic, molecular, or ionic structure. The many different techniques that have been developed to measure nuclear moments can be grossly grouped in three categories: the conventional techniques based mostly on spectroscopy of energy levels, the methods based on the detection of nuclear radiation from aligned excited nuclei, and techniques involving the interactions of fast ions with matter or of fast ions with laser beams.

Hyperfine structure of spectral lines. The hyperfine interaction causes a splitting of the electronic energy levels which is proportional to the magnitude of the nuclear moments, to the angular momenta I and J of the nuclei and their electronic environment, and to the magnetic field or electric field gradient at the nucleus. The magnitude of the splitting is determined by the nuclear moments, and the multiplicity of levels is given by the relevant angular momenta I or J

involved in the interaction. The energy levels are identified either by optical or microwave spectroscopy.

Optical spectroscopy (in the visible and ultraviolet) has the advantage of allowing the study of atomic excited states and of atoms in different states of ionization. Furthermore, optical spectra provide a direct measure of the monopole moments, which are manifested as shifts in the energy levels of atoms of different nuclear isotopes exhibiting different nuclear radii and charge distributions. Optical spectroscopy has a special advantage over other methods in that the intensity of the lines often yields the sign of the interaction constant a. SEE ISOTOPE SHIFT.

Microwave spectroscopy is a high-resolution technique involving the attenuation of a signal in a waveguide containing the absorber in the form of a low-pressure gas. The states are identified by the observation of electric dipole transitions of the order of 20,000 MHz. The levels are split by quadrupole interactions of the order of 100 MHz. Very precise quadrupole couplings are obtained, as well as vibrational and rotational constants of molecules, and nuclear spins. SEE MICROWAVE SPECTROSCOPY.

Atomic and molecular beams and nuclear resonance. Atomic and molecular beams passing through inhomogeneous magnetic fields are deflected by an amount depending on the nuclear moment. However, because of the small size of the nuclear moment, the observable effect is very small. The addition of a radio-frequency magnetic field at the frequency corresponding to the energy difference between hyperfine electronic states has vastly extended the scope of the technique. For nuclei in solids, liquids, or gases, the internal magnetic fields and gradients of the electric fields may be quenched if the pairing of electrons and the interaction between the nuclear magnetic moment and the external field dominate. The molecular beam apparatus is designed to detect the change in orientation of the nuclei, while the nuclear magnetic resonance system is designed to detect absorbed power (resonance absorption) or a signal induced at resonance in a pick-up coil around the sample (nuclear induction). The required frequencies for fields of about 0.5 T are of the order of 1–5 MHz. The principal calibration of the field is accomplished in relation to the resonant frequency for the proton whose g-factor is accurately known. Sensitivities of 1 part in 10^8 are possible under optimum experimental conditions. The constant a for ^{133}Cs has been measured to 1 part in 10^{10}, and this isotope is used as a time standard. SEE MOLECULAR BEAMS.

The existence of quadrupole interactions produces a broadening of the resonance line above the natural width and a definite structure determined by the value of the nuclear spin.

In some crystals the electric field gradient at the nucleus is large enough to split the energy levels without the need for an external field, and pure quadrupole resonance spectra are observed. This technique allows very accurate comparison of quadrupole moments of isotopes.

Atomic and molecular beams with radioactive nuclei. The conventional atomic and molecular beam investigations can be applied to radioactive nuclei if the beam current measurement is replaced by the much more sensitive detectors of radiations emitted in a radioactive decay. Moments of nuclei with half-lives down to the order of minutes have been determined.

Perturbed angular correlations. The angular distribution and the polarization of radiation emitted by nuclei depend on the angle between the nuclear spin axis and the direction of emissions. In radioactive sources in which the nuclei have been somewhat oriented by a nuclear reaction or by static or dynamic polarization techniques at low temperatures, the ensuing non-isotropic angular correlation of the decay radiation can be perturbed by the application of external magnetic fields, or by the hyperfine interaction between the nuclear moment and the electronic or crystalline fields acting at the nuclear site. Magnetic dipole and electric quadrupole moments of ground and excited nuclear states with half-lives as short as 10^{-9} have been measured.

Techniques involving interactions of fast ions. Techniques involving the interaction of intense light beams from tuned lasars with fast ion beams have extended the realm of resonance spectroscopy to the study of exotic species, such as nuclei far from stability, fission isomers, and ground-state nuclei with half-lives shorter than minutes. The hyperfine interactions in beams of fast ions traversing magnetic materials result from the coupling between the nuclear moments and unpaired polarized s-electrons, and are strong enough (H_e is of the order of 2×10^3 T) to extend the moment measurements to excited states with lifetimes as short as 10^{-12} s. Progress in atomic and nuclear technology has contributed to the production of hyperfine interactions of increasing strength, thus allowing for the observation of nuclear moments of nuclei and nuclear states of increasing rarity.

Bibliography. P. Averbuch (ed.), *Magnetic Resonance and Radio-frequency Spectroscopy*, 1969; W. E. Burcham, *Elements of Nuclear Physics*, 1986; H. Kopferman, *Nuclear Moments*, 1958; *Proceedings of the International Conference on Nuclear Moments and Nuclear Structure*, vol. 43, Physics Society, Japan, 1973; E. Segrè, *Nuclei and Particles*, 2d ed., 1977.

MOLECULAR STRUCTURE AND SPECTRA
ROBERT S. MULLIKEN AND U. FANO

Until the advent of quantum theory, ideas about the structure of molecules evolved gradually from analysis and interpretation of the facts of chemistry. Chemists developed the concept of molecules as built from atoms in definite proportions, and identified and constructed (synthesized) a great variety of molecules. Later, when the structure of atoms as built from nuclei and electrons began to be understood with the help of quantum theory, a beginning was made in seeing why atoms can combine in definite ways to form molecules; also, infrared spectra began to be used to obtain information about the dimensions and the nuclear motions (vibrations) in molecules. However, a fundamental understanding of chemical binding and molecular structure became possible only by application of the present form of quantum theory, called quantum mechanics. This theory makes it possible to obtain from the spectra of molecules a great deal of information about the nature of molecules in their normal as well as excited states, and about dissociation energies and other characteristics of molecules.

Molecular sizes. The size of a molecule varies approximately in proportion to the numbers and sizes of the atoms in the molecule. Simplest are diatomic molecules. These may be thought of as built of two spherical atoms of radii r and r', flattened where they are joined. The equilibrium value R_e of the distance R between their nuclei is then smaller than the sum of the atomic radii (**Fig. 1**). However, the nuclei of atoms in two different molecules cannot normally approach more closely than a distance $r + r'$; r and r' are called the van der Waals radii of the atoms. The smallest molecule is hydrogen (H_2), with two electrons whose negative charges equal the positive charges of the two nuclei. Here r is about 0.12 nanometer giving $r + r' = 0.24$ nm, but R_e is only 0.074 nm. In HCl $r = 0.12$ nm for H and 0.18 nm for Cl, but R_e is only 0.127 nm.

To describe a polyatomic molecule, one must specify not merely its size but also its shape or configuration. For example, carbon dioxide (CO_2) is a linear symmetrical molecule, the O—C—O angle being 180°. The H—O—H angle in the nonlinear water (H_2O) molecule is 105°. Many molecules which are essential for life contain thousands or even millions of atoms. Proteins are often coiled or twisted and cross-linked in curious ways which are important for their biological functioning.

Dipole moments. Most molecules have an electric dipole moment. In atoms, the electron cloud surrounds the nucleus so symmetrically that its electrical center coincides with the nucleus, giving zero dipole moment; in a molecule, however, these coincidences are disturbed, and a dipole moment usually results.

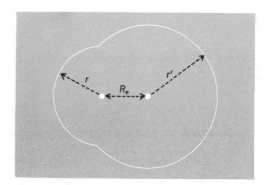

Fig. 1. Diatomic molecule with nuclei at distance R_e apart, built from atoms of radii r and r'.

Thus, when the atoms of HCl come together, there is some shifting of the H-atom electron toward the Cl. A complete shift would give H^+Cl^-, which would constitute an electric dipole of magnitude eR_e, where e is the electronic charge. But in fact the dipole moment is only 0.17 eR_e. This is because the actual electronic shift is only fractional.

Although in molecules such as H_2, N_2, and CO_2 partial shifts of electronic charge from the original atoms do take place, these necessarily occur so symmetrically that no dipole moment results. Many larger molecules also have zero dipole moments by virtue of high symmetry. Examples are methane (CH_4), uranium hexafluoride (UF_6), and benzene (C_6H_6).

In general, the dipole moment of a neutral molecule is defined as the vector sum of quantities $+Q\mathbf{S}$ for the nuclei and $-e\mathbf{s}$ for the electrons. Here Q is the charge on any nucleus and $\mathbf{S}$ its vector distance from any fixed point in the molecule; $\mathbf{s}$ is the average vector distance of any electron from the same point. To calculate a dipole moment with these definitions, quantum mechanics must be used.

However, a study of what is known experimentally about molecular dipole moments has led to useful semiempirical generalizations. Bond moments and group moments have been obtained for various types of chemical bonds and of chemical groups or radicals. By adding these vectorially, the actual dipole moment of a large molecule can often be reproduced fairly accurately. In CH_4 or CO_2, one can assume a moment for each C—H or C=O bond, even though these cancel out vectorially to give a zero resultant. In the linear molecule OCS, the unequal moments of the C=O and C=S bonds give a nonzero resultant. Because of the zero moment of CH_4, the CH bond moment and the CH_3 group moment must be equal and opposite. In CH_3Cl, the total moment can be thought of as the vector sum of the H_3C group moment and the C—Cl bond moment.

When molecules vibrate, their dipole moments usually vary. **Figure 2** shows how the dipole moment μ in a diatomic molecule may vary with R; the quantity previously discussed is μ_e, the value of μ at R_e. When μ_e is zero because of symmetry, it remains zero for symmetrical vibrations but, in polyatomic molecules, varies during unsymmetrical vibrations.

Molecules may possess magnetic as well as electric dipole moments. SEE MICROWAVE SPECTROSCOPY.

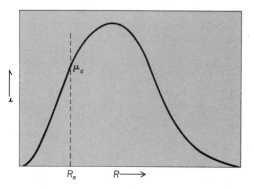

Fig. 2. Electric dipole moment μ of typical diatomic molecule as function of internuclear distance R; μ_e is the dipole moment at the equilibrium distance R_e.

Molecular polarizability. In the preceding consideration of dipole moments, the discussion has been in terms of atoms and molecules free from external forces. An atom field pulls the electrons of an atom or molecule toward it and pushes the nuclei away, or vice versa. This action creates a small induced dipole moment, whose magnitude per unit strength of the field is called the polarizability.

Molecular polarizabilities can be expressed as sums of atomic polarizabilities, plus corrections depending on the types of bonds present. Polarizabilities increase rather rapidly in such series as F, Cl, Br, I, and also from HF to HI, or F_2 to I_2.

Molecular polarizabilities can also be expressed approximately as sums of bond polarizabilities. These polarizabilities are anisotropic, being greater along bonds than perpendicular to bonds.

Molecular energy levels. The stationary states of motion of nuclei and electrons in a molecule, or of electrons in an atom, are restricted by quantum mechanics to special forms with definite energies. (Nonstationary states, which vary in the course of time, are constructed by mixing stationary states of different energies.) The state of lowest energy is called the ground state; all others are excited states. In analogy to water levels, the energies of the stationary states are called energy levels. Excited states exist only momentarily, following an electrical or other stimulus. *See* ENERGY LEVEL.

Energy levels are either discrete or continuous. The levels of a self-contained atom or molecule are restricted to special, sharply defined values (discrete levels). When an atom or molecule is ionized, that is, when one of its electrons has enough energy to escape completely, the energy can take on any value exceeding the minimum escape energy. This range of energies is called a continuous level or an ionization continuum. Molecules also have dissociation continua, which are discussed below.

Excitation of an atom consists of a change in the state of motion of its electrons. Electronic excitation of molecules can also occur, but alternatively or additionally, molecules can be excited to discrete states of vibration and rotation.

In a diatomic vibration, R varies periodically above and below R_e. The possible vibration energies E_v are given by Eq. (1), where $c\omega_e$ is just the small-amplitude vibration frequency, and h

$$E_v = hc\omega_e[(v + 1/2) - x_e(v + 1/2)^2] + \ldots \quad (1)$$

is Planck's constant (6.62×10^{-34} joule · s); x_e is a small quantity which is nearly always positive. The vibrational quantum number v can take whole number values 0, 1, 2, etc. The + . . . in Eq. (1) indicates small correction terms. The zero-point vibration energy $\frac{1}{2}hc\omega_e(1 - \frac{1}{2}x_e)$ present in the ground vibration state ($v = 0$) is a characteristic manifestation of quantum phenomena.

The value of $c\omega_e$ depends on the masses m_1 and m_2 of the atoms and the force constant k, as shown in Eq. (2). The frequency $c\omega_e$ (c = speed of light) is written in this manner for reasons

$$c\omega_e = \sqrt{k[(1/m_1) + (1/m_2)]} \quad (2)$$

of convenience in spectroscopic work, where the factor c is usually dropped.

The quantities R_e, k, and the dissociation energy D are the most important properties of a potential curve, which shows how the energy of attraction $U(R)$ of the atoms varies with R; k is d^2U/dR^2 taken at R_e. The $U(R)$ curve and vibrational levels for the ground electronic state of H_2 are shown in **Fig. 3**. Similar curves, but with other R_e, k, and D values, exist for other electronic states and other molecules. Molecules have also repulsive electronic states, whose $U(R)$ curves rise steadily with decreasing R. These are often important for spectroscopy and in atomic collisions. For stable (attractive) $U(R)$ curves, the vibrational levels decrease in spacing as v increases, until finally, as the spacing approaches zero, a maximum v is reached; in Fig. 3 this is 14. After a small gap, a dissociation continuum of energy levels then sets in. Here the atoms have enough mutual kinetic energy to fly apart. For repulsive states, there is only a dissociation continuum,

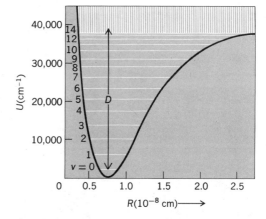

Fig. 3. $U(R)$ curve of ground electronic state of H_2 with vibrational levels and dissociation continuum. D indicates the dissociation energy. Maximum v here is 14. (*After G. Herzberg, Molecular Spectra and Molecular Structure, vol. 1, 2d. ed., Van Nostrand, 1950*)

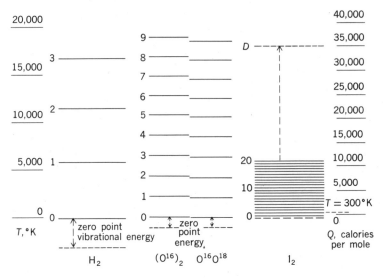

Fig. 4. Lowest vibrational levels of H_2, O_2, and I_2, numbered by vibrational quantum number v. Vibration level spacings decrease with increasing v. Where spacing reaches zero, the molecule dissociates; dissociation level D is indicated for I_2. Energies are given by the scale at right. The scale at left shows the average energy of vibration at various temperatures. °F = (K × 1.8) − 459.67.

with no vibrational levels. **Figure 4** illustrates how strongly vibration level spacings can vary: both k and $1/m$, and therefore $c\omega_e$, decrease from H_2 to O_2 to I_2. Figure 4 likewise illustrates the effect of mass in isotopic molecules.

The total energy of any molecule can be written as Eq. (3). Both the electronic energy E_{el}

$$E = E_{el} + E_v + (E_r + E_{fs} + E_{hfs} + E_{ext}) \tag{3}$$

and vibration energy E_v can be discrete or continuous. The quantities E_r, E_{fs}, and E_{hfs} denote rotational, fine-structure, and hyperfine-structure energies. The last two appear as small or minute splittings of the rotation levels. The spacings ΔE of adjacent discrete levels of each type are usually in the order given in notation (4).

$$\Delta E_{el} \gg \Delta E_v \gg \Delta E_r \gg \Delta E_{fs} \gg \Delta E_{hfs} \tag{4}$$

The fine structures of rotational levels differ strongly for different types of electronic states. The simplest diatomic electronic states are called $^1\Sigma$ states, and include $^1\Sigma^+$ and $^1\Sigma^-$ types for heteropolar and $^1\Sigma_g^+$, $^1\Sigma_u^+$, $^1\Sigma_g^-$, and $^1\Sigma_u^-$ for homopolar molecules. Most even-electron diatomic and linear polyatomic molecule ground states are $^1\Sigma^+$ states ($^1\Sigma_g^+$ if homopolar). The rotational levels of $^1\Sigma$ states have no fine structure; hyperfine structure, because of interaction with nuclear spins, is usually on too small a scale to detect by optical spectroscopy, to which the present article is limited. The E_{ext} term in Eq. (3) refers to additional fine structure which appears on subjecting molecules to external magnetic fields (Zeeman effect) or electric fields (Stark effect). SEE FINE STRUCTURE; HYPERFINE STRUCTURE; STARK EFFECT; ZEEMAN EFFECT.

The rotational levels of any $^1\Sigma$ state are given by Eq. (5). The quantity B_v is related to the

$$E_r = hcB_vJ(J + 1) + \ldots \tag{5}$$

moment of inertia I [$I = m_1m_2R^2/(m_1 + m_2)$], and to v, by Eq. (6). The rotational quantum number

$$B_v = (h/8\pi^2c)\,\overline{(1/I)}_v = B_e - \alpha_e(v + 1/2) + \ldots = B_0 - \alpha_e v + \ldots \tag{6}$$

J can have any whole number value from 0 up, and corresponds to an angular momentum $(h/2\pi)\sqrt{J(J + 1)}$. The averaging of $1/I$ in Eq. (6) normally results in a slow decrease of B with increasing v (α_e is usually a small positive quantity). The quantity B_e refers to a hypothetical nonvibrating molecule ($R = R_e$).

Figure 5 illustrates how enormously rotational level spacings can vary because of differences in m and R_e (both are much greater for I_2 than H_2). The effect of mass for isotopic molecules is illustrated for O_2. Comparison with Fig. 4 illustrates the relation $\Delta E_v \gg \Delta E_r$ mentioned earlier.

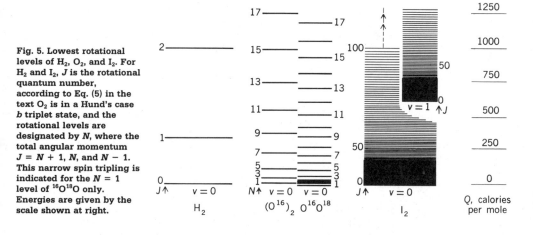

Fig. 5. Lowest rotational levels of H_2, O_2, and I_2. For H_2 and I_2, J is the rotational quantum number, according to Eq. (5) in the text O_2 is in a Hund's case b triplet state, and the rotational levels are designated by N, where the total angular momentum $J = N + 1$, N, and $N - 1$. This narrow spin tripling is indicated for the $N = 1$ level of $^{16}O^{18}O$ only. Energies are given by the scale shown at right.

Polyatomic molecules have much more complicated patterns of vibrational and (usually) of rotational energy levels than diatomic molecules. The number of normal modes (independent forms) of vibration for a molecule with n atoms is $3n - 6$ for nonlinear molecules, and $3n - 5$ for linear molecules. Each normal mode is a cooperative vibration of some or all the atoms moving with the same frequency, characteristic of the mode. Sometimes two or even three modes are so related in form that their frequencies are identical. These are called degenerate vibrations.

Figure 6 depicts the normal modes of H_2O and CO_2. They are labeled by symbols which also denote their frequencies. The arrows indicate the directions of motion of the atoms during one phase of vibration. The CO_2 frequency v_2 is twofold degenerate: there are two independent modes corresponding to motion in either of two planes at right angles. The other two CO_2 modes, and all three H_2O modes, are nondegenerate.

Molecular spectra. The frequencies cv of electromagnetic spectra obey the Einstein-Bohr equation (7). The quantities v, in waves per centimeter, or wave numbers (cm^{-1}), will

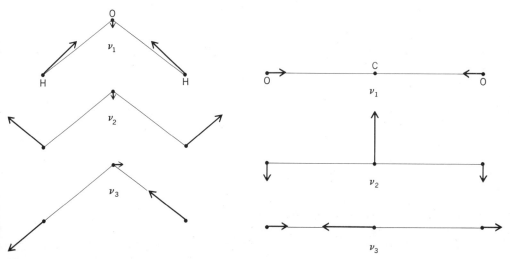

Fig. 6. Normal vibration modes of H_2O and CO_2. Synchronized displacements of atoms occur in proportion to lengths of the arrows. Diagram corresponds to snapshot taken at one phase of vibration.

$$hc\nu = E' - E'' \tag{7}$$

hereafter be called frequencies, as is usual in spectroscopy, although properly only the $c\nu$ are frequencies. Molecular emission spectra accompany jumps in energy from higher to lower levels; absorption spectra accompany transitions from lower to higher levels. Both E' and E'' can be either discrete or continuous levels. If both are discrete, they give a spectrum of discrete frequencies; otherwise, they give a continuous spectrum. Discrete spectra are the main type considered here. Discrete frequencies are usually called spectrum lines because of their appearance when recorded by an optical spectrograph. The wealth and precision of spectroscopic observations have been increased by orders of magnitude by the advent of laser sources and techniques. *See Laser Spectroscopy.*

Besides its frequency, the intensity and width of a spectrum line are important. Intensities vary over wide ranges. In the extreme case of nearly zero intensity for a spectroscopic transition, the transition is called forbidden. Only a small minority of all pairs of levels yield allowed transitions. These are governed by selection rules derivable from quantum theory.

Under disturbing influences, however, some lines are seen, weakly, which violate these rules. Further, the usual selection rules are electric dipole rules, and additional transitions become very weakly allowed if magnetic dipole, electric quadrupole, and other selection rules are also considered. The following discussion is confined to spectra which obey the electric dipole rules.

Molecular spectra can be classified as fine-structure or low-frequency spectra, rotation spectra, vibration-rotation spectra, and electronic spectra. Low-frequency spectra are discussed elsewhere. *See Electron paramagnetic resonance (EPR) spectroscopy; Magnetic resonance; Microwave spectroscopy; Molecular beams.*

Pure rotation spectra. Transitions between energy levels differing only in rotational state give rise to pure rotation spectra. For diatomic molecules in $^1\Sigma$ states, Eq. (5), the relation is given by Eq. (8). The transitions obey the selection rule $\Delta J = 1$ (ΔJ means $J' - J''$). Putting $J' = J'' + 1$, Eq. (9) is obtained. Equation (9) represents a sequence of lines spaced almost equidistantly

$$hc\nu = E'_r - E''_r = hcB_v[J'(J' + 1) - J''(J'' + 1)] + \ldots \tag{8}$$

$$\nu = 2B_v(J'' + 1) + \ldots \tag{9}$$

($2B_v$, $4B_v$, $6B_v$, ...), and lying in the far infrared or (for small B or low J'') the microwave region. Their intensities are proportional to μ_e^2, where μ_e is the electric moment at R_e (Fig. 2); hence homopolar molecules (H_2, N_2, and so on) show no pure rotation spectra. The intensities are proportional also to the lower-state (v'', J'') level population and to ν (for absorption) or ν^4 (for emission).

Pure rotation spectra of linear polyatomic molecules are like those of diatomic molecules. Polyatomic molecules having $\mu_e = 0$, whatever their shape (examples are CO_2, CH_4, C_6H_6), have no pure rotation spectra. In other cases, the spectra can be obtained using $hc\nu = E'_r - E''_r$ with appropriate E_r expressions and selection rules.

Vibration-rotation bands. Spectra involving only vibrational and rotational state changes lie mainly in the infrared. For a $^1\Sigma$ diatomic state, using Eqs. (1), (5), and (7), Eq. (10) is obtained, with ν_0 defined in Eq. (11). In Eq. (10) B' and B'' mean B_v for v' and v'', respectively. Each band

$$\nu = \nu_0(v',v'') + [B'J'(J' + 1) - B''J''(J'' + 1)] + \ldots \tag{10}$$

$$\nu_0 = \omega_e(1 - x_e)(v' - v'') - x_e\omega_e(v'^2 - v''^2) \tag{11}$$

consists of two sets of rotational lines, one on each side of its ν_0. Each line corresponds to a particular rotational transition conforming to $\Delta J = \pm 1$. The two series (branches) have frequencies defined in Eq. (12) for R or positive branch ($J' = J'' + 1$), and in Eq. (13), for P or negative branch ($J' = J'' - 1$). Both can be represented by a single equation (14) by letting $M = J'$

$$\nu = \nu_0 + 2B''(J'' + 1) + (B' - B'')(J'' + 1)(J'' + 2) \tag{12}$$

$$\nu = \nu_0 - 2B''J + (B' - B'')J''(J'' - 1) \tag{13}$$

$$\nu = \nu_0 + (B' + B'')M + (B' - B'')M^2 + \ldots \tag{14}$$

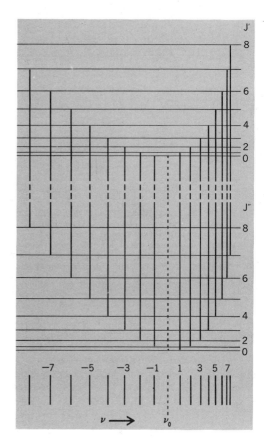

Fig. 7. Relation of band lines (lower part) [see Eqs. (8) and (9)] to rotational levels [see Eq. (6)] for a vibration-rotation band or an electronic band. In the former case, the upper and lower sets of rotational levels belong to two vibrational levels of a $^1\Sigma$ electronic state. In the latter case, they belong to two different $^1\Sigma$ states. Positive M values, R branch; negative M values, P branch.

$+ 1$ for the R and $M = -J''$ for the P branch. Neglecting the term in M^2, Eq. (14) represents a series of equidistant lines with one missing ($M = 0$) at ν_0. **Figure 7** shows how the line positions are related to the upper (v') and lower (v'') sets of rotational levels.

Since $B' - B''$ is a small negative quantity [see Eq. (6), noting that $v' > v''$], the M^2 term makes the P line spacing increase and the R line spacing decrease slowly as M increases. This is shown, exaggerated, in Fig. 7. At some large M value, the R branch turns back on itself, but usually the lines have become weak before this value is reached.

The relative intensities of band lines depend primarily on the initial rotational distribution of molecules. More precisely, Eq. (15) holds. Here B_{in}, J_{in}, and n in ν_n are B', J', and 4, respectively,

$$\text{Intensity} = C(v',v'')\nu_n(J' + J'' + 1)e^{-B_{in}J_{in}(J_{in}+1)hc/kT} \qquad (15)$$

for an emission band, and B'', J'', and 1, respectively, for an absorption band. **Figure 8** shows diagrammatically how the values of B and T affect the appearance of a typical absorption band ($B' = B''$ has been assumed for simplicity in Fig. 8). **Figure 9** shows the appearance of an actual HCl band. The weaker HCl^{37} lines are at slightly lower frequencies than the HCl^{35} lines, mainly because ω_e is smaller [see Eqs. (2) and (11)].

The factor $C(v',v'')$ is largest by far for fundamental bands ($\Delta v = 1$), and falls rapidly with increasing Δv in the overtone bands or harmonics (Δv is $v' - v''$). For fundamental bands, C depends on the slope of the $\mu(R)$ curve (Fig. 2), being approximately proportional to $(d\mu/dR)^2$ taken at R_e. For overtone bands, C depends on the detailed shapes of both the $\mu(R)$ and $U(R)$ curves. Fundamental or overtone bands arising from $v'' > 0$ are called hot bands.

Vibration-rotation absorption bands of liquids and solutions are widely used in chemical analysis. Here the rotational structure is blurred out, and only an "envelope" is seen. For many

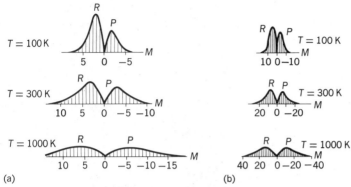

Fig. 8. Intensity distribution at several temperatures for a diatomic absorption band. Line positions are based on Eq. (9) assuming $B' = B''$ for simplicity; frequency increases toward the left (opposite to Fig. 7). (a) and (b) correspond respectively to B values of HCl ($B = 10.44$ cm^{-1}) and of 2 cm^{-1} (approximately the value for CO, for which $B = 1.93$ cm^{-1}). 100 K = $-280°$F; 300 K = 80°F; 1000 K = 1340°F. (*After G. Herzberg, Molecular Spectra and Molecular Structure, vol. 1, 2d ed., Van Nostrand, 1950*)

purposes, it is sufficient to know empirically the spectrum of each molecule which may be present. Also, groups of atoms which recur in many molecules often have nearly constant frequencies, of use for identification and in determining molecular structure. *See Infrared spectroscopy.*

Electronic band spectra. These are the most general type of molecular spectra. The characteristic feature is a change of electronic state. From Eqs. (3) and (7), neglecting fine structure, Eqs. (16) and (17) are obtained. Diatomic electronic spectra are often observed in emission,

$$\nu = \frac{(E'_{el} - E''_{el}) + (E'_v - E''_v) + (E'_r - E''_r)}{hc} \quad (16) \qquad \nu = \nu_{el} + \nu_v + \nu_r = \nu_0 + \nu_r \quad (17)$$

while the electronic spectra of polyatomic molecules are usually absorption spectra. Depending mainly on the magnitude of ν_{el}, electronic spectra occur in the infrared, visible, ultraviolet, or vacuum ultraviolet.

For any one electronic transition, the spectrum consists typically of many bands. These lie in general at frequencies both above and below ν_{el}, since ν_v can be positive or negative. They constitute a band system. Each band consists of numerous rotational lines arranged in two or more branches and lying on both sides of a central position ν_0.

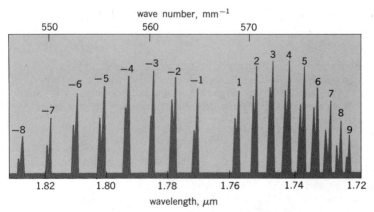

Fig. 9. First harmonic (2,0) vibration-rotation band of HCl in absorption. R branch to right, P branch to left, showing intensity distribution. The stronger lines are ^{35}HCl; the weaker companions, at lower frequencies, are ^{37}HCl. (*After C. F. Meyer and A. A. Levin, Phys. Rev., 34:44, 1929*)

For diatomic molecules, ν_0 depends on a single v' and v'' and, using Eq. (1) for each electronic state, is given by Eq. (18). Since ω_e and $x_e\omega_e$ are now different (often strongly) in the upper and lower states, $\nu_0(v',v'')$ cannot be reduced to as simple an expression as the corresponding Eq. (11) for vibration-rotation bands. Eq. (18) is more convenient when rewritten as Eq. (19), where Eqs. (20) apply. The relative intensities of the bands depend on (1) the initial distribution of mol-

$$\nu_0(v',v'') = \nu_{el} + [\omega'_e(v' + 1/2) - x'_e\omega'_e(v' + 1/2)^2 + \ldots]$$
$$- [\omega''_e(v'' + 1/2) - x''_e\omega''_e(v'' + 1/2)^2 + \ldots] \quad (18)$$

$$\nu_0(v',v'') = \nu_{00} + (\omega'_0 v' - a'v'^2) - (\omega''_0 v'' - a''v''^2) + \ldots \quad (19)$$

$$\nu_{00} = \nu_{el} + \frac{1}{2}(\omega'_e - \omega''_e) - \frac{1}{4}(x'_e\omega'_e - x''_e\omega''_e) \qquad \omega'_0 = \omega'_e(1 - x'_e) \qquad a' = x'_e\omega'_e, \text{ etc.} \quad (20)$$

ecules among vibrational levels, and (2) the relative transition probabilities from any initial to various final vibrational levels.

The simplest example is the absorption spectrum of a cool gas of low molecular weight, for which all molecules initially have $v'' = 0$. The spectrum then consists of one "v' progression," a single series of bands with various values of v'; the frequencies are given by $\nu = \nu_{00} + \omega_0 v' - a'v'^2$. For a hot or a heavy gas, additional weaker v' progressions with $v'' > 0$ also appear.

In emission spectra, the initial population usually ranges over a number of v' values, from each of which transitions occur to a number of v'' values, so that the system contains many bands on both sides of ν_{00}. In the special case of fluorescence spectra, the molecule is excited to various v' values by absorbing light; it then emits light belonging to the same (or sometimes another) electronic transition. From each v', it can descend not only to the original v'' but also to various other, mainly larger, values. Hence fluorescence bands lie mainly at lower frequencies than the absorption bands used to excite them. *See* FLUORESCENCE.

Relative transition probabilities are governed by the Franck-Condon principle. This takes note of the very great rapidity of electronic motions as compared with those of the far more massive nuclei, and concludes that during the extremely brief time for an electronic transition, the nuclei tend to remain unchanged in their positions and momenta. It is applicable to both polyatomic and diatomic spectra. Consider first a diatomic molecule starting from the $v'' = 0$ level of a ground state $U(R)$ curve like the lower curves in **Fig. 10**. A vertical line drawn from the bottom point A ($v'' = 0$ if zero-point vibration is neglected) to point B on any one of the upper curves corresponds to an electronic absorption transition in which the nuclei have not moved.

In the case of Fig. 10a, point B corresponds to $v' = 0$, and the conclusion is that this is the most probable v' for $v'' = 0$. In the case of Fig. 10b, point B corresponds to an excited molecule at the inner turning point of a vibration with a v' of possibly 3 or 4, in a typical case. One then concludes (with J. Franck) that the strongest absorption bands for $v'' = 0$ have $v' = 3.0$ and 4.0. To obtain more exact information, a quantum-mechanical calculation (first carried out by E. U. Condon) is necessary.

In the case of Fig. 10c, point B corresponds to an energy level in the dissociation contin-

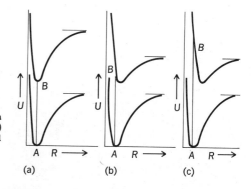

Fig. 10. Diatomic $U(R)$ curves for three cases to explain the vibrational intensity distribution according to the Franck-Condon principle. The asymptote of each curve for large R corresponds to dissociation into atoms, with one or both atoms excited in the case of the upper curves. Starting in each case from the bottom of the lower curve (essentially $v'' = 0$), the most probable transition in absorption is (a) to $v' = 0$, (b) to $v' = 3$ or 4, and (c) to the dissociation continuum, as shown by vertical lines. (*After G. Herzberg, Molecular Spectra and Molecular Structure, vol. 1, 2d ed., Van Nostrand, 1950*)

uum above the asymptote of the upper $U(R)$ curve. According to the Franck-Condon principle, the absorption spectrum will have maximum intensity in a continuous range of frequencies, with $hc\nu$ about equal to the energy difference AB. The quantum-mechanical calculation shows that the actual spectrum will extend with appreciable intensity over a range of both higher and lower frequencies than this, including, on the lower-frequency side, a number of high-v' bands. Actual examples of such spectra (a long v' progression followed by a strong continuum) are the far-ultraviolet Schumann-Runge bands of oxygen and the visible bands of iodine. By measuring the frequency at which the continuum begins, one obtains an exact value of the dissociation energy of these molecules. In so doing, any excitation energy of the atomic dissociation products to which the upper $U(R)$ curve leads is subtracted.

The Franck-Condon method is useful in understanding intensity distributions and structure in emission as well as absorption band systems. For diatomic spectra, various patterns of intensity as functions of v' and v'' occur, depending largely on the R_e values of the two $U(R)$ curves and, of course, also on the initial distribution among v' levels. Sometimes the upper-state $U(R)$ curve is stable (has a minimum) but the lower state is repulsive. Continuous emission spectra then occur, with the atoms flying apart on reaching the lower state. The H_2 molecule shows such a spectrum, as do rare gas molecules such as He_2 and Kr_2, which are stable only in excited or ionized states.

Molecular electronic states. Before discussing the structures of electronic bands, one must consider the nature of molecular electronic states. Each electronic state has orbital and spin characteristics. The spin quantum number S has a whole-number value if the number of electrons is even, a half-integral value if it is odd. Electronic states with $S = 0$ are called singlet states, all others multiplet states. The orbital characteristics differ sharply for linear (including diatomic) and nonlinear molecules.

For linear molecules only, there is a quantum number Λ such that $\pm \Lambda h/2\pi$ is the component of angular momentum around the line of nuclear centers. Linear-molecule electronic states can be discussed under three headings: (1) singlet states; (2) multiplet states with strong spin coupling (Hund's case a); and (3) multiplet states with weak spin coupling (Hund's case b). Strictly speaking, actual multiplet states are intermediate between cases a and b, or between these and certain other cases called c and d. The discussion to follow is largely restricted to singlet electronic states.

Singlet states with $\Lambda = 0$ include $^1\Sigma^+$ and $^1\Sigma^-$ states: states with $\Lambda = 1, 2, \ldots$ are called $^1\Pi$, $^1\Delta$, and so on. In linear molecules with a center of symmetry (H_2, CO_2 and so on), one must further distinguish even and odd (g and u) states: $^1\Sigma^+_g$, $^1\Sigma^+_u$, $^1\Sigma^-_g$, $^1\Sigma^-_u$, $^1\Pi_u$, $^1\Pi_g$, $^1\Delta_g$, $^1\Delta_u$, etc. The rotational levels of singlet states obey the symmetric rotor formula, Eq. (21). Here J is

$$E_r = hc[BJ(J + 1) - \Lambda^2] + \ldots \qquad (21)$$

restricted to integral values equal to or greater than Λ.

For $\Lambda > 0$, each rotational level is a narrow doublet (Λ-doubling). Corresponding fine structure [see E_{fs} in Eq. (3)] can usually be detected in electronic bands, but (for ground states only) it can be much more accurately studied in low-frequency spectra. Hyperfine structure [see E_{hfs} in Eq. (3)] is usually on too small a scale to be detected in electronic band lines, but has been found in a few cases. Hyperfine structure is best studied in low-frequency spectra.

Electronic band structures. The simplest electronic bands occur for transitions between singlet electronic states. The possible types of electronic transitions are limited by the selection rule $\Delta\Lambda = 0, \pm 1$. The structures of $^1\Sigma - ^1\Sigma$ bands are essentially the same as for the $^1\Sigma$-state vibration-rotation bands described earlier. Equations (12) to (15) and Fig. 7, also Fig. 8, for the intensities in absorption are still applicable if Eq. (18) instead of Eq. (11) is used for ν_0, and it is recognized that B' and B'' now belong to two different electronic states.

The quantity $B' - B''$ in Eq. (14), instead of always being a small negative quantity, may now be either positive or negative, and $(B' - B'')/(B' + B'')$ is often fairly large (although it can also be nearly zero). As a result, it is usual in electronic bands to find so-called heads. A head is a position of maximum or minimum frequency; by using Eq. (14) to obtain $d\nu/dM = 0$, one finds $M_{head} = (B' + B'')/2(B' - B'')$. Then, on inserting M_{head} into Eq. (14), one obtains $\nu_{head} = \nu_0 - (B' + B'')^2/4(B' - B'')$. [Since $(B' + B'')/2(B' - B'')$ is not usually a whole number, the actual M_{head} is the nearest whole-number M to that calculated.] According to whether $B' - B''$ is negative or positive, the positive (R) or the negative (P) branch forms the head. Figure 7, if continued to somewhat larger M values, illustrates the formation of an R-branch head at a calculated M of 10.5; the actual head is formed by the two coincident lines $M = 10$ and 11.

Although homopolar molecules (H_2, N_2, and so on) have no pure rotation or vibration-rotation spectra, they do have electronic spectra. For homonuclear homopolar molecules, the band lines show alternating intensities. The lines in each branch are alternately stronger and weaker as M increases, this effect being superposed on the otherwise smoothly varying intensity distribution. The alternation ratio depends on the nuclear spin I and has been, in several cases, the means of determining I. When $I = 0$, alternate lines are completely missing. Heteronuclear molecules, even if homopolar (for example, HD or $^{16}O^{18}O$) do not show alternating intensities.

Polyatomic electronic spectra. These differ from diatomic electronic spectra because several initial and final vibration quantum numbers are involved, and because the rotational structure (except for linear molecules) is usually much more complicated. However, the detailed structures of the electronic spectra of a number of simple molecules and radicals in the vapor state in emission and in absorption have been studied. Nevertheless, for the most part, the spectra of polyatomic molecules are examined as absorption spectra in solution. The rotational structure is then completely blurred out, but the vibrational structure can be seen.

The Franck-Condon principle is here a useful guide. One of its corollaries, which amounts almost to a selection rule, is that only totally symmetric vibrations (vibrations during which the equilibrium symmetry of the molecule is preserved) undergo quantum number changes. This greatly simplifies the vibrational structure, especially of absorption spectra where most molecules are initially mainly in the $v''=0$ state of all vibrations. One finds then mostly v' progressions of one or a very few totally symmetric vibrations, and combinations of these.

Rather often, polyatomic band systems do not even show obvious vibrational structure. This can happen for any of several reasons: The upper state may involve dissociation; in CH_3I, for example, the first ultraviolet absorption region yields $CH_3 + I$; there may be so many low-frequency, upper-state vibrations that the spectrum looks continuous; or there may be a combination of these and other reasons. Such continuous or pseudocontinuous band systems are often loosely referred to as bands. For complicated molecules, the spectra of several different electronic transitions often overlap strongly so that it is difficult even to separate one system from another. SEE ATOMIC STRUCTURE AND SPECTRA; NEUTRON SPECTROMETRY; RAMAN EFFECT.

Bibliography. C. N. Banwell, *Fundamentals of Molecular Spectroscopy*, 3d ed., 1983; E. F. Brittain et al., *Introduction to Molecular Spectroscopy: Theory and Experiment*, 1978; P. R. Bunker, *Molecular Symmetry and Spectroscopy*, 1979; P. A. Gorry, *Basic Molecular Spectroscopy*, 1986; W. H. Flygare, *Molecular Structure and Dynamics*, 1978; I. N. Levine, *Molecular Spectroscopy*, 1975; W. G. Richards and P. R. Scott, *Structure and Spectra of Molecules*, 1985; J. Steinfeld, *Molecules and Radiation: An Introduction to Modern Molecular Spectroscopy*, 2d ed., 1985.

BAND SPECTRUM
W. F. MEGGERS AND W. W. WATSON

A spectrum consisting of groups or bands of closely spaced lines. Band spectra are characteristic of molecular gases or chemical compounds. When the light emitted or absorbed by molecules is viewed through a spectroscope with small dispersion, the spectrum appears to consist of very wide asymmetrical lines called bands. These bands usually have a maximum intensity near one edge, called a band head, and a gradually decreasing intensity on the other side. In some band systems the intensity shading is toward shorter waves, in others toward longer waves. Each band system consists of a series of nearly equally spaced bands called progressions; corresponding bands of different progressions form groups called sequences.

Six spectra of diatomic molecular fragments are shown in the **illustration**. The spectrum of a discharge tube containing air at low pressure is shown in *a*. It has four band systems: the γ-bands of nitrogen oxide (NO, 230–270 nanometers), negative nitrogen bands (N_2^+, 290–350 nm), second-positive nitrogen bands (N_2, 290–500 nm), and first-positive nitrogen bands (N_2, 550–700 nm). The spectrum of high-frequency discharge in lead fluoride vapor in *b* has bands in prominent sequences. The spectrum in *c* shows part of one band system of SbF, and was obtained by

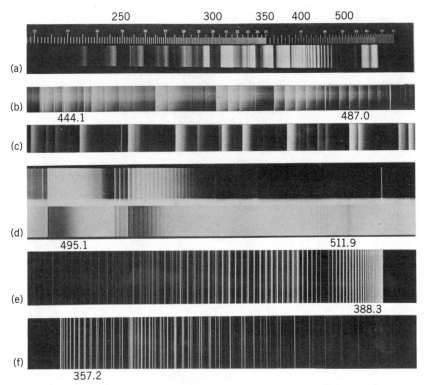

Photographs of band spectra of (a) a discharge tube containing air at low pressure; (b) high-frequency discharge in lead fluoride vapor; (c) SbF (b and c taken with large quartz spectrograph, after Rochester); (d) BaF emission and absorption; (e) CN; and (f) NO. The measurements are in nanometers. (*From F. A. Jenkins and H. E. White, Fundamentals of Optics, 3d ed., McGraw-Hill, 1957*)

vaporizing SbF into active nitrogen. Emission from a carbon arc cored with barium fluoride (BaF$_2$) and absorption of BaF vapor in an evacuated steel furnace are illustrated in d. These spectra were obtained in the second order of a diffraction grating, as were the spectra in e and f. The photograph e is that of the CN band at 388.3 nm from an argon discharge tube containing carbon and nitrogen impurities, and f is a band in ultraviolet spectrum of NO, obtained from glowing active nitrogen containing a small amount of oxygen.

When spectroscopes with adequate dispersion and resolving power are used, it is seen that most of the bands obtained from gaseous molecules actually consist of a very large number of lines whose spacing and relative intensities, if unresolved, explain the appearance of bands of continua (parts e and f of the figure). For the quantum-mechanical explanations of the details of band spectra SEE MOLECULAR STRUCTURE AND SPECTRA.

CRYSTAL ABSORPTION SPECTRA
DAVID ASPNES

The wavelength or energy dependence of the attenuation of electromagnetic radiation as it passes through a crystal, due to its conversion to an equivalent amount of energy in the crystal.

When atoms are grouped into a regular array to form a crystal, their interaction with electromagnetic radiation is greatly modified. Absorption spectra of free atoms consist of a series of sharp lines at well-defined energies, owing to excitation of electrons between the discrete energy levels of the atoms. In a crystal, the sharp energy levels interact and are broadened into energy bands, and the cores of the atoms vibrate about their equilibrium positions. The ability of electromagnetic radiation to interact with these and other energy-storing processes of a crystal leads to a broad absorption spectrum that bears little resemblance to the spectra of the free parent atoms. *See* ATOMIC STRUCTURE AND SPECTRA.

The absorption spectrum of a crystal includes a number of distinct features, as indicated in the **illustration**. These absorption types are called lattice, intrinsic, extrinsic, or free-carrier, according to the physical process by which energy is extracted from the electromagnetic radiation. The illustration shows the absorption spectrum of a typical semiconducting crystal, gallium arsenide (GaAs). Absorption spectra of insulating crystals are similar, except that free-carrier absorption is negligible, and the absorption edge moves to higher energies. For metal crystals, the free-carrier contribution dominates, and the intrinsic absorption processes are much less pronounced.

General properties. Absorption is measured in terms of the absorption coefficient, defined as the rate of attenuation of radiation per unit length, and it is expressed typically in units of cm^{-1}. This quantity varies from less than $3 \times 10^{-5} cm^{-1}$ in the near infrared, for the exceedingly highly refined glasses used in fiber optics applications, to greater than $10^6\ cm^{-1}$, for certain intrinsic absorption processes in crystals. Absorption is related to reflection through a more fundamental quantity called the dielectric function. The absorption coefficient is also related to the index of refraction by means of a frequency integral known as the Kramers-Kronig transform.

Lattice absorption. Lattice absorption arises from the excitation of lattice vibrations (phonons) and is the equivalent of vibrational absorption in molecules. It occurs in the infrared and is responsible for most structure in the absorption spectra of crystals at energies below about 0.1 eV (wavelengths above about 10 nanometers). In a diatomic, partially ionic crystal such as GaAs, an electric field that forces the Ga cores one way simultaneously forces the As cores the other way. Near the frequency corresponding to a wavelength of 37 nm in GaAs, a resonance occurs which results in the generation of transverse optical phonons and a large increase in absorption. This is seen as the reststrahl peak in the spectrum of illus. *b*. Since the lattice restoring forces are anharmonic, several phonons can be created at the same time at the higher multiples

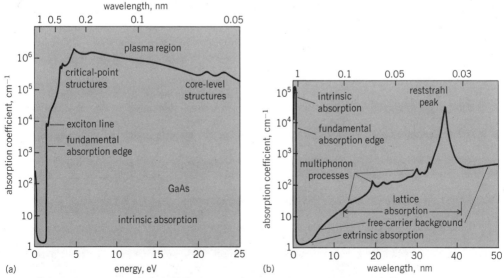

Absorption spectrum of GaAs. The region in which each type of absorption process is important is indicated. (*a*) Spectrum at higher energies. (*b*) Spectrum at lower energies.

of the reststrahl frequency. These multiphonon processes are responsible for the structures shown from 10 to 30 nm in the illustration. SEE MOLECULAR STRUCTURE AND SPECTRA.

Free-carrier absorption. If unbound charges are available to carry a current, the electromagnetic field causes motion of the charges directly, with a net energy transfer from the field to the medium as a result of charge-charge and charge-lattice collisions. Unbound charges are free carriers such as conduction (non-bonding-state) electrons and valence (bonding-state) vacancies or holes. The decreasing efficiency of energy dissipation by collision processes at high frequencies causes free-carrier absorption to fall off roughly as $1/E^2$, where $E = \hbar\omega$ is the energy of the electromagnetic radiation ($\hbar$ is Planck's constant divided by 2π, and ω is the frequency of radiation). For crystals with free-carrier concentrations less than about 10^{17} cm^{-3}, free-carrier absorption is significant only in the infrared. Since metallic free-carrier concentrations are much larger, of the order of 10^{23} cm^{-3}, free-carrier concentrations dominate the absorption properties of most metals throughout the visible and near-ultraviolet, producing their lustrous neutral gray appearance. But in a few spectacular exceptions, such as copper and gold, both free-carrier and intrinsic absorption processes are of comparable importance, and the absorption spectra in the visible are modified significantly.

Intrinsic absorption. Electronic transitions between the bonding (filled-valence) energy bands and the antibonding (empty-conduction) energy bands produce the crystal equivalent of the line-absorption spectra of free atoms. These intrinsic absorption processes dominate the optical behavior of semiconductors and insulators in the visible and ultraviolet spectral regions. The variation of electron energy with electron momentum within the energy banks gives rise to structure in the absorption spectra of crystals in the visible and near ultraviolet. The most readily apparent intrinsic absorption feature is the fundamental absorption edge, which marks the boundary between the range of transparency at lower energies and the strong absorption that occurs at higher energies. The energy of the fundamental absorption edge is determined by the forbidden gap, the energy difference between the highest valence and lowest conduction-bank state of the crystal. The absorption edge or forbidden gap may be nonexistent (in metals such as lead, white tin; in semimetals such as antimony, bismuth), zero (in semiconductors such as gray tin, one of the mercury-cadmium-tellurium alloys), fall in the infrared [in semiconductors such as germanium, silicon, gallium arsenide (GaAs)], or occur in the visible [in semiconductors such as gallium phosphide (GaP), cadmium sulfide (CdS)] or well into the ultraviolet [in insulators such as carbon, diamond; sodium chloride (NaCl); silicon dioxide (SiO$_2$); lithium fluoride (LiF)].

Intrinsic absorption spectra are of two types, indirect or direct, according to whether or not a phonon participates in the absorption process. Indirect absorption processes, which are much less probable events, give rise to absorption coefficients typically 100–1000 times less than those for direct transitions. They are important in certain crystals (germanium, silicon, gallium phosphide) in which the highest valence and lowest conduction states have different values for the electron momentum, and in which an absorption process requires a phonon for momentum conservation. In other crystals (GaAs, ZnO) the highest valence and lowest conduction states occur at the same electron momentum, and the fundamental edge is direct.

For either direct or indirect transitions, the excited electron is attracted by the Coulomb interaction to the vacancy, or hole, that was created in the valence band by its excitation. There is a strong tendency for the electron and hole to bind together to form a hydrogenlike state called an exciton. Excitons greatly affect the shape of absorption spectra near the fundamental edge. The positive binding energy means that the energy needed to create an exciton is somewhat less than that of the forbidden gap. Creating excitons makes it possible to absorb radiation at energies less than that of the forbidden gap. The lowest possible exciton absorption process results in a single line seen at the absorption edges of most semiconductors and insulators. In addition, excitonic effects cause a strong enhancement of the absorption process immediately above the fundamental edge.

Absorption structures above the fundamental edge are due to critical points, which—like the fundamental edge itself—represent energies at which new valence-conduction band-pair transitions become possible (an M_0 critical point at which the electron energy reaches a relative minimum as a function of electron momentum) or where formerly available band-pair transitions become no longer possible (an M_3 critical point at which electron energy reaches a relative maximum). Such singularities also occur at saddle points, where a relative maximum of electron energy is reached in one direction of momentum, and relative minima in the other two (an M_1 critical point), or where a relative maxima are reached in two directions, and a relative minimum in the remainder (an M_2 critical point).

Above approximately 10 eV, the conduction-band states become more free-electron-like, and structure in absorption spectra dies out. This is the plasma region, as indicated in the illustration. Absorption structures beyond this region, in the far-ultraviolet and x-ray spectral regions, originate from core valence bands. Since the crystal potential is a minor perturbation on deep core levels, and the deep core levels contribute negligibly to bonding, they are very narrow in energy and can be considered more properly as atomic levels. Consequently, the source of any structure in core-level absorption spectra is primarily conduction bands. Core-level absorption spectra begin in the 20–25 eV region for Ga-V compounds [gallium phosphide, gallium arsenide, gallium selenide (GaSe)], near 100 eV for silicon, and near 340 eV for carbon.

Extrinsic absorption. Extrinsic absorption processes involve states associated with deviations from crystal perfection, such as vacancies, interstitials, and impurities. Since these states occur in low (approximately 10^{14}–10^{18} cm^{-3}) concentrations relative to the host atoms (approximately 10^{23} cm^{-3}), extrinsic absorption is weak relative to intrinsic processes and is typically important only where the crystal is otherwise transparent in the forbidden gap. Nevertheless, the ability to obtain macroscopic lengths of crystals (on the order of a centimeter) can make extrinsic absorption readily apparent, as, for example, in the color centers of alkali halides, or in the poor optical quality of industrial diamonds. Although extrinsic processes are not so important in absorption, they are vitally important in luminescence, where the presence or absence of radiation may depend entirely on the types and concentrations of impurities in a crystal. SEE LUMINESCENCE.

Bibliography. V. M. Agaronovich and V. Ginzburg, *Crystal Optics with Spatial Dispersion and Excitons*, 2d ed., 1984; J. N. Hodgson, *Optical Absorption and Dispersion in Solids*, 1970; T. S. Moss et al., *Semiconductor Opto-electronics*, 1973; J. I. Pankove, *Optical Processes in Semiconductors*, 2d ed., 1976.

LUMINESCENCE
CLIFFORD C. KLICK AND JAMES H. SCHULMAN

Light emission that cannot be attributed merely to the temperature of the emitting body. Various types of luminescence are often distinguished according to the source of the energy which excites the emission. When the light energy emitted results from a chemical reaction, such as in the slow oxidation of phosphorus at ordinary temperatures, the emission is called chemiluminescence. When the luminescent chemical reaction occurs in a living system, such as in the glow of the firefly, the emission is called bioluminescence. In the foregoing two examples part of the energy of a chemical reaction is converted into light. There are also types of luminescence that are initiated by the flow of some form of energy into the body from the outside. According to the source of the exciting energy, these luminescences are designated as cathodoluminescence if the energy comes from electron bombardment; radioluminescence or roentgenoluminescence if the energy comes from x-rays or from gamma rays; photoluminescence if the energy comes from ultraviolet, visible, or infrared radiation; and electroluminescence if the energy comes from the application of an electric field. By attaching a suitable prefix to the word luminescence, similar designations may be coined to characterize luminescence excited by other agents. Since a given substance can frequently be made to luminesce by a number of different external exciting agents, and since the atomic and electronic phenomena that cause luminescence are basically the same regardless of the mode of excitation, the classification of luminescence phenomena into the foregoing categories is essentially only a matter of convenience, not of fundamental distinction.

When a luminescent system provided with a special configuration is excited, or "pumped," with sufficient intensity of excitation to cause an excess of excited atoms over unexcited atoms (a so-called population inversion), it can produce laser action. (Laser is an acronym for light amplification by stimulated emission of radiation.) This laser emission is a coherent stimulated luminescence, in contrast to the incoherent spontaneous emission from most luminescent systems as they are ordinarily excited and used.

Fluorescence and phosphorescence. A second basis frequently used for characterizing luminescence is its persistence after the source of exciting energy is removed. Many substances continue to luminesce for extended periods after the exciting energy is shut off. The delayed light emission (afterglow) is generally called phosphorescence; the light emitted during the period of excitation is generally called fluorescence. In an exact sense, this classification, based on persistence of the afterglow, is not meaningful because it depends on the properties of the detector used to observe the luminescence. With appropriate instruments one can detect afterglows lasting on the order of a few thousandths of a microsecond, which would be imperceptible to the human eye. The characterization of such a luminescence, based on its persistence, as either fluorescence or phosphorescence would therefore depend upon whether the observation was made by eye or by instrumental means. These terms are nevertheless commonly used in the approximate sense defined here, and are convenient for many practical purposes. However, they can be given a more precise meaning. For example, fluorescence may be defined as a luminescence emission having an afterglow duration which is temperature-independent, while phosphorescence may be defined as a luminescence with an afterglow duration which becomes shorter with increasing temperature. For details on various types of luminescence SEE FLUORESCENCE; PHOSPHORESCENCE.

Type of radiation emitted. Because of their many practical applications, materials that give a visible luminescence have been studied and developed more intensively than those which emit in other spectral regions. Luminescence, however, may consist of radiation in any region of the electromagnetic spectrum. The production of x-radiation by the bombardment of a metal target by a fast electron beam is an example of luminescence. Certain fluorescent lamps, called black-light lamps, are coated with a luminescent powder chosen for its ability to emit ultraviolet light of approximately 360 nanometers rather than visible light. A number of luminescent solids have been developed which luminesce in the near infrared under excitation by visible light, by a cathode-ray beam, or by electric fields.

Luminescent substances. The ability to luminesce is not confined to any particular state of matter. Luminescence is observed in gases, liquids, and amorphous and crystalline solids. The passage of an electrical discharge through a gas will excite the gaseous atoms or molecules to luminesce under certain conditions. An example of such a gaseous luminescence is the mercury-vapor lamp, in which an electrical discharge excites the mercury vapor to emit both visible and ultraviolet light. Many liquids, such as oil or solutions of certain dyestuffs in various solvents, luminesce very strongly under ultraviolet light. A large number of solids, including many natural minerals as well as thousands of synthetic inorganic and organic compounds, luminesce under various types of excitation. Applications of gas luminescence, formerly confined to advertising signs (neon signs) and fluorescent lamps, have multiplied greatly with the invention of the laser, many atomic and molecular gases providing the laser-active materials. The same is true of dye solutions and inorganic glasses. The major nonlaser applications of luminescence involve solid luminescent materials.

The term phosphor, originally applied to certain solids that exhibit long afterglows, has been extended to include any luminescent solid regardless of its afterglow properties. Other terms sometimes used synonymously with phosphor are luminophor, fluor, or fluorphor. The term luminophor is preferable, since it carries no connotation that afterglow times are long or short. In conformity with current usage, however, the term phosphor is used in the succeeding discussion of solid materials.

Comparatively few pure solids luminesce efficiently, at least at normal temperatures. In the category of organic solids, the pure aromatic hydrocarbons, such as naphthalene and anthracene, which consist exclusively of condensed phenyl rings, are luminescent, as are many heterocyclic closed-ring compounds. However, the closed phenyl ring structure is not in itself a guarantee of efficient luminescence, since certain substituents for hydrogen in the structure, particularly halogens, tend to reduce luminescence efficiency. This quenching of efficiency is called internal conversion. Among the pure inorganic solids that luminesce efficiently at room temperature are the tungstates, uranyl salts, platinocyanids, and a number of salts of the rare-earth elements.

Activators and poisons. The development of a large number of inorganic phosphors is due to the discovery that certain impurities, called activators, when present in amounts ranging from a few parts per million to several percent, can confer luminescent properties on the compounds (host or matrix compounds) in which they are incorporated. The activator and its nearby atoms are often referred to as the luminescent center.

By the same token, small amounts of other impurities or imperfections, called poisons or quenchers, can inhibit or destroy the luminescence, evidently by providing alternative mechanisms for the radiationless dissipation of the energy imparted to the phosphor or by emitting radiation characteristic of the poison itself which may not be in a spectral region of interest for the purpose at hand.

Manganese is a particularly effective activator in a wide variety of matrices when incorporated in amounts ranging from a small trace up to the order of several percent. The emission spectrum of these manganese-activated phosphors generally lies in the green, yellow, or orange spectral regions. Other frequently used activators are copper, silver, thallium, lead, cerium, chromium, titanium, antimony, tin, and the rare-earth elements. Poisons in inorganic phosphors generally are from the iron-nickel-cobalt group. The most important host materials are silicates, phosphates, aluminates, sulfides, selenides, the alkali halides, and oxides of calcium, magnesium, barium, and zinc. Preparation of phosphors and the incorporation of the activators is generally done by high-temperature reaction of well-mixed, finely ground powders of the components. A number of semiconducting compounds such as gallium phosphide (GaP) and gallium arsenide (GaAs) can be made to luminesce and display laser action when prepared so that charges can be injected into the semiconductor.

In order for ultraviolet light to provoke luminescence in a substance, the substance must be able to absorb it. Because of variations in its ability to absorb ultraviolet light of different wavelengths, a material may be nonluminescent under ultraviolet light of one wavelength and strongly luminescent under a different wavelength in the ultraviolet region. For a similar reason a substance may not luminesce under ultraviolet light at all and yet be strongly luminescent under x-ray irradiation.

LUMINESCENCE IN ATOMIC GASES

The processes that occur in luminescence may be most simply illustrated in the case of an atom in a gas. The atom can exist only in certain specific states of energy, some of which are shown schematically in **Fig. 1**. The lowest energy level of the atom, E_1, corresponds to the atom in its unexcited, or ground, state, and the higher energy levels, E_2 and E_3, represent electronically excited states of the atom. The excitation of the atom from state E_1 to state E_2 requires the absorption of an amount of energy $\Delta E = E_2 - E_1$. If this excitation is to be produced by the absorption of light, the energy of the acting light photon, $E_{absorbed}$, must equal ΔE. The frequency of the exciting light must therefore be $\nu = E_{absorbed}/h = (E_2 - E_1)/h$, and the wavelength of the exciting light must be $\lambda_{absorbed} = hc/(E_2 - E_1)$, where h is Planck's constant and c is the velocity of light. In an isolated atom this extra energy cannot be dissipated and is emitted as radiation when the atom eventually returns to its ground state. The emitted light will therefore again correspond in energy to ΔE and it will have a wavelength $\lambda_{emitted} = \lambda_{absorbed}$. When $\lambda_{emitted} = \lambda_{absorbed}$, the emitted light is sometimes referred to as resonance luminescence or resonance radiation. SEE ATOMIC STRUCTURE AND SPECTRA.

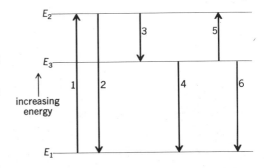

Fig. 1. Schematic representation of energy levels and electronic transitions in an atomic gas. E_1, ground state; E_2 and E_3, excited states; 1, excitation; 2, emission of resonance luminescence; 3, radiationless transition to lower excited state; 4, luminescence emission, if transition $E_3 \to E_1$ is allowed (if it is not allowed, 4 does not occur and E_3 is called a metastable state); 5, stimulation of atom back to emitting state; 6, radiationless transition of atom back to ground state (quenching).

If a large number of atoms are excited and the excitation then removed, the luminescence intensity will decrease with time exponentially according to the equation $I_t = I_0 e^{-t/\tau_l}$, where I_t is the intensity of luminescence at a time t after removal of the excitation, I_0 is the intensity at $t = 0$, and τ_l is the average time required by an atom to make a spontaneous luminescent transition. The quantity τ_l, called the radiative lifetime, is independent of temperature, and it is this temperature independence that is emphasized in the more precise definition of fluorescence given earlier. If the transition between the energy levels E_1 and E_2 is highly probable (a so-called permitted or allowed transition), τ_l is very small, of the order of 10^{-8} s for transitions involving visible light.

The excited atom can also lose a certain amount of energy and fall to an energy state E_3, of intermediate energy between E_1 and E_2. This can happen, for example, if the excited atom collides with another atom. If the transition from state E_3 to the ground state E_1 can occur with a reasonably high probability, fluorescence will occur starting from this intermediate excited state. In this case the fluorescent wavelength $\lambda_{emitted} = hc/(E_3 - E_1)$. Since $(E_3 - E_1)$ is smaller than $(E_2 - E_1)$, the fluorescence in this case will be of longer wavelength than the resonance radiation. Although the quantum efficiency of luminescence is unity (one photon of emitted light per photon of absorbed light), the energy efficiency is less than unity.

If, however, a transition between state E_3 and state E_1 is highly improbable (a so-called forbidden transition), state E_3 is known as a metastable state. The atom can remain in this state for long periods of time and cannot return to the ground state with the emission of radiation. Luminescence can occur under these circumstances only if the atom regains the energy ($E_2 - E_3$) by a collision with another atom or by some other process. Once the atom has been brought back to state E_2, a transition to the ground state is again allowed, and luminescence corresponding to ($E_2 - E_1$) will be emitted. The existence of metastable states like E_3 explains the delayed emission termed phosphorescence. The atom may spend a considerable amount of time in such a state before some external influence causes it to return to an emitting state such as E_2, in which case the luminescence is correspondingly delayed and appears as an after-glow. In order for the atom to get from E_3 to E_2 it must absorb energy somehow. The rate of return to the emitting state, and hence the duration of the afterglow, will therefore depend to a very large extent on temperature. At high temperatures the atoms will be excited back to the emitting state rapidly, and there will be a bright afterglow of short duration. At lower temperatures the atoms will be raised back to the emitting state very slowly, and the afterglow will be of long duration but of low intensity. This temperature dependence is the basis for the more precise definition of phosphorescence that was given earlier. As an alternative to regaining the energy ($E_2 - E_3$), the atom may lose the energy ($E_3 - E_1$) by a competing radiationless process, for example, by collision with another atom. The energy of excitation is thus dissipated without luminescence, and the quantum yield of luminescence is zero (quenching).

The principles illustrated in the foregoing discussion may be extended with little modification to the case where the primary absorption or excitation act completely removes an electron from an atom (that is, ionizes the atom) instead of merely raising the atom to an excited state. Under these conditions the electron can be trapped temporarily by other atoms, and its return to the parent atom can also be delayed by this mechanism.

The same principles are also operative in the case of complex configurations of atoms, such as in organic molecules or solids. However, the forbiddenness of radiative transitions can be modified in these cases, and efficient luminescent emissions can consequently be observed due to radiative transitions from metastable states, albeit with longer afterglows.

When nonlinear optics conditions exist, luminescence can be excited in certain systems by multiphoton absorption, the resulting luminescence being of higher frequency than the exciting light because the energy of two or more photons of the latter are combined to give one photon of emitted light. There are also cases where a single photon is absorbed by a pair of activator atoms or ions, each of which then emits a luminescence photon, leading to a quantum efficiency of 2. However, the preceding discussion and the exposition of those principles which follow deal primarily with the more usual case of single-photon absorption and emission.

CONFIGURATION COORDINATE CURVE MODEL

Luminescence in atomic gases is adequately described by the concepts of atomic spectroscopy, but luminescence in molecular gases, in liquids, and in solids introduces two major new effects which need special explanation. One is that the emission band appears on the long wave-

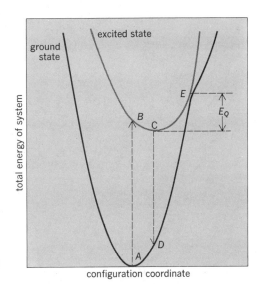

Fig. 2. Configuration coordinate curves for a simple luminescence center. Transition $A \rightarrow B$ shows absorption, and transition $C \rightarrow D$ shows emission. Energy E_Q is that necessary for the excited state to reach point E, from which a transition to the ground state can be made without luminescence.

length (low-energy) side of the absorption band; the other is that emission and absorption often show as bands tens of nanometers wide instead of as the lines found in atomic gases.

Both of these effects may be explained by using the concept of configuration coordinate curves illustrated in **Fig. 2**. As in the case of atomic gases, the ground and excited states represent different electronic states of the luminescent center, that is, the region containing the atoms, or electrons, or both, involved in the luminescent transition. On these curves the energy of the ground and excited state is shown to vary parabolically as some configuration coordinate, usually the distance from the activator to its nearest neighbors, is changed. There is a value of the coordinate for which the energy is a minimum, but this value is different for the ground and excited states because of the different interactions of the activator with its neighbors. Absorption of light gives rise to the transition from A to B. This transition occurs so rapidly that the ions in the luminescent center do not have time to rearrange. Once the system is at B it gives up heat energy to its surroundings by means of lattice vibrations and reaches the new equilibrium position at C. Emission occurs when the system makes the transition from C to D, and once again heat energy is given up when the system goes from D back down to A. This loss of energy in the form of heat causes the energy associated with the emission $C \rightarrow D$ to be less than that associated with the absorption $A \rightarrow B$.

When the system is at an equilibrium position, such as C of the excited-state curve, it is not at rest but migrates over a small region around C because of the thermal energy of the system. At higher temperatures these fluctuations cover a wider range of the configuration coordinate. As a result, the emission transition is not just to point D on the ground state curve but covers a region around D. In the vicinity of D the ground state curve shows a rapid change of energy, so that even a small range of values for the configuration coordinate leads to a large range of energies in the optical transition. This explains the broad emission and absorption bands that are observed. An analysis of this sort predicts that the widths of the band (usually measured in energy units between the points at which the emission or absorption is half its maximum value) should vary as the square root of the temperature. For many systems this relationship is valid for temperatures near and above room temperature. At low temperatures, quantum-mechanical effects, described below, become dominant.

Two other phenomena which can be explained on the basis of the model described in Fig. 2 are temperature quenching of luminescence and the variation of the decay time of luminescence with temperature. In **Fig. 3** a curve of temperature quenching for the emission in thallium-activated potassium chloride is shown. At low temperatures there is very little change in brightness with temperature, but at elevated temperatures the luminescence efficiency decreases rapidly (so-

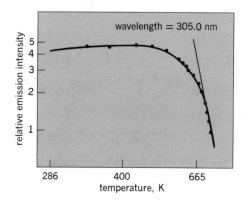

Fig. 3. Variation of the brightness of a thallium-activated potassium chloride phosphor with temperature. °F = (K × 1.8) − 459.67.

called thermal quenching). On the scheme of Fig. 2 this is interpreted as meaning that the thermal vibrations become sufficiently intense to raise the system to point E. From point E the system can fall to the ground state by emitting a small amount of heat, or infrared radiation. If point E is at an energy E_Q above the minimum of the excited state curve, it may be shown that the quantum efficiency, η, of luminescence is given by Eq. (1), where C is a constant, k is Boltzmann's con-

$$\eta = 1 + C \exp\left(-E_Q/kT\right)^{-1} \tag{1}$$

stant, and T is the temperature on the Kelvin scale. By fitting an expression of this form to the data of Fig. 3, a value of 0.60 eV is obtained for E_Q.

Another result of the onset of thermal quenching is that the luminescence decays faster, since the excited state is now depopulated by two processes simultaneously—a dissipative process in parallel with the luminescent process.

Quantum-mechanical corrections. Although the configuration coordinate curve model of Fig. 2 is successful in describing many aspects of luminescence in solids, it predicts that emission and absorption bands should become narrow lines as the temperature is reduced to absolute zero (0 K). This is not the case, as is shown in **Fig. 4**, which gives the width of the absorption band of the F-center in potassium chloride as a function of the square root of the temperature. (An F-center is the simplest type of color center, a color center being a lattice defect which can absorb light.) At high temperatures the previously quoted results are valid, but at low temperatures the width of the band is constant.

This phenomenon may be explained by treating the configuration coordinate curves quantum mechanically. The curves have the energy versus displacement characteristics of the simple harmonic oscillator. For this case the quantum-mechanical analysis shows that there is a series of equally spaced energy levels separated by an energy of $h\nu$, where h is Planck's constant and ν is the frequency of vibration. The lowest of these levels occurs at a value of $1/2\, h\nu$ above the minimum of the classical curve, and this energy is called the zero point energy. Its importance is that even at absolute zero the system is not at rest but varies over a range of configuration coordinates characteristic of this lowest vibrational level. Analysis shows that under these conditions the widths of the bands, ΔE, vary as in Eq. (2), where A is a constant. A curve of this form is drawn

$$\Delta E = A[\coth\left(h\nu/2kT\right)]^{1/2} \tag{2}$$

through the experimentally obtained points of Fig. 4 and shows satisfactory agreement.

One other result of the introduction of quantum mechanics is that this simple model predicts that both absorption and emission bands should be gaussian in shape at all temperatures; that is, they should be of the form of Eq. (3), where I is the emission intensity or absorption

$$I = I_0 \exp\left[-A(E - E_0)^2\right] \tag{3}$$

strength for light of energy E, and I_0 and E_0 refer to corresponding quantities at the maximum of the curve. **Figure 5** shows the emission spectrum of thallium-activated potassium chloride plotted

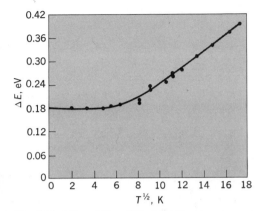

Fig. 4. Variation of the width of the F-center absorption band in potassium chloride at its half-maximum points, ΔE, as a function of the square root of the temperature. °F = (K × 1.8) − 459.67.

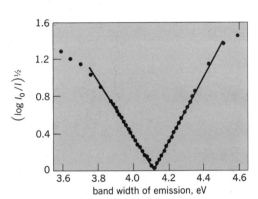

Fig. 5. Emission of thallium-activated potassium chloride at 4 K (−452°F). In a plot of this particular form, a gaussian curve would consist of two straight lines which make equal angles with the abscissa.

in such a way that the expression of Eq. (3) would give two straight lines making equal angles with the horizontal. Although these is some disagreement in the wings of the emission spectrum, and although the lines are not quite at the same angle, there still is fairly good agreement with the predictions of Eq. (3).

High dielectric constant materials. The use of configuration coordinate curves is justifiable only when the electron taking part in an optical transition is tightly bound to a luminescent center and interacts primarily with its nearest neighbors. This appears to be generally the case in materials with low dielectric constants such as the alkali halides. For high dielectric constant materials the situation is very different. It has been estimated that for boron in silicon the electron is spread over about 500 atoms so that its interaction with the nearest neighbors is small. In these cases the center interacts with the lattice during an optical transition through the creation of many vibrational phonons (sound quanta) at relatively large distances from the center. A case of this sort is illustrated in **Fig. 6**, which shows edge emission in cadmium sulfide near the absorption edge of the material. The major peaks correspond to an optical transition with simultaneous emission of 0, 1, 2, and 3 phonons, respectively, starting with the highest peak. The peaks are equidistant in energy, and the spacing is just that to be expected for the creation of phonons. Although both short- and long-range interactions of a center with its environment occur in all

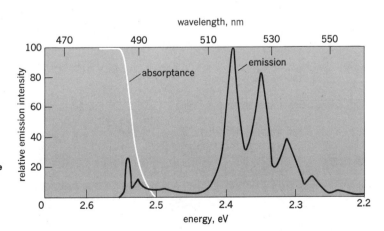

Fig. 6. Graph showing the absorptance (100 minus percent of transmission) and edge emission of cadmium sulfide at 4 K (−452°F).

cases, the short-range interaction dominates for tightly bound electrons in low dielectric constant materials, while the long-range interaction dominates for loosely bound electrons in high dielectric constant materials.

SENSITIZED LUMINESCENCE

A process of considerable interest occurs in systems where one type of activator absorbs the exciting light and transfers its energy to a second type of activator which then emits. The transfer does not involve motion of electrons. This process is often called sensitized luminescence, and it is widespread in both inorganic and organic luminescent systems. The principal phosphors used in fluorescent lamps are of this type. The results on calcium carbonate ($CaCO_3$) shown in **Fig. 7** illustrate this kind of system. If divalent manganese (Mn^{2+}) alone is incorporated as an activator into $CaCO_3$, the system gives only a very weak Mn^{2+} emission for all wavelengths of exciting ultraviolet light. This is because the Mn^{2+} ion has only forbidden transitions in this spectral region, as shown by its very low absorption strength and slow decay of luminescence. In other words, the Mn^{2+} does not absorb very much ultraviolet light. Bright phosphors may be prepared by incorporating divalent lead (Pb^{2+}), monovalent thallium (Tl^+), or trivalent cerium (Ce^{3+}) along with Mn^{2+} in $CaCO_3$. In each case the Mn^{2+} emission is the same, but the wavelengths which excite these phosphors vary as the other activator (called the sensitizer) is changed. All these absorptions occur for energies well below the energy at which free electrons are formed in the solid. The ultraviolet light is absorbed by the Pb^{2+}, Tl^+, or Ce^{3+} sensitizers, which have allowed transitions leading to strong absorption of the light; this absorbed energy is then passed on to nearby Mn^{2+} activators and excite the Mn^{2+} emission. It is important to know over what distances the energy may be transferred from absorber to emitter. Experiments on calcium silicate with Pb^{2+} and Mn^{2+} as added impurities have shown that if a Mn^{2+} ion is in any one of the 28 nearest lattice sites around the Pb^{2+} ion, the energy transfer may take place.

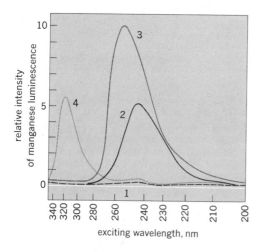

Fig. 7. Excitation spectra for emission of calcium carbonate phosphors, using manganese as an activator, with different impurity sensitizers: 1, no sensitizer; 2, thallium; 3, lead; and 4, cerium.

Resonant transfer. The transfer of energy has been treated theoretically using quantum mechanics and assuming a model of resonance between coupled systems. A number of cases have been examined; in each case the sensitizer has an allowed transition since only in this case will it absorb the exciting light appreciably.

The first case is one in which the activator undergoes an allowed transition. The probability of energy transfer, P (defined as the reciprocal of the time required for a transition to occur) varies as in expression (4). Here R is the distance between sensitizer and activator; n is the index of

$$P \propto \frac{1}{R^6 n^4} \int f_s F_A dE \qquad (4)$$

refraction of the host material, which may be liquid or solid; f_s is a function describing the emission band of the sensitizer as a function of energy, and is normalized so that the area under the curve is unity; and F_A is a function describing the absorption band of the activator, also normalized to unity. The integral of expression (4) measures the overlap of these bands and determines the resonance transfer. In typical cases the sensitizer will transfer energy to an activator if the activator occupies any one of the 1000–10,000 nearest available sites around the sensitizer. Another case is that of a forbidden quadrupole transition in the activator; in this case the number of sites for transfer would be about 100. If the transition in the activator is even more strongly forbidden, quantum-mechanical "exchange" effects predominate and the number of available sites for transfer should be reduced to 40 or less. In both of the cases just mentioned the integral measuring the overlap enters in the same way as it does in expression (4). From this theoretical treatment it appears that phosphors with Mn^{2+} as the activator probably receive their energy by exchange interactions.

Concentration quenching. Another phenomenon related to the resonant transfer of energy is that of concentration quenching. If phosphors are prepared with increasing concentrations of activators, the brightness will first increase but eventually will be quenched at high concentrations. It is believed that at high concentrations the absorbed energy is able to move from one activator to a nearby one by resonant transfer and thus migrate through the solid or liquid. If there are "poisons," or quenching sites, distributed in the material, the migrating energy may reach one of these, be transferred to it, and dissipate without luminescence. Impurity atoms, vacancies, jogs at dislocations, normal lattice ions near dislocations, and even a small fraction of the activator ions themselves, when associated in pairs or higher aggregates, can act as poisons. As the concentration is increased, the speed of migration is increased, and the quenching process becomes increasingly important.

LUMINESCENCE INVOLVING ELECTRON MOTION

In an important group of luminescent materials the transfer of energy to the luminescent center is brought about by the motion of electrons. Many oxides, sulfides, selenides, tellurides, arsenides, and phosphides are of this type, and of these, zinc sulfide has been most widely studied. It is frequently used as the luminescent material in cathode-ray tubes and electroluminescent lamps.

Electronic processes in insulating materials are described by a band model such as that illustrated in **Fig. 8**. The electron energy increases vertically and the horizontal dimension shows position in a crystal. In the shaded areas, called the filled band, or the valence band, all the energy levels are filled with electrons. No electron can be accelerated or moved to higher energies within this band since the higher levels are already filled. Thus the material is an insulator. Above the

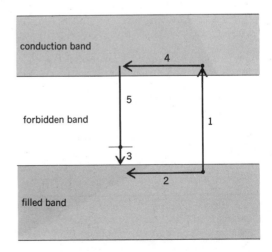

Fig. 8. Energy-band model for luminescence processes in zinc sulfide. Electron energy increases vertically.

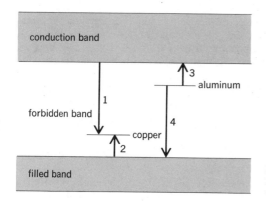

Fig. 9. Energy-band model for luminescence processes in zinc sulfide which uses copper as the activator and aluminum as the coactivator.

valence band is an energy region called the forbidden band, which has no energy levels in it for pure materials. However, imperfections may introduce a local energy level in this region, as illustrated by the short line above 3 in Fig. 8. Above the forbidden band is an energy region called the conduction band. Here there are energy levels but, in an insulator, no electrons. If an electron in the filled band absorbs light of sufficiently high energy, it may jump up into the conduction band, as shown in 1 of Fig. 8. The empty position left behind in the filled band, called a hole, has properties which allow it to be described as an electron except that it has a positive charge. Both the electron in the conduction band and the hole in the filled band are free to move and gain energy. As a result, current can flow when a voltage is applied externally. The electron and hole can also diffuse far from their origin and can thus transport energy to a distant luminescent center.

A simple luminescent transition is illustrated in Fig. 8. Assume that the impurity level (black dot in the forbidden band) is due to a luminescent center and that there is an electron in the level at the beginning of the process. Transition (1) shows the creation of a free electron and hole due to the absorption of light. The hole migrates to the center (2), and the electron in the impurity level falls into the hole (3), thus destroying it. The free electron now migrates toward the center (4) and falls into it (5), giving off luminescence. The cycle is complete and the center once again has an electron. It is important to note that in this process both electrons and holes must be assumed to move.

In zinc sulfide the normal activators are monovalent metals such as Ag^+ or Cu^+. They replace Zn^{2+} ions. To maintain electrical neutrality it is necessary also to incorporate into the lattice ions such as Cl^- for S^{2-} or Al^{3+} for Zn^{2+}. These additional ions are called coactivators. The situation is now quite complex, and is illustrated for ZnS with Cu and Al as added impurities in **Fig. 9**. The various arrows in this figure show some of the electronic transitions that give rise to observable effects. Arrows 1 and 4 show two different luminescent transitions. Arrow 2 shows the excitation of an electron into the luminescent center, which may occur by absorption of infrared light or by thermal fluctuations. In either case the center is prevented from luminescing and thus is quenched. Transition 3 is the excitation of an electron into the conduction band. The electron may be excited by thermal energy and its appearance may be detected by the appearance of transition 1. This process is called thermoluminescence.

The wide variety of luminescence phenomena, the insight they give into the constitution of very different classes of materials, and their many important applications to light production, imaging devices, and radiation detectors make luminescence a perennially challenging field of study.

Bibliography. H. J. Cantow et al. (eds.), *Luminescence*, 1981; J. H. Crawford, Jr., and L. M. Slifkin (eds), *Point Defects in Solids*, 3 vols., 1972–1978; B. Di Bartolo (ed.), *Luminescence in Solids*, 1978; J. D. Dunitz et al. (eds.), *Luminescence and Energy Transfer*, 1981; R. J. Elliott and A. F. Gibson, *An Introduction to Solid State Physics*, 1974; M. Karas, *Luminescence*, 1983; S. G. Schulmar (ed.), *Molecular Luminescence Spectroscopy: Methods and Applications*, 1985; S. Shionoya et al. (eds.), *Proceedings of the 1975 International Conference on Luminescence, Tokyo*, 1976.

FLUORESCENCE

James H. Schulman and Clifford C. Klick

Fluorescence is generally defined as a luminescence emission that is caused by the flow of some form of energy into the emitting body, this emission ceasing abruptly when the exciting energy is shut off. In attempts to make this definition more meaningful it is often stated, somewhat arbitrarily, that the decay time, or afterglow, of the emission must be of the order of the natural lifetime for allowed radiative transitions in an atom or a molecule, which is about 10^{-8} s for transitions involving visible light. Perhaps a better distinction between fluorescence and its counterpart, phosphorescence, rests not on the magnitude of the decay time per se, but on the criterion that the fluorescence decay is temperature-independent. If this latter criterion is adopted, the luminescence emission from such materials as the uranyl compounds and rare-earth-activated solids would be called slow fluorescence rather than phosphorescence. The decay of their luminescence takes place in milliseconds to seconds, rather than in 10^{-8} s, showing that the optical transitions are somewhat "forbidden"; but the decay is temperature-independent over a considerable range of temperature, and it follows an exponential decay law, shown in the equation below,

$$I = I_0 \exp(-t/\tau_1)$$

that is to be expected for spontaneous transitions of electrons from an excited state of an atom to the ground state when the atom has a transition probability per unit time $1/\tau_1$. (τ_1 is called the natural radiative lifetime or fluorescence lifetime.) In this equation I is the luminescence intensity at a time t, and I_0 is the intensity when $t = 0$.

In applying this criterion, one should take note of the following restriction which arises because all luminescent systems ultimately lose efficiency or are "quenched" at elevated temperatures, each system having its own characteristic temperature for the onset of this so-called thermal quenching. Quenching sets in because increase of temperature makes available other competing atomic or molecular transitions that can depopulate the excited state and dissipate the excitation energy nonradiatively. Hence, even for a fluorescence, a temperature dependence of decay time will be observed at temperatures where thermal quenching becomes operative, because $1/\tau_{obs} = 1/\tau_1 + 1/\tau_{diss}$, where τ_{obs} is the observed fluorescence lifetime and $1/\tau_{diss}$ is the probability for a competing dissipative transition.

In the literature of organic luminescence, the term fluorescence is used exclusively to denote a luminescence which occurs when a molecule makes an allowed optical transition. Luminescence with a longer exponential decay time, corresponding to an optically forbidden transition, is called phosphorescence, and it has a different spectral distribution from the fluorescence. See Phosphorescence.

The decay time of fluorescent materials varies widely, from the order of 5×10^{-9} s for many organic crystalline materials up to 2 s for the europium-activated strontium silicate phosphor. Fluorescent materials with decay times between 10^{-9} and 10^{-7} s are used to detect and measure high-energy radiations, such as x-rays and gamma rays, and high-energy particles such as alpha particles, beta particles, and neutrons. These agents produce light flashes (scintillations) in certain crystalline solids, in solutions of many polynuclear aromatic hydrocarbons, or in plastics impregnated with these hydrocarbons.

PHOSPHORESCENCE

Clifford C. Klick and James H. Schulman

A delayed luminescence, that is, a luminescence that persists after removal of the exciting source. It is sometimes called afterglow.

This original definition is rather imprecise, because the properties of the detector used will determine whether or not there is an observable persistence. There is no generally accepted rigorous definition or uniform usage of the term phosphorescence. In the literature of inorganic luminescent systems, some authors define phosphorescence as delayed luminescence whose persistence time decreases with increasing temperature. According to this usage, luminescence whose persistence time is independent of temperature is called fluorescence regardless of the length of the afterglow; a temperature-independent afterglow of long duration is called simply a slow fluorescence, which implies that the atomic or molecular transition involved is forbidden to a greater or lesser degree by the spectroscopic selection rules. In nonphotoconductive inorganic systems, phosphorescence arises when some excitation process has placed an atom (or ion or molecule) in a metastable energy state M (from which transitions to the state of lowest energy, or ground state G, are highly improbable or forbidden), and energy from the thermal vibrations of the system subsequently raises the atom to a higher energy state E from which luminescent transitions are highly probable or allowed (see **illus**.). The most common mechanism of phosphorescence in photoconductive inorganic systems, however, occurs when electrons or holes, set free by the excitation process and trapped at lattice defects, are expelled from their traps by the thermal energy in the system and recombine with oppositely charged carriers with the emission of light. In these cases, the level M represents the state of the system with the electron or hole trapped.

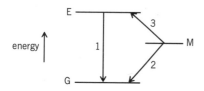

1 = allowed transition
2 = forbidden transition
3 = thermally excited (radiationless) transition

process	usage 1	usage 2	usage 3 followed by 1
inorganic	fluorescence	fluorescence	phosphorescence
organic	fluorescence	phosphorescence	delayed fluorescence

Atomic and molecular transitions involved in luminescence, and corresponding terminology.

In the study of organic systems much attention has been given to the G→E→M process, which is the excitation of a molecule from its ground state to an excited state, both usually spectroscopic singlet states, followed by a radiationless transition (called an intersystem crossing) from the singlet excited state to the metastable triplet state M. In the organic literature the term phosphorescence is reserved for the forbidden luminescent transition M → G, while the afterglow corresponding to the M→E→G process is called delayed fluorescence. The spectrum (color) of organic "phosphorescence," defined in this way, is necessarily different from the spectrum of the ordinary fluorescence, because the emitting states (E and M) in the two cases are different and the final (ground) state G is the same. Conversely, the spectra of ordinary fluorescence and delayed fluorescence in organic systems are the same, because the luminescent transition takes place between the same emitting state E and the ground state G in both cases.

The temperature-dependent (M→E→G process) luminescence of a given system can exhibit a wide range of persistence times. A very low temperature where there is insufficient energy available to raise atoms from metastable to emitting states, or to expel electrons from traps, little or no afterglow is observed. At some higher temperature a low-intensity, long-lived afterglow will be observed; at a still higher temperature the afterglow will be brighter but of shorter duration. Finally, at some high temperature where rate of expulsion of atoms from metastable states or rate of expulsion of electrons from traps is very rapid, afterglow can become immeasurably short.

The time dependence of the luminescence intensity (the decay law) can be extremely complex, depending on the number and energies of the metastable states or electron traps involved. Phosphors which give a phosphorescent emission visible to the eye for about half a day at normal temperatures have been synthesized. SEE ABSORPTION OF ELECTROMAGNETIC RADIATION; FLUORESCENCE; LUMINESCENCE.

OPTICAL ACTIVITY

Vincent Madison

The effect of asymmetric compounds on polarized light. To exhibit this effect, a molecule must be nonsuperimposable on its mirror image, that is, must be related to its mirror image as the right hand is to the left hand. An optically active compound and its mirror image are called enantiomers or optical isomers. Enantiomers differ only in their geometric arrangements; they have identical chemical and physical properties. The right-handed and left-handed forms of a molecule can be distinguished only by their optical activity or by their interactions with other asymmetric molecules. Optical activity can be used to probe other aspects of molecular geometry, as well as to identify which enantiomer is present and its purity.

As an example of optical isomers, consider tartaric acid (**Fig. 1**), which was one of the first synthetic molecules to be separated into its enantiomers. In this case the asymmetry of each isomer is magnified when trillions of molecules form a crystal; two types of asymmetric crystals are formed.

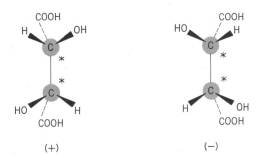

Fig. 1. Enantiomers of tartaric acid.

The physical basis of optical activity is the differential interaction of asymmetric substances with left versus right circularly polarized light. If solids and substances in strong magnetic fields are excluded, optical activity is an intrinsic property of the molecular structure and is one of the best methods of obtaining structural information from a sample in which the molecules are randomly oriented. The relationship between optical activity and molecular structure results from the interaction of polarized light with electrons in the molecule. Thus the molecular groups that contribute most directly to optical activity are those that have mobile electrons which can interact with light. Such groups are called chromophores, since their absorption of light is responsible for the color of objects. For example, the chlorophyll chromophore makes plants green.

Methods of measurement. Optical activity is measured by two methods, optical rotation and circular dichroism.

Optical rotation. This method depends on the different velocities of left and right circularly polarized light beams in the sample. The velocities are not measured directly, but both beams are passed through the sample simultaneously. This is equivalent to using plane-polarized light. The differing velocities of the left and right circularly polarized components yield a rotation of the plane of polarization. A polarimeter for observing optical rotation consists of a light source, a fixed polarizer, a sample compartment, and a rotatable polarizer. A cell containing solvent is placed between the polarizers, and one of them is adjusted to be perpendicular to the other, excluding the passage of light. The solvent in the cell is then replaced by a solution of the sample, and the polarizer is rotated to again exclude passage of light. The optical rotation a is the number of degrees the polarizer was rotated. A positive or negative sign indicates the direction of rotation. Enantiomers have rotations of equal magnitude, but opposite signs. The optical rotation depends on the substance, solvent, concentration, cell path length, wavelength of the light, and temperature. Standardized specific rotations [α] are reported as defined in Eq. (1), where T is the temper-

$$[\alpha]_\lambda^T = \frac{a}{cl} \qquad (1)$$

ature (°C), λ the wavelength (often the orange sodium D line), l the cell path length in decimeters, and c the concentration in grams per milliliter. Alternatively, M_ϕ is defined by normalizing to the rotation for a 1-molar solution, Eq. (2), where M_ϕ is the molar rotation and MW the molecular

$$M_\phi = [\alpha]_\lambda^T \text{MW}/100 \qquad (2)$$

weight. For polymers, the mean residue rotation, m_ϕ, may be defined by the right side of Eq. (2) by using the mean residue (monomer unit) weight for MW. The variation of optical rotation with wavelength is known as optical rotatory dispersion (ORD).

Circular dichroism. Circular dichroism (CD) is the difference in absorption of left and right circularly polarized light. Since this difference is about a millionth of the absorption of either polarization, special techniques are needed to determine it accurately. Circular dichroism spectrometers consist of a light source, a monochromator to select a single wavelength, a modulator to produce circularly polarized light, a sample compartment, a phototube to detect transmitted light, and associated electronic components. The modulator rapidly switches (typically 50,000 times per second) between left and right circular polarization of the light beam. The absorption of an optically inactive sample is independent of polarization, so that the light intensity at the phototube is constant; thus a constant direct current is generated. The absorption of an optically active sample depends on the polarization, so that the light intensity at the phototube varies at the frequency of the modulator; thus an alternating current is generated. The circular dichroism is proportional to the amplitude of the alternating current. The proportionality constant is determined through calibration by using a compound of known circular dichroism.

Circular dichroism is reported as a difference in absorption, Eq. (3), or as an ellipticity (a measure of the elliptical polarization of the emergent beam), Eq. (4), for a 1-molar solution, where

$$\Delta\epsilon = \epsilon_L - \epsilon_R = (A_L - A_R)/(c'l') \qquad (3) \qquad\qquad M_\theta = 3300\Delta\epsilon \qquad (4)$$

ϵ is the extinction coefficient, A is the absorbance [log (I_0/I)], subscripts L and R indicate left or right circular polarization, c' is the concentration in moles per liter, l' is the path length in centimeters, I_0 and I are the light intensities in the absence and presence of the sample, respectively, and M_θ is the molar ellipticity. Either $\Delta\epsilon$ or ellipticity, m_θ, may be expressed per residue by making c' the concentration of residues (monomer units). As in the case of optical rotation, enantiomers have circular dichroism spectra of equal magnitude but opposite signs.

Variation with wavelength. Optical rotation and circular dichroism are two manifestations of the same interactions between polarized light and molecules. They are related by a mathematical transformation. An important difference between the two measurements is the way in which they vary with wavelength. Optical rotation extends to wavelengths far from any absorption of light. Thus colorless substances still have significant optical rotation at the sodium line. However, all groups which absorb light (chromophores) contribute at all wavelengths, and it can be difficult to extract the contribution of a single group. On the other hand, circular dichroism is confined to the narrow absorption band of each chromophore. Thus it is easier to determine the contribution of individual chromophores, information vital to structural analysis. SEE COTTON EFFECT; ROTATORY DISPERSION.

Correlation with molecular structure. In synthesizing enantiomers, chemists focus on an asymmetric center, that is, a locus which imparts asymmetry to the whole molecule. A common asymmetric center is a tetrahedral carbon atom with four different groups attached, such as the carbons marked with asterisks in tartaric acid (Fig. 1). However, in correlating optical activity with molecular structure, the focus is on the three-dimensional arrangement of the chromophores which interact most strongly with light.

As examples, consider the nucleoside adenosine and its dimer (**Fig. 2**). The most mobile electrons are in the aromatic ring system, the chromophore (Fig. 2a). The electrons in the sugar ribose are more tightly bound and interact less strongly with visible and ultraviolet light. However, all the asymmetric centers are in the ribose part of the molecule. For adenosine, light interacting

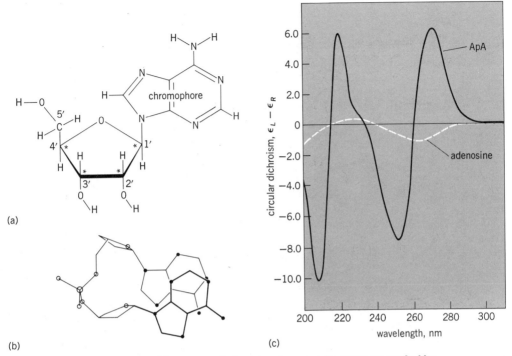

Fig. 2. Adenosine and its dimer. (a) Structure of adenosine. Asymmetric centers are marked by an asterisk. (b) Stacked arrangement of adenosine dimer (ApA). The 3' carbon of one adenosine is linked to the 5' carbon of the other by a phosphate group. (c) Circular dichroism spectra of ApA and adenosine at neutral pH in aqueous solution at room temperature.

with the aromatic chromophore is only weakly influenced by the asymmetric centers in ribose, so that small circular dichroism bands are observed (Fig. 2c).

In the covalently linked dimer of adenosine, the observed circular dichroism bands are about 10 times larger than those of the monomer. In the 240- to 300-nm region of the spectra (Fig. 2c), two bands are observed for the dimer, but only one for the monomer. This indicates strong interaction of the two aromatic chromophores, and hence their close proximity in the dimer. Analysis of the circular dichroism spectra expected for various arrangements of the two chromophores, as well as other types of experimental data, indicates that the aromatic rings are stacked (Fig. 2b). The asymmetric centers in ribose cause the formation of the stacked arrangement shown rather than its mirror image.

Stacking of aromatic rings, as exemplified by the adenosine dimer, is a common feature of nucleic acid polymers (deoxyribonucleic acid and ribonucleic acid) isolated from biological sources. Slight differences in the stacking geometry gives each of these polymers a characteristic circular dichroism spectrum. Alterations in the stacking arrangement caused by some pharmacologically active agents can be detected through alterations in the circular dichroism spectra. These structural changes may in turn be related to the pharmacological action.

A derivative of the amino acid proline (**Fig. 3**) can be used to illustrate another way in which optical activity depends on molecular structure. In this molecule only the OCN group (amide chromophore) which is in the horizontal plane of the drawing and the hydrogen which is marked H^* need be considered. By forming the N-H bond, H^* acquires a charge of about $+\frac{1}{3}$ electron. It has been predicted that such a positive charge will perturb the motion of the electrons in the amide chromophore in a manner which will produce a negative circular dichroism band when the charge is above the plane of the amide group and to the right of the oxygen. Only for

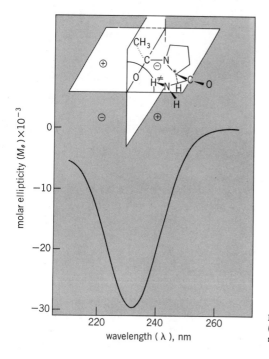

Fig. 3. Folded arrangement of *L*-proline derivative (*N*-acetyl-*L*-proline-amide) and its circular dichroism spectrum in *p*-dioxane solution at room temperature.

the arrangement shown is the magnitude of the circular dichroism band expected to be as large as observed. Furthermore, it has been shown that there will be no circular dichroism if $H^{\neq}$ is in either of the two planes shown, and that for $H^{\neq}$ in adjacent quadrants the sign of the circular dichroism band alternates (Fig. 3). For this compound, reflection through the horizontal plane will generate the enantiomer. This would place $H^{\neq}$ in the lower right quadrant and generate a positive circular dichroism band with magnitude equal to that of Fig. 3.

Bibliography. L. D. Barron, *Molecular Light Scattering and Optical Activity*, 1983; S. F. Mason, *Molecular Optical Activity and the Chiral Discriminations*, 1982; J. A. Schellman, Symmetry rules for optical rotation, *Accounts Chem. Res.*, 1:144–155, 1968.

ROTATORY DISPERSION
Bruce H. Billings

A term used to describe the change in rotation as a function of wavelength experienced by linearly polarized light as it passes through an optically active substance. *See* Optical activity.

Optically active materials. Substances that are optically active can be grouped into two classes. In the first the substances are crystalline and the optical activity depends on the arrangement of nonoptically active molecular units. When these crystals are dissolved or melted, the resulting liquid is not optically active. In the second class the optical activity is a characteristic of the molecular units themselves. Such materials are optically active as liquids or solids. A typical substance in the first category is quartz. This crystal is optically active and rotates the plane of polarization by an amount which depends on the direction in which the light is propagated with respect to the optic axis. Along the axis the rotation is 29.73°/mm (755°/in.) for light of wavelength 508.6 nanometers. At other angles the rotation is less and is obscured by the crystal's linear birefringence. Molten quartz, or fused quartz, is isotropic. Turpentine is a typical material of the second class. It gives rotation of $-37°$ in a 10-cm (3.94-in.) length for the sodium D lines.

Reasons for variation. In all materials the rotation varies with wavelength. The variation is caused by two quite different phenomena. The first accounts in most cases for the majority of the variation in rotation and should not strictly be termed rotatory dispersion. It depends on the fact that optical activity is actually circular birefringence. In other words, a substance which is optically active transmits right circularly polarized light with a different velocity from left circularly polarized light.

Any type of polarized light can be broken down into right and left components. Let these components be R and L. The lengths of the rotating light vectors will then be $R/\sqrt{2}$ and $L/\sqrt{2}$. At $t=0$, the R vector may be at an angle ψ_r with the x axis and the L vector at an angle ψ_l. Since the vectors are rotating at the same velocity, they will coincide at an angle β which bisects the difference as in Eq. (1). If $R=L$, the sum of these two waves will be linearly polarized light vibrating at an angle α to the axes given by Eq. (2).

$$\beta = \frac{\psi_r + \psi_l}{2} \quad (1) \qquad \alpha = \frac{\psi_r - \psi_l}{2} \quad (2)$$

If, in passing through a material, one of the circularly polarized beams is propagated at a different velocity, the relative phase between the beams will change in accordance with Eq. (3),

$$\psi'_r - \psi'_l = \frac{2\pi d}{\lambda}(n_r - n_l) + \psi_r - \psi_l \quad (3)$$

where d is the thickness of the material, λ is the wavelength, and n_r and n_l are the indices of refraction for right and left circularly polarized light. The polarized light incident at an angle α has, according to this equation, been rotated an angle given by Eq. (4).

$$\gamma = \frac{\pi d}{\lambda}(n_r - n_l) \quad (4)$$

This shows that the rotation would depend on wavelength, even in a material in which n_r and n_l were constant and which thus had no circular dipersion. It is for this reason that the term rotatory dispersion is perhaps ill-defined in much of the literature.

In addition to this pseudodispersion which depends on the material thickness, there is a true rotatory dispersion which depends on the variation with wavelength of n_r and n_l.

From Eq. (4) it is possible to compute the circular birefringence for various materials. This quantity is of the order of magnitude of 10^{-8} for many solutions and 10^{-5} for crystals. It is 10^{-1} for linear birefringent crystals.

Bibliography. L. D. Barron, *Molecular Light Scattering and Optical Activity*, 1983; P. Crabble, *ORD and CD in Chemistry and Biochemistry*, 1972; T. M. Lowry, *Optical Rotatory Power*, 1935, reprint 1964; S. F. Mason, *Molecular Optical Activity and the Chiral Discriminations*, 1982; G. Snatzke (ed.), *Optical Rotatory Dispersion and Circular Dichroism in Organic Chemistry*, 1976.

COTTON EFFECT
Albert Moscowitz

The characteristic wavelength dependence of the optical rotatory dispersion curve or the circular dichroism curve or both in the vicinity of an absorption band.

When an initially plane-polarized light wave traverses an optically active medium, two principal effects are manifested: a change from planar to elliptic polarization, and a rotation of the major axis of the ellipse through an angle relative to the initial direction of polarization. Both effects are wavelength dependent. The first effect is known as circular dichroism, and a plot of its wavelength (or frequency) dependence is referred to as a circular dichroism (CD) curve. The second effect is called optical rotation and, when plotted as a function of wavelength, is known as an optical rotatory dispersion (ORD) curve. In the vicinity of absorption bands, both curves take on characteristic shapes, and this behavior is known as the Cotton effect, which may be either positive or negative (**Fig. 1**). There is a Cotton effect associated with each absorption process,

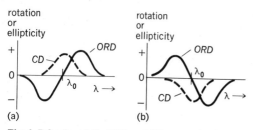

Fig. 1. Behavior of the ORD and CD curves in the vicinity of an absorption band at wavelength λ_0 (idealized). (a) Positive Cotton effect. (b) Negative Cotton effect.

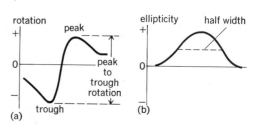

Fig. 2. Curves used to determine relative rotatory intensities. (a) Partial ORD curve. (b) Partial CD curve.

and hence a partial CD curve or partial ORD curve is associated with each particular absorption band or process. *See* ROTATORY DISPERSION.

Measurements. Experimental results are commonly reported in either of two sets of units, termed specific and molar (or molecular). The specific rotation $[\alpha]$ is the rotation in degrees produced by a 1-decimeter path length of material containing 1 g/ml of optically active substance, and the specific ellipticity θ is the ellipticity in degrees for the same path length and same concentration. Molar rotation $[\varphi]$ (sometimes $[M]$) and molar ellipticity $[\theta]$ are defined by Eqs. (1) and (2). For comparisons among different compounds, the molar quantities are more useful, since they

$$[\varphi] = [\alpha]M/100 \qquad (1) \qquad\qquad [\theta] = \theta M/100 \qquad (2)$$

allow direct comparison on a mole-for-mole basis.

The ratio of the area under the associated partial CD curve to the wavelength of the CD maximum is a measure of the rotatory intensity of the absorption process. Moreover, for bands appearing in roughly the same spectral region and having roughly the same half-width (**Fig. 2**), the peak-to-trough rotation of the partial ORD curve is roughly proportional to the wavelength-weighted area under the corresponding partial CD curve. In other words, relative rotatory intensities can be gaged from either the pertinent partial ORD curves or pertinent partial CD curves. A convenient quantitative measure of the rotatory intensity of an absorption process is the rotational strength. The rotational strength R_i of the ith transition, whose partial molar CD curve is $[\theta_i(\lambda)]$, is given by relation (3).

$$R_i \approx 6.96 \times 10^{-43} \int_0^\infty \frac{[\theta_i(\lambda)]}{\lambda} d\lambda \qquad (3)$$

Molecular structure. The rotational strengths actually observed in practice vary over quite a few orders of magnitude, from $\sim 10^{-38}$ down to 10^{-42} cgs and less; this variation in magnitude is amenable to stereochemical interpretation. In this connection it is useful to classify optically active chromophores, which are necessarily dissymmetric, in terms of two limiting types: the inherently dissymmetric chromophore, and the inherently symmetric but dissymmetrically perturbed chromophore. *See* OPTICAL ACTIVITY.

A symmetric chromophore is one whose inherent geometry has sufficiently high symmetry so that the isolated chromophoric group is superimposable on its mirror image, for example, the carbonyl group $>C=O$. The transitions of such a chromophore can become optically active, that is, exhibit a Cotton effect, only when placed in a dissymmetric molecular environment. Thus, in symmetrical formaldehyde, $H_2C=O$, the carbonyl transitions are optically inactive; in ketosteroids, where the extrachromophoric portion of the molecule is dissymmetrically disposed relative to the symmetry planes of the $>C=O$ group, the transitions of the carbonyl group exhibit Cotton effects. In such instances the signed magnitude of the rotational strength will depend both upon the chemical nature of the extrachromophoric perturbing atoms and their geometry relative to that of the inherently symmetric chromophore. In a sense, the chromophore functions as a molecular probe for searching out the chemical dissymmetries in the extrachromophoric portion of the molecule.

The type of optical activity just described is associated with the presence of an asymmetric carbon (or other) atom in a molecule. The asymmetric atom serves notice to the effect that, if an inherently symmetric chromophore is present in the molecule, it is almost assuredly in a dissymmetric environment, and hence it may be anticipated that its erstwhile optically inactive transitions will exhibit Cotton effects. Moreover, the signed magnitude of the associated rotational strengths may be interpreted in terms of the stereochemistry of the extrachromophoric environment, as compared with that of the chromophore. But an asymmetric atom is not essential for the appearance of optical activity. The inherent geometry of the chromophore may be of sufficiently low symmetry so tht the isolated chromophore itself is chiral, that is, not superimposable on its mirror image, for example, in hexahelicene.

In such instances the transitions of the chromophore can manifest optical activity even in the absence of a dissymmetric environment. In addition, it is very often true that the magnitudes of the rotational strengths associated with inherently dissymmetric chromophores will be one or more orders of magnitude greater ($\sim 10^{-38}$ cgs, as opposed to $< 10^{-39}$ cgs) than those associated with inherently symmetric chromophores. Hence, in the spectral regions of the transitions of the inherently dissymmetric chromophore, it will be the sense of handedness of the inherently dissymmetric chromophore itself that will determine the sign of the rotational strength, rather than perturbations due to any dissymmetric environment in which the inherently dissymmetric chromophore may be situated.

The sense of handedness of an inherently dissymmetric chromophore may be of considerable significance in determining the absolute configuration or conformations of the entire molecule containing that chromophore. Accordingly, the absolute configuration or conformation can often be found by focusing attention solely on the handedness of the chromophore itself. For example, in the chiral molecule shown in **Fig. 3** there is a one-to-one correspondence between the sense of helicity of the nonplanar diene chromophore present and the absolute configuration at the asymmetric carbon atoms. Hence there exists a one-to-one correspondence between the handedness of the diene and the absolute configuration of the molecule. Since it is known that a right-handed diene helix (**Fig. 4**) associates a positive rotational strength with the lowest diene singlet transition in the vicinity of 260 nanometers, by examination of the pertinent experimental Cotton effect (positive), the absolute configuration of the molecule is concluded to be as shown.

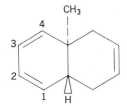

Fig. 3. Structural formula of (+)-*trans*-9-methyl-1,4,9,10-tetrahydronaphthalene.

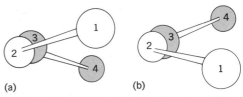

Fig. 4. Schematic representation of the twisted diene chromophore showing the two possible handednesses; the numbering is as indicated in Fig. 3. (*a*) Right-handed. (*b*) Left-handed.

Other examples of inherently dissymmetric chromophores are provided by the helical secondary structures of proteins and polypeptides. Here the inherent dissymmetry of the chromophoric system arises through a coupling of the inherently symmetric monomers, which are held in a comparatively fixed dissymmetric disposition relative to each other through internal hydrogen bonding. The sense of helicity is then related to the signs of the rotational strengths of the coupled chromophoric system. The destruction of the hydrogen bonding destroys the ordered dissymmetric secondary structure, and there is a concomitant decrease in the magnitude of the observed rotational strengths.

Bibliography. L. D. Barron, *Molecular Light Scattering and Optical Activity*, 1983; S. F. Mason, *Molecular Optical Activity and the Chiral Discriminations*, 1982; *Proc. Roy. Soc. London*, ser. A, 297:1–172, 1976; L. Velluz et al., *Optical Circular Dichroism: Principles, Measurements, and Applications*, 1969.

INSTRUMENTATION AND TECHNIQUES

Arc discharge	80
Synchrotron radiation	81
Wavelength measurement	92
Wavelength standards	93
Optical prism	94
Diffraction grating	95
Interferometry	98
Spectrohelioscope	106
Spectrography	107
Gamma-ray detectors	107
Matrix isolation	110

ARC DISCHARGE

GLENN H. MILLER

A type of electrical conduction in gases characterized by high current density and low potential drop. It is closely related to the glow discharge but has a much lower potential drop in the cathode region, as well as greater current density.

There are may arc devices, and they operate under a wide range of conditions. For example, an arch discharge may be sustained at either high pressure (of the order of atmospheres) or low pressure. The cathode may or may not be heated from an external source. Furthermore, the applied potential difference may be either direct or alternating. Numerous applications have been made of such devices, some having large commercial value. A few of these applications are illuminating devices, high-current rectifiers, high-current switches, welding devices, and ion sources of nuclear accelators and thermonuclear devices.

Arc production. Although an arc may be initiated in several ways, it is instructive to consider the transition from a cold-cathode glow discharge (see **illus**.). In the normal condition, the glow partially covers the cathode, and the voltage drop remains nearly constant as the current is increased. In this condition. The ionization is produced primarily by electron impact. As the current is increased, the cathode glow spreads out, eventually covering the cathode completely. Further increase in current can now be obtained only by an increase in the potential drop across the discharge. The cathode temperature increases in this process. This is the abnormal glow region. As the cathode temperature is increased further, thermionic emission becomes an important factor. At this point the discharge characteristic may acquire a negative slope; that is, a further increase in current may increase the cathode temperature, and the resulting thermionic emission, enough actually to reduce the potential drop. Unless the external resistance is sufficiently great to make the overall resistance positive, the discharge will change suddenly to the arc mode. Typical values in this region are a potential drop of a few tens of volts and a current ranging from amperes to thousands of amperes.

Regions of an arc. There are three geometrical regions in an arc. These are the cathode fall, the anode rise, and the arc body.

Cathode fall. The cathode region is characterized by a high potential gradient. There will be a large positive ion current to the cathode, but the electron current from the cathode may be greater than this. It is not difficult to understand the large thermionic emission for a cathode of refractory metal having a very high melting temperature. However, for a metal with a low melting temperature, such as copper, the situation is not quite so clear. The best explanation appears to be that advanced first by J. J. Thomson and later by L. B. Loeb, who pointed out that the cathode must be viewed on a microscopic basis, and that the local temperature may be vastly greater than the macroscopic cathode temperature. Under these conditions it is not unreasonable to be-

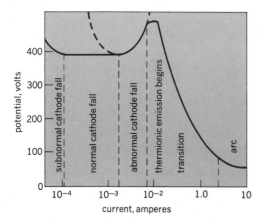

Transition from glow discharge to arc with increase of current. (*After L. B. Loeb, Basic Processes of Gaseous Electronics, University of Caliornia Press, 1955*)

lieve that there may be extensive thermionic emission. As in the glow discharge, the cathode is surrounded by a positive ion sheath, resulting in a large space charge.

Anode rise. The anode region requires considerable explanation. There is to this electrode a large electron current which may raise the anode to a temperature even greater than that of the cathode. Thus it is also a thermionic emitter, although the electrons are returned to the anode. Hence there is a large negative space charge and a resulting anode rise. Cooling the anode may result in a reduction in anode rise, indicating a decrease in thermionic emission. Again it is necessary to use the microscopic viewpoint.

Arc body. The main body of the arc is characterized by secondary ionization. This is predominantly a temperature effect. The electrons produce very little ionization by impact, because the electron energy is generally low. However, energy can be imparted to the gas molecules in the form of atomic and molecular excitation. With the many intermolecular collisions that occur, this energy is readily degraded to thermal energy of the gas. Thus a high temperature may be achieved and ionization occur by virtue of intermolecular collisions. This is expressed quantitatively by the equation, shown below, derived by M. N. Saha. Here n_i and n_0 are the ionic and

$$\log_{10}\left(\frac{n_i^2}{n_0}\right) = -5040\frac{v_i}{T} + \tfrac{3}{2}\log_{10}(T) + 15.385$$

neutral densities, respectively, v_i the ionization potential in volts, and T the Kelvin temperature. This equation refers to the equilibrium condition.

It is difficult to make accurate and meaningful measurements because of the high temperature and large current. Thus there is much that is not understood or substantiated.

An interesting situation obtains in connection with an externally heated cathode. Here may exist an arc in which the potential drop is less than the ionization potential of the gas. Several factors are important in this case. First, there is a positive space charge in the arc body which results in a maximum potential there, and there is actually a potential drop from the point to the anode. Second, there may be many excited or metastable atoms or molecules in the discharge. It seems likely that ionization may take place in several steps rather than one. The most important mechanism, however, appears to be that for removal of electrons that have lost so much energy by inelastic collisions that they are trapped near the potential maximum. A fast electron may give up enough energy by coulomb interaction with one of these so that both electrons may reach the anode. If this were not so, neutralization would occur and the arc would be quenched.

Finally, it should be pointed out that in any arc there will be vaporization of the electrodes, and that the gas will have in it molecules of the electrode material. For this reason the electrode material may have a profound effect on the arc characteristics.

Bibliography. E. Kuhnhardt and L. H. Luessen (eds.), *Electrical Breakdown and Discharges in Gases*, 1983; P. Llewellyn-Jones, *The Glow Discharge and an Introduction to Plasma Physics*, 1966; L. B. Loeb, *Basic Processes of Gaseous Electronics*, 1955; F. A. Maxfield and R. R. Benedict, *Theory of Gaseous Conduction and Electronics*, 1941; J. Millman and S. Seely, *Electronics*, 2d ed., 1951.

SYNCHROTRON RADIATION
Arthur Bienenstock and Herman Winick

Electromagnetic radiation emitted by relativistic charged particles curving in magnetic fields. Such emission of polarized white light by electrons guided by celestial magnetic fields has long been known to astronomers and is responsible, for example, for the beautiful background light in the Crab Nebula.

Radiation by electrons in circular orbits in early synchrotrons and betatrons was predicted by J. P. Blewett and others before 1945 and was first (accidentally) observed on a 70-MeV synchrotron in 1947. A detailed theoretical treatment of the properties of synchrotron radiation was given by J. Schwinger in 1949, showing, for example, that for particles of different mass the power radiated varies inversely as the fourth power of mass. This means that synchrotron radiation from protons is 13 orders of magnitude weaker than from electrons (at the same energy and bending

Storage ring synchrotron radiation sources*

Location	Ring (laboratory)	Electron energy GeV	Type of operation
China			
Beijing	BEPC (HEP)	2.2–2.8	Parasitic[†]
Hefei	HESYRL (USTC)	0.8	Dedicated[†]
England			
Daresbury	SRS (Daresbury)	2.0	Dedicated
France			
Orsay	ACO (LURE)	0.54	Dedicated
	DCI (LURE)	1.8	Partly dedicated
	SuperACO (LURE)	0.8	Dedicated[†]
Germany			
Hamburg	DORIS (HASYLAB)	5.5	Partly dedicated
West Berlin	BESSY	0.8	Dedicated
Italy			
Frascati	ADONE (Frascati)	1.5	Partly dedicated
Japan			
Tsukuba	Photon Factory (KEK)	2.5	Dedicated
	Accumulator (KEK)	6–8	Partly dedicated
	TRISTAN (KEK)	30	Parasitic[†]
Tokyo	SOR (ISSP)	0.4	Dedicated
Okasaki	UVSOR (IMS)	0.6	Dedicated
Tsukuba	TERAS (ETL)	0.6	Dedicated
Sweden			
Lund	Max (LTH)	0.55	Dedicated
Taiwan			
Hsinchu	TLS (SRRC)	1.0	Dedicated[†]
United States			
Gaithersburg, MD	SURF (NBS)	0.28	Dedicated
Ithaca, NY	CESR (CHESS)	5.5–8	Parasitic
Stanford, CA	SPEAR (SSRL)	4.0	Partly dedicated
	PEP (SSRL)	15.0	Parasitic[†]
	SXRL	1.0	Partly dedicated[†]
Stoughton, WI	Tantalus (SRC)	0.24	Dedicated
	Aladdin (SRC)	1.0	Dedicated
Upton, NY	NSLS I (BNL)	0.75	Dedicated
	NSLS II (BNL)	2.5	Dedicated
Soviet Union			
Karkhov	N-100 (KPI)	0.10	Dedicated
Moscow	Siberia I (Kurchatov)	0.45	Dedicated
Novosibirsk	VEPP-2M (INP)	0.7	Partly dedicated
	VEPP-3 (INP)	2.2	Partly dedicated
	VEPP-4 (INP)	5–7	Parasitic

*As of May 1985.
[†]In construction as of May 1985.

radius). In a multi-GeV electron storage ring, the rate of energy loss is so high that a circulating electron would lose all its energy in a few milliseconds. To store beams with decay lifetimes of a few hours (limited by collisions with the residual gas even at pressures at 10^{-9} torr or 10^{-7} pascal), powerful radio-frequency systems replenish the lost energy. By contrast, the synchrotron radiation from the 28-GeV proton storage rings that operated for many years at CERN in Geneva was so weak that protons could have circulated for years with negligible loss of energy.

Experimental facilities. The radiation has many features (natural collimation, high intensity and brilliance, broad spectral bandwidth, high polarization, pulsed time structure, small source size, and high-vacuum environment) which make it ideal for a wide variety of applications in experimental science and technology. Very powerful sources of synchrotron radiation in the ultraviolet and x-ray parts of the spectrum became available in many countries when high-energy physicists began operating electron synchrotrons in the 1950s. Although synchrotrons produce large amounts of radiation, their cyclic nature results in pulse-to-pulse intensity changes and variations in spectrum and source shape during each cycle. By contrast, the electron-positron storage rings developed for colliding-beam experiments starting in the 1960s offered a constant spectrum and much better stability. Beam lines were constructed on both synchrotrons and storage rings to allow the radiation produced in the bending magnets of these machines to leave the ring vacuum system and reach experimental stations. In most cases the research programs were pursued on a parasitic basis, secondary to the high-energy physics programs.

Even with the limitations of parasitic operation, these first facilities offered radiation that was vastly superior to conventional sources such as gas discharge lamps and electron impact x-ray tubes. In many cases the number of photons per second delivered to the sample within a given narrow bandwidth was three to five orders of magnitude higher than the flux delivered by conventional sources. As was the case with other sources such as lasers, this dramatic improvement in source intensity and brilliance opened up new research opportunities, and many basic and applied scientists sought access to the radiation. Important results were achieved in many disciplines such as biochemistry, materials science, surface science, and crystallography through the use of various techniques such as extended x-ray absorption fine structure (EXAFS), photoemission spectroscopy, x-ray scattering, x-ray topography, and x-ray microscopy.

The demonstrated capabilities of synchrotron radiation and the shortage of experimental stations led to decisions in many countries in the mid and late 1970s to expand existing facilities and make them at least partly dedicated to the production of radiation and also to construct new fully dedicated storage ring sources. Since about 1980 new dedicated facilities have been completed in England, Germany, Japan, and the United States and are in construction in the People's Republic of China and the Soviet Union (see **table** and **Figs. 1** and **2**).

Starting in the late 1970s, advances were also made in the development of special magnets which may be inserted into the straight sections between ring bending magnets to produce beams with extended spectral range or with higher flux and brilliance than is possible with the

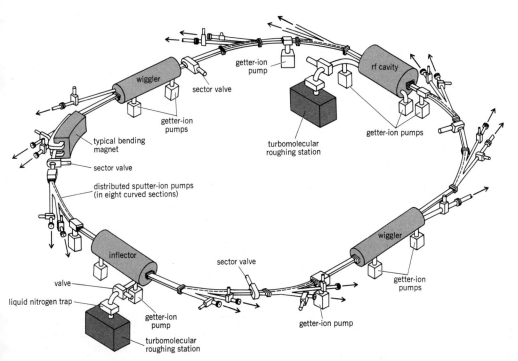

Fig. 1. Electron storage ring designed as a synchrotron radiation source. Only one bending magnet is shown. Not shown is the injector accelerator. (*After J. Godell, Brookhaven National Laboratory*)

ring bending magnets. These devices, called wiggler and undulator magnets, utilize periodic transverse magnetic fields to produce transverse oscillations of the electron beam with no net deflection or displacement. They have already provided another order-of-magnitude or more improvement in flux and brilliance over ring bending magnets, again opening up new research opportunities. The potential of these devices goes well beyond their present performance levels, particularly in the brilliance that undulators could produce in specially designed rings. Such rings would have many more straight sections for wiggler and undulator insertions than those now in use.

Even with the expansion of existing facilities and the completion of new rings, there is still a shortage of beam time and experimental stations at many facilities. This is particularly true for applications that require the high flux and brilliance provided by wigglers and undulators. The total number of insertion device possibilities on present rings is quite small, and most rings are not optimized for insertion devices, particularly undulators. Because of this, plans have been made in the United States and other parts of the world to construct new storage rings which will be optimized to serve a large number of wiggler and undulator beam lines. In addition to providing more experimental stations, these rings will open up new research opportunities because they will provide radiation with higher flux and brilliance than existing rings.

Properties of radiation. The radiation produced by an electron in circular motion at low energy (speed much less than the speed of light) is weak and rather nondirectional (**Fig. 3**a). At relativistic energies (speed close to the speed of light) the radiated power increases markedly, and the emission pattern is folded forward into a cone (Fig. 3b) with a half-opening angle in radians given approximately by $\gamma^{-1} = mc^2/E$, where mc^2 is the rest-mass energy of the electron (0.51 MeV) and E is the total energy. Thus, at electron energies of the order of 1 GeV, much of the very strong radiation produced is confined to a forward cone with an instantaneous opening angle of about 1 mrad. At higher electron energies this cone is even smaller. The large amount of radiation produced combined with the natural collimation gives synchrotron radiation its intrinsic high brightness and brilliance. Brightness is the photon flux per unit solid angle within a unit bandwidth. Brilliance is brightness per unit source area. The brilliance is further enhanced by the small

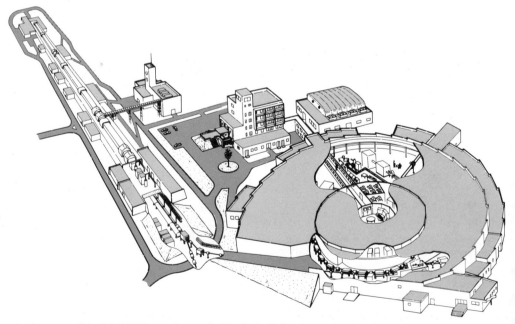

Fig. 2. Layout of the 2.5-GeV Photon Factory facility in Tsukuba, Japan, showing the storage ring, linac injector, and office and support buildings.

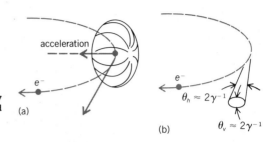

Fig. 3. Instantaneous radiation emission patterns by electrons in circular motion (**a**) at low energy (speed much less than speed of light) and (**b**) at relativistic energy (speed close to speed of light).

size of the electron beam which typically has a cross-sectional area of about 10^{-3} in.^{2} (1 mm^2) or less in existing rings and could be even smaller in new rings.

Bending magnet sources. As an electron moves in an arc through a bending magnet (**Fig. 4a**), the instantaneous forward cone sweeps out a horizontal fan of continuum, polarized radiation. The small vertical opening angle, $2\gamma^{-1}$, is preserved, but the horizontal angle is determined by the acceptance of the beam pipe or optical elements (for example, mirrors, gratings, or crystals). By accepting horizontal angles larger than γ^{-1}, more flux can be utilized. The brightness and brilliance are, however, not increased.

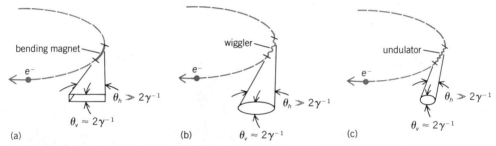

Fig. 4. Spatial characteristics of beams from (**a**) bending magnets, (**b**) wiggler magnets, and (**c**) undulator magnets.

Wiggler magnet sources. In a wiggler magnet the electron beam executes transverse oscillations with an angular excursion much larger than γ^{-1}. A fan of radiation is produced (Fig. 4b) which is similar to that produced by a bending magnet but with an enhancement of intensity, brightness, and brilliance due to the superposition of radiation from each pole. Because a wiggler produces no net deflection or displacement of the electron beam, there is great flexibility in its design. It can produce a wide fan (one or more degrees) which can be shared by several experiments (**Fig. 5**), or a narrow fan which directs all the flux to a single experiment. The magnetic field can be higher than the ring bending magnet field, resulting in a spectrum that extends to higher photon energy. Superconducting wigglers with fields up to 6 teslas have been used to extend the spectrum deeper into the hard x-ray region. Although most wigglers have vertical magnetic fields, as do ring bending magnets, devices with horizontal fields have also been made. These produce vertically polarized radiation. Wigglers (and undulators) can be electromagnets (**Fig. 6**) or permanent magnets (**Fig. 7**). Permanent magnets are increasingly used for these applications, particularly for devices with short periods and many poles.

Undulator magnet sources. An undulator magnet is designed to cause the electron beam to execute transverse oscillations with an angular excursion of the order of γ^{-1}. By keeping the electron beam angular range down to the same order as the natural emission angle of synchrotron radiation, the radiation emerging from the device is concentrated into the smallest possible opening angle and the intrinsic high brightness of the radiation is preserved in both planes (Fig. 4c). Undulators thus have the highest brightness and brilliance of any magnet sources. Fur-

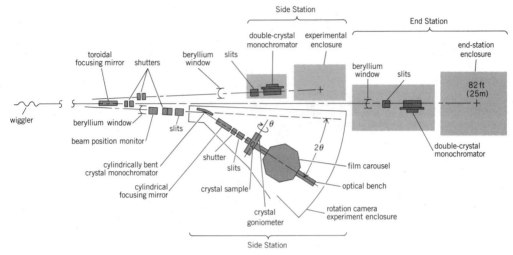

Fig. 5. Layout drawing of a multistation wiggler beam line at SPEAR. The wiggler source is shown in Fig. 6.

Fig. 6. Lower half of a 1.8-tesla electromagnetic wiggler used in SPEAR. The magnet has seven full poles and two end half-poles. Each pole is powered by a coil, one of which is shown. The higher field of the magnet shifts the spectrum to harder x-ray energies, and the overall intensity is enhanced by the number of poles. (*SLAC photo by J. Faust*)

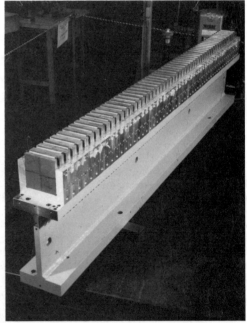

Fig. 7. Lower half of a 1.2-tesla, 54-pole permanent-magnet wiggler, now in use in the SPEAR ring. The samarium cobalt blocks seen at the end alternate with vanadium permendur poles. (*Lawrence Berkeley Laboratory, photo by Steve Adams*)

thermore, the spectral distribution of radiation from an undulator is qualitatively different from a bending magnet or wiggler as explained below.

Spectrum. The main spectral features of the radiation produced by bending magnets, wigglers, and undulators are compared in **Figs. 8** and **9**. Figure 8 shows the spectrum of the bending magnets of the SPEAR ring compared to two wigglers now operating in that ring, and clearly indicates the higher flux provided by the wigglers. These spectra are characterized by a single parameter, the critical energy ϵ_c, given (in keV) by $2.2E^3/R$ or $0.665BE^2$, where E is the electron energy in GeV, R is the radius of curvature in meters, and B is the magnetic induction in teslas. Half the power is radiated above the critical energy and half below. Typically, useful flux is available out to four or five times the critical energy. In use, the broadband radiation is monochromatized by a grating or crystal so that the bandwidth ($\Delta\epsilon/\epsilon$ or $\Delta\lambda/\lambda$, where ϵ and λ are the energy and wavelength of the radiation) is 10^{-3} or smaller. Thus, highly monochromatic radiation at any energy within the range of the synchrotron radiation source is readily available.

In an undulator, interference effects in the radiation produced at a large number of essentially colinear source points result in a spectrum with quasimonochromatic peaks at wavelengths given by Eq. (1), where λ_u is the period length of the undulator, θ is the angle of observation relative to the average electron direction, n is the harmonic number, and K is the undulator strength parameter given by Eq. (2). Here, B is in teslas, λ_u is in centimeters, and 2δ is the full

$$\lambda = \frac{\lambda_u}{2n\gamma^2}\left[1 + \frac{K^2}{2} + \gamma^2\theta^2\right] \quad (1) \qquad K = 0.934\, B_{max}\, \lambda_u = \gamma\delta \quad (2)$$

angular excursion of the electron beam traversing the magnet, in radians.

For $K \ll 1$ only the fundamental peak ($n = 1$) is important. For $K \approx 1$ the power in the fundamental is a maximum and the first few harmonics have appreciable intensity. For $K > 1$ the wavelength of the fundamental becomes longer [according to Eq. (1)] and more harmonics appear. For $K \gg 1$ the fundamental has very long wavelength and there are many closely spaced harmon-

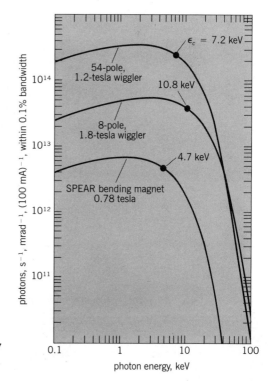

Fig. 8. Photon flux as a function of photon energy from the SPEAR bending magnets and from the wiggler magnets shown in Figs. 6 and 7.

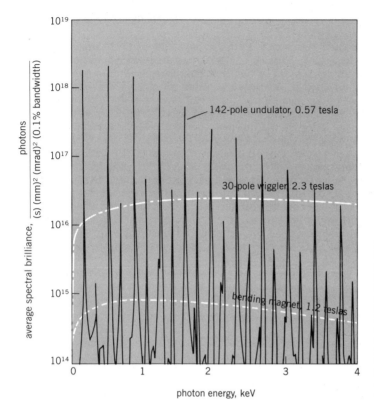

Fig. 9. Comparison of the brilliance from the bending magnets, a wiggler magnet, and an undulator magnet in the proposed Advanced Light Source at the Lawrence Berkeley Laboratory, with electron energy of 1.3 GeV and current of 400 mA. The period length of the undulator λ_u is 3.5 cm (1.4 in.). (*After K. J. Kim, Lawrence Berkeley Laboratory*)

ics. In this limit, the device is a wiggler, with the envelope of the spectrum approaching the continuous spectrum of that device. Figure 9 compares the spectral brilliance of the radiation from a bending magnet, a wiggler, and an undulator in the Advanced Light Source (ALS) proposed by the Lawrence Berkeley Laboratory, and shows the extremely high brilliance offered by the undulator at discrete photon energies covering a range of values.

The spectral width of undulator peaks is determined by the number of periods in the undulator, the transverse size and divergence of the electron beam, and the angular acceptance of the detector. For an electron beam of negligible size and divergence and for a detector with very small acceptance (that is, a pinhole), the fractional bandwidth of the peaks is given by $\Delta\lambda/\lambda \approx 1/(nN)$, where N is the number of periods. In this case the brightness increases as N^2. A device with 100 periods could therefore produce a brightness that is four orders of magnitude greater than that produced by ring bending magnets. When the finite beam size and divergence of most present rings is taken into account, the brightness gain is less than N^2, but is still large. New rings can be designed with reduced electron beam size and divergence, resulting in much higher brightness and brilliance from undulators. These are the low emittance rings proposed as the next-generation synchrotron radiation sources.

Power. The total power radiated (in kilowatts) by the bending magnets is given by $88E^4I/R$, where E is the electron energy in GeV, I is the electron current in amperes, and R is the radius of curvature in meters. The largest storage rings now used routinely for synchrotron radiation research (SPEAR, DORIS, and the Photon Factory) radiate about 100 kW, or about 15 W/mrad. The higher-energy rings now used for colliding beam research at Stanford and Hamburg (PEP and Petra) radiate about 1 MW. The power radiated by a wiggler of undulator magnet is given by $1.267\langle B^2\rangle E^2IL$, where $\langle B^2\rangle$ is the average value of the square of the magnetic field in teslas taken over the length of the device L in meters. As above, E is in GeV, I is in amperes, and the power is in kilowatts. Wigglers now in use produce total power up to several kilowatts and power densities up to about 1 kW/mrad. The difficulties in handling such high power densities on beam line optical elements has led to increased interest in undulators, which produce high brightness and flux in the spectral region to which they are tuned, but little flux elsewhere, so that the total power radiated is less than that of wigglers.

Polarization. In the plane of the electron orbit, the radiation from bending magnets and from wigglers and undulators is nearly 100% linearly polarized, with the electric vector parallel to the electron acceleration. Out of the plane of the orbit, bending magnets produce elliptically polarized radiation. Undulators with helical magnetic fields have been built, producing circularly polarized radiation. It should be possible to use a pair of crossed linear undulators to produce radiation with polarization that can be flexibly selected by the experimenter and rapidly switchable, for example, from right to left circular polarization.

Pulsed time structure. Due to the bunching of the electron beam by the radio-frequency accelerating system of the ring, the radiation is produced in pulses as each electron bunch sweeps by the observation point. The maximum number of bunches, given by the ratio of the radio frequency to the orbital frequency, ranges from a few (1 to 10) for small rings to hundreds or thousands for large rings. The interval between pulses can be adjusted by only filling certain bunches with electrons. The pulse duration can be as short as 100 picoseconds and, in large rings, the interval between pulses can be 1 microsecond or more. This pulse structure facilitates study of time-dependent phenomena and is also used to enhance the data collection rate in some experiments (for example, by measuring photoelectron velocities using time-of-flight techniques).

Research applications. Ultraviolet and x-radiation from conventional sources has been employed extensively for decades in research utilizing a great variety of experimental techniques in many disciplines, such as physics, chemistry, and biology, as well as in a variety of technological processes. Many of these techniques and processes have experienced considerable improvements as a result of one or more of the special properties of synchrotron radiation described above.

Extended x-ray absorption fine structure (EXAFS). A plot of x-ray absorption versus photon energy (**Fig. 10**) shows the steep rise in the absorption at the absorption-edge energy, where the incident photon is just able to excite core electrons into empty states. Just above this energy, the observed oscillatory structure on the absorption is due to interference between the outgoing photoelectron waves and the electron waves backscattered from atoms adjacent to the absorbing atoms. This interference modulates the probability of exciting the electron, as explained by R. de L. Kronig in 1931. Because the structure arises from an interference phenomenon and the absorption-edge energies of different atomic species are usually quite different, analysis of this structure provides considerable information about the average atomic environment of each atomic species in complex, polyatomic solids, liquids, and gases.

The highly intense, continuous synchrotron radiation spectrum from multi-GeV storage rings is ideally suited for such measurements as a function of photon energy. Consequently, it has been used extensively for structural studies of amorphous materials, heterogeneous catalysts, and noncrystalline metalloproteins whose atomic arrangements are not ordinarily determinable by

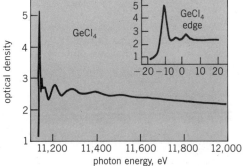

Fig. 10. X-ray absorption spectrum above the Ge K-edge in GeCl$_4$ vapor. The inset shows the edge region with a shifted energy scale. (After B. M. Kincaid and P. Eisenberger, Synchrotron radiation studies of the K-edge photo-absorption spectra of Kr, Br$_2$, and GeCl$_2$: A comparison of theory and experiment, Phys. Rev. Lett., 34:1361–1364, 1975.

other structural techniques. These structural studies have led, in turn, to increased understanding of the properties of these materials. EXAFS patterns have also been measured in fractions of a second, so that time-resolved studies of structural changes may be anticipated.

Similar modulations of the x-ray fluorescence yield, the Auger electron yield, and the desorbed atom yield versus photon energy are caused by the same interference phenomenon. The fluorescence yield is used extensively to study the atomic environments of extremely dilute (less than 100 parts per million) constituents of complex systems such as alloys, semiconductors, and proteins in solution. The Auger electron yield is used for the determination of the atomic environments of atoms at the surfaces of materials. The desorbed atom yield is used to determine the surroundings of atoms adjacent to specific atoms on a surface. For example, it is used to determine the environments of those metal atoms adjacent to oxygen atoms on a partially oxidized metal surface. SEE AUGER EFFECT; SURFACE PHYSICS; X-RAY FLUORESCENCE ANALYSIS.

Photoemission spectroscopy. Energy and angular distributions of electrons ejected from solids by photons of varying incident energy provide basic information about band structure and core levels of both the surface and bulk of the sample. The extremely high intensity and tunability of synchrotron radiation have made possible high-resolution studies which delineate the band structures of crystalline solids. In addition, electrons of energy between 30 and 100 eV have such a large probability for interaction with other atoms that they have an escape depth of only about 0.5 nanometer in most solids. As a result, the detected electrons reveal the properties of surface electronic energy levels of clean materials and the extrinsic surface states of chemisorbed material, opening a route to the study of the phenomena of oxidation, corrosion, and catalysis. Such studies have led to major revisions of models of the metal-semiconductor interface determines many properties of solid-state electron devices. Again, the tunability, when coupled with the high intensity of synchrotron radiation, makes possible the separate delineation of surface and bulk electronic states.

In addition, the radiation facilitates photoemission studies in two ways: the high-vacuum environment of the storage ring minimizes contamination of clean surfaces prepared in place; and the sharply pulsed time structure of the radiation makes it possible to measure photoelectron energies by time of flight.

Figure 11 shows photoemission spectra from clean and heavily oxidized gallium arsenide surfaces, each taken in less than 10 min by using synchrotron radiation from SPEAR. The curves taken at 100 eV incident photon energy show clear differences due to oxidation. The new shifted peak in the spectrum of the oxidized surface, to the left of the arsenic peak, shows that the oxygen primarily attaches to the arsenic rather than the gallium (although the reverse was expected from thermodynamic arguments based on bulk chemistry). At 100 eV photon energy, essentially only two atomic layers at the surface are being probed. At higher energies, contributions from the bulk become dominant, as can be seen from the change in relative heights of the shifted and unshifted arsenic peaks. The strength of emission from valence states is also seen to depend strongly on the state of oxidation and photon energy. SEE ELECTRON SPECTROSCOPY.

Surface and surface layer structure. The extreme brightness of x-ray synchrotron radiation has made possible the determination of atomic arrangements on two-dimensional layers, surfaces, and surface layers by x-ray diffraction. Among the most exciting studies have been those of melting and "freezing" in two-dimensional systems, as well as the changes in these processes as the system is changed from two- to three-dimensional through the addition of layers. It has also been possible to determine atomic arrangements in amorphous surface layers less than 25 nm thick by x-ray diffraction.

X-ray diffraction topography. The availability of highly intense white radiation has also accelerated markedly the development of time-resolved studies of defects in nearly perfect crystals through x-ray topography. Such studies include the effects of magnetic fields on domains and of temperature on defects. Studies may be anticipated of defect motion due to stress and of corrosion due to toxic environments on the highly perfect single crystals required in solid-state electronics. The highly collimated monochromatic radiation has also been used to study the relationship between defects in a substrate and those in an epitaxial layer, which are important in many solid-state devices. **Figure 12** is an example of an x-ray diffraction topograph taken with synchrotron radiation.

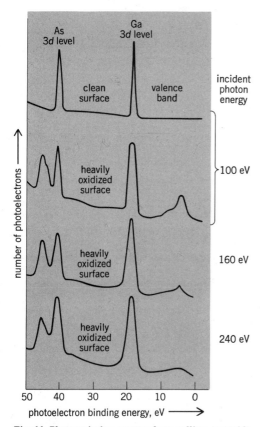

Fig. 11. Photoemission spectra from gallium arsenide surfaces. (*After I. Lindau et al., Determination of the oxygen binding site on GaAs (110) using soft x-ray photoemission spectroscopy, Phys. Rev. Lett., 35:1356, 1976*)

Fig. 12. Diffraction topograph of a lithium fluoride crystal taken in 5 min with SPEAR operating at 1.89 GeV and 7.2 mA. The smallest features are 1 μm. (*W. Parrish, IBM, San Jose, California*)

Time-resolved fluorescence spectroscopy. Synchrotron radiation is utilized extensively for time-resolved spectroscopy in those temporal domains in which 50–200-picosecond pulses at 1-microsecond intervals are appropriate. These include studies of an electron-hole recombination in semiconductors and rotational times of tryptophan side chains on proteins. The high photon energies available make possible experiments which cannot be performed in the more restricted spectral domain of pulsed lasers.

X-ray fluorescence trace-element analysis. The high intensity, tunability, high collimation, and high linear polarization of synchrotron radiation make it valuable for trace-element analysis through x-ray fluorescence. In a search for superheavy elements, a sensitivity sufficient to observe about 5×10^8 atoms of superheavy elements per sample was demonstrated. With improvements in technique, concentrations as low as 10^6 atoms per sample may be observable. SEE TRACE ANALYSIS.

Bibliography. E. E. Koch (ed.), *Handbook on Synchrotron Radiation*, vol. 1, 1983; C. Kunz (ed.), *Topics in Current Physics—Synchrotron Radiation: Techniques and Applications*, 1979; H. Winick and A. Bienenstock, Synchrotron radiation research, *Annu. Rev. Nucl. Part. Sci.*, 28:33–113, 1978; H. Winick and S. Doniach (eds.), *Synchrotron Radiation Research*, 1980.

WAVELENGTH MEASUREMENT
W. R. C. ROWLEY

Determination of the distance between successive wavefronts of equal phase of a wave. The wavelength of an oscillating electromagnetic wave depends upon the frequency of the oscillation and the velocity of propagation in the medium or in the transmission system in which the wave is propagating. This article discusses wavelength measurement of electromagnetic waves in the infrared, and optical regions.

As a consequence of the 1983 definition of the meter as the length of the path traveled by light in vacuum during a time interval of 1/299,792,458 of a second, absolute wavelength values are derived directly from the corresponding frequency values by using Eq. (1), where the speed of

$$\lambda_0 = c/f \tag{1}$$

light c is exactly 299,792,458 m/s. In practice, the frequencies of a small number of stablized laser radiations have been precisely determined by reference to the cesium-133 primary frequency standard. These infrared and visible radiations form a set of basic frequency and wavelength standards, with uncertainties between 1 part in 10^9 and 1 part in 10^{10}. SEE WAVELENGTH STANDARDS.

Dispersion methods. Wavelength values to an accuracy of 1 part in 10^5 can be determined with a spectrometer, spectrograph, or monochromator, in which a prism or diffraction grating is used as a dispersive element. Each wavelength forms a line image of the entrance slit at a particular angle. An unknown wavelength can be determined by interpolation with the pattern formed by a lamp emitting the tabulated characteristic wavelengths of a particular element. In particular, the spectra of the iron arc and the thorium lamp are recommended by the International Astronomical Union as standards for this purpose. Care must be taken in using tabulated wavelength values because the values may be either for vacuum or for the refractive index of standard air. (Air wavelengths are smaller than vacuum values by about 1 part in 3000.) SEE DIFFRACTION GRATING; OPTICAL PRISM.

Use of interferometers. The most precise wavelength measurements use an interferometer to compare the unknown wavelength λ_1 with a standard wavelength λ_2. Usually either the two-beam Michelson form or the multiple-beam Fabry-Perot form of interferometer is used. The general equation, applicable to both forms, is Eq. (2), in which θ is the angle of incidence, t is the

$$(m_1 + f_1)\lambda_1 = (m_2 + f_2)\lambda_2 = 2nt \cos \theta \tag{2}$$

real or virtual separation of the reflectors, and n is the refractive index of the medium between the reflectors. Thus at any arbitrary reference point in the interference pattern, Eq. (3) holds,

$$\lambda_1 = \frac{\lambda_2 (m_2 + f_2)}{m_1 + f_1} \tag{3}$$

where m_1 and m_2 are integers, usually called the orders of interference, and f_1 and f_2 are fractions. Maximum transmission of light of wavelength λ occurs when its corresponding fractional order f is zero. With visible light, by accurate interpolation to determine the fractional orders, measurements precise to 10^{-10} are possible with an evacuated interferometer.

Several forms of interferometric "wavemeter" are commercially available for use with laser radiations. One of their principal uses is to measure the wavelength emitted by a dye laser to better than 1 part in 10^6, so that it may be tuned into coincidence with a desired spectral transition. One form of wavemeter has a retroreflector that moves to and fro along a track. This reflector forms part of a two-beam interferometer so that sinusoidal intensity signals are generated. The signals for the unknown and standard radiations are counted electronically, giving totals that correspond to the order numbers m_1 and m_2. SEE LASER SPECTROSCOPY.

Fourier transform method. When a number of wavelengths are mixed together in the input to a moving-carriage two-beam interferometer, the output signal is the summation of the many separate sine-wave signals having different periods. A Fourier analysis of this composite signal enables the separate wavelengths to be identified. This Fourier transform method is particularly useful for the measurement of complex spectra in the infrared. SEE INFRARED SPECTROSCOPY.

WAVELENGTH STANDARDS
Donald A. Jennings

Accurately known wavelengths of spectral radiation emitted from specified sources that are used to measure the wavelengths of other spectra. In the past, the radiation from the standard source and the source under study were superimposed on the slit of a spectrometer (prism or grating) and then the unknown wavelengths could be determined from the standard wavelengths by using interpolation. This technique has evolved into the modern computer-controlled photoelectric recording spectrometer. Accuracy of many more orders of magnitude can be obtained by the use of interferometric techniques, of which Fabry-Perot and Michelson interferometers are two of the most common. *See* INTERFEROMETRY.

The newest definition of the meter is in terms of the second. The wavelength of radiation from the cesium atomic clock is not used to realize length because diffraction problems at this wavelength are severe. Instead, lasers at shorter wavelengths whose frequencies have been measured are used. Frequency measurements can now be made even into the visible spectral region with great accuracy. Hence, when the 1983 Conférence Général des Poids et Mesures redefined the meter, it also gave a list of very accurate wavelengths of selected stabilized lasers which may be used as wavelength standards; these are shown in the **table**. Nearly ten times better accuracy can be achieved by using these wavelengths than by using the radiation from the krypton lamp which provided the previous standard. *See* HYPERFINE STRUCTURE; LASER SPECTROSCOPY; MOLECULAR STRUCTURE AND SPECTRA.

In the past, mainly for visible spectroscopic applications, other wavelength standards such as the mercury and iron lines have been used. These will continue to serve in good stead, and improved values have been reported in the International Astronomical Union Transactions. These old wavelength standards are less accurate by two orders of magnitude and more than the new ones realized from stabilized lasers. However, they do serve a useful purpose as they are less costly and easier to use. In many cases where the spectral lines under investigation are broad or the resolution of the spectrometer is small, the use of these less accurate standards is completely justified. The laser wavelength standards would be used only when the ultimate in accuracy is needed. *See* ASTRONOMICAL SPECTROSCOPY.

The progress in laser frequency measurements since 1974 has established wavelength standards throughout the infrared spectral region. This has been accomplished with the accurate frequency measurement of rotational-vibrational transitions of selected molecules. The OCS molecule is used in the 5-micrometer spectral region. At 9–10 μm, the carbon dioxide (CO_2) laser itself with over 300 accurately known lines is used. From 10 to 100 μm, rotational transitions of various molecules are used; most are optically pumped laser transitions. The increased accuracy

Wavelength standards from stabilized lasers

Absorbing molecule	Transition	Hyperfine component	Wavelength
CH_4	ν_3,P(7)	$F_2(2)$	3 392.231 397 0 nm
$^{127}I_2$	17-1,P(62)	o	576.294 760 27 nm
$^{127}I_2$	11-5,R(127)	i	632.991 398 1 nm
$^{127}I_2$	9-2,R(47)	o	611.970 769 8 nm
$^{127}I_2$	43-0,P(13)	a_3	514.673 466 2 nm

of frequency measurements makes this technique mandatory where ultimate accuracy is needed.
Bibliography. Bureau International des Poids et Mesures, *Comité Consultatif pour la Définition du Mètre*, 7th Session, 1982; International Astronomical Union, *Transactions*, 1962; C. E. Moore, *Atomic Energy Levels*, 3 vols., National Standard Data Reference Systems, 1971.

OPTICAL PRISM
Max Herzberger

An optical system consisting of two or more usually plane surfaces of a transparent solid or embedded liquid at an angle with each other. Prisms are used for deviating light. Since the amount of deviation depends on the refractive index of the prism, which varies with wavelength, prisms can also be used for dispersing light.

Reflecting prisms. Prisms can be used instead of mirrors for deviating light, with the added advantage that the reflecting surfaces are protected against corrosion. In this case there is at least one internal reflection. When the angles of incidence and emergence are zero, there is no dispersion. The overall dispersion is also zero when the geometry of the prism is such that the dispersion at the entering surface is compensated by dispersion in the opposite sense at the emergent surface.

Dispersing prisms. Dispersing prisms deviate light of different wavelengths by different amounts, and they can therefore be used to separate white light into its monochromatic parts. A parallel beam of light entering the prism leaves the prism as a parallel beam of light but its diameter may be changed. The ratio of its diameter after refraction to its diameter before refraction can be considered as the magnification of the prism.

The prism magnification for a bundle parallel to the prism edge is always equal to unity, whereas it varies in the meridional plane (normal to the edge) as the angle of incidence is varied. It is equal to unity in this plane only if the prism is traversed at minimum deviation. If α is the prism angle, n the refractive index, and δ the deviation, it can be shown that, for minimum deviation, the equation below holds.

$$\sin(\delta + \alpha)/2 = n \sin \alpha/2$$

To increase the dispersion, several prisms with their refracting edges parallel can be used. The Rayleigh prism, shown in **Fig. 1**a, is an example of such a system. By using a prism made of a material, such as flint glass, that has a high dispersion, and adding one or more prisms made of a material having a low dispersion, such as crown glass (Fig. 1b), the deviation can be neutralized without neutralizing the dispersion to give a direct-vision prism system. The arrangement shown in Fig. 1b is known as the Amici prism system. By using a similar arrangement but adjusting the angle so that the dispersion, but not the deviation, is neutralized, it is possible to make a prism system that is achromatic over a small part of the spectrum, like an achromatic lens.

An achromatic prism in front of an optical system with its refracting edge normal to the meridional plane can be used to change the magnification of the optical system in that plane. This amount can be varied by rotating prism A in **Fig. 2**a about an axis normal to the meridional plane. A second achromatic prism, B, with its edge parallel to the meridional plane, can be used to adjust the sagittal magnification of the optical system. Therefore, an arrangement of two such

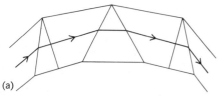
(a)

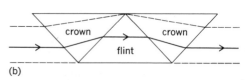
(b)

Fig. 1. Two types of dispersing prisms. (*a*) Rayleigh prism system. (*b*) Amici direct-vision system consisting of a flint-glass prism and two crown-glass prisms.

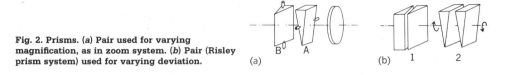

Fig. 2. Prisms. (*a*) Pair used for varying magnification, as in zoom system. (*b*) Pair (Risley prism system) used for varying deviation.

prisms with their motions linked together can be used for the purpose of forming a variable-focal-length lens system, which is commonly called a zoom lens.

A thin prism is one whose angle is so small that the prism angle expressed in radians is practically equal to the tangent. Such prisms are used in ophthalmology, and their powers are usually expressed in prism diopters. The Risley prism system, used for testing ocular convergence, consists of two thin prisms mounted so that they can be rotated simultaneously in opposite directions, as shown in Fig. 2*b*. When they are in the orientation sketched at 1, their combined deviation is zero; when both have been rotated by 90° in opposite directions, as shown at 2, their combined deviation is a maximum; at intermediate positions, their combined deviation lies between zero and the maximum, but the plane of deviation is constant. A similar pair of rotating wedges is used in certain types of rangefinders.

Bibliography. G. A. Boutry, *Instrumental Optics*, 1962; E. U. Condon and H. Odishaw, *Handbook of Physics*, 1967; D. F. Horne, *Optical Instruments and Their Applications*, 1980; F. A. Jenkins and H. E. White, *Fundamentals of Optics*, 4th ed., 1976; Optical Society of America, *Handbook of Optics*, 1978; D. C. O'Shea, *Elements of Modern Optical Design*, 1985.

DIFFRACTION GRATING
GEORGE R. HARRISON

An optical device consisting of an assembly of narrow slits or grooves, which by diffracting light produces a large number of beams which can interfere in such a way as to produce spectra. Since the angles at which constructive interference patterns are produced by a grating depend on the lengths of the waves being diffracted, the waves of various lengths in a beam of light striking the grating will be separated into a number of spectra, produced in various orders of interference on either side of an undeviated central image. By controlling the shape and size of the diffracting grooves when producing a grating and by illuminating the grating at suitable angles, a beam of light can be thrown into a single spectrum whose purity and brightness may exceed that produced by a prism. Gratings can now be made with much larger apertures than prisms, and in such form that they waste less light and give higher intrinsic dispersion and resolving power. A single grating can be used over a much broader range of spectrum than can any single prism, and its dispersion will vary less rapidly with wavelength. Gratings are used in large spectrographs and for highly precise spectroscopic work, as well as in monochromators and analytical spectrographs. *See* OPTICAL PRISM.

Transmission gratings consist of a large number of narrow transparent and opaque slits alternating side by side in regular order and with uniform separation, through which a beam of light will appear as a series of spectra in various orders of interference. Such gratings are conveniently used in small spectroscopes and spectrometers, but only for visible light, since they are usually not transparent to ultraviolet or infrared radiation. They are commonly made by contact molding from a master grating.

Reflection gratings, either plane or concave, are used in most spectrographs. Such a grating may consist of an original ruling or of a metal-coated replica from an original. Large grating replicas practically indistinguishable in performance or permanence from an original can now be made.

Production of gratings. Gratings are engraved by highly precise ruling engines, which use a diamond tool to press into a highly polished mirror surface a series of many thousands of fine shallow burnished grooves. Gratings for the range 150–1000 nanometers are commonly ruled with 5000–30,000 grooves per inch or 200–1200 grooves per millimeter (the usual value is near 15,000 per inch or 600 per millimeter), on a thin layer of aluminum deposited on glass by evaporation in vacuum. Gratings for the infrared region are often ruled on gold, silver, copper, lead, or tin mirrors, with coarser groove spacings.

If a grating is to give resolution approaching the theoretical limit, its grooves must be ruled straight, parallel, and equally spaced to within a few tenths of the shortest incident wavelength. The proper overall spacing of grooves must also be maintained if changes in focal properties are not to result. Scattered light and false images may arise from local spacing error and groove shape variations of only a few hundredths of the diffracted wavelength.

Among the false lines produced by imperfect gratings are Rowland ghosts, which arise from periodic errors in groove position; Lyman ghosts, which come from a combination of periodicities in ruling; and satellites, caused by sets of irregularly placed grooves, which may seriously reduce resolution. Target pattern arises from unequal contribution of light from all parts of a grating, and is especially prevalent in concave gratings, in which the shape of the grooves may change as the cutting angle of the diamond changes.

Gratings of 2, 4, or 6 in. (5, 10, or 15 cm) ruled width are commonly used in commercial spectrographs, with projection distances of 20–180 in. (50–450 cm). In large research instruments, gratings of 6 to 10 in. (15 to 25 cm) ruled width are used with projection distances of 10–50 ft (3–15 m) or more. The largest modern gratings, used in their highest orders, show resolving power $\lambda/\delta\lambda$ in excess of 900,000 in the green region of the spectrum, and in excess of 1.5×10^6 at shorter wavelengths. Here λ is the mean wavelength of two closely spaced, just resolvable spectral lines, and $\delta\lambda$ is their wavelength difference. Such gratings give resolution equal to that of most interferometers, and in addition provide greater photographic speed, are easier to adjust, follow more simple laws of wavelength distribution, and permit a wider range of wavelengths to be photographed at one time without crossed dispersion. SEE INTERFEROMETRY.

Properties of gratings. A grating illuminated at angle α (measured from the normal) will direct wavelength λ toward angle β in accordance with the formula $m\lambda = d(\sin \alpha \pm \sin \beta)$, where m is an integral order of interference, d is the grating constant, or distance between consecutive grooves, and the + and − signs refer to orders on opposite sides of the normal. The linear dispersion produced by a grating on a photographic plate depends on its intrinsic angular dispersion multiplied by the distance P from grating to plate. The intrinsic angular dispersion is given by the formula shown below. Theoretically the resolving power of a grating is mN, where N

$$\frac{d\beta}{d\lambda} = \frac{1}{\lambda}\left(\frac{\sin \alpha}{\cos \beta} + \tan \beta\right)$$

is the number of grooves in the grating. Resolving power is not directly dependent on the number of grooves, since for gratings of a given size m and N are inversely related. It is basically dependent on the number of wavelengths of optical retardation the grating introduces between the extreme rays leaving it. Another useful concept is the resolving limit $d\sigma$, the smallest wave number difference the grating can resolve, which remains essentially constant for a given angle of illumination of a given grating at all wavelengths, except as errors in groove spacing become more important for shorter wavelengths.

The manner in which incident light will be distributed among the various orders of interference depends upon the shape and orientation of the groove sides and on the relation of wavelength to groove separation. When $d \lesssim \lambda$, diffraction effects predominate in controlling the intensity distribution among orders, but when $d > \lambda$, optical reflection from the sides of the grooves is more strongly involved. It is possible to "blaze" a grating by ruling its grooves so that their sides reflect a large fraction of the incoming light of suitably short wavelengths in one general direction. Controlled groove shape is especially important in the gratings known as echelettes and echelles, in which as much as 80% of the incoming light may be sent into one particular order for a given wavelength. Many ordinary gratings are blazed.

Grating spectroscopes. These consist usually of a slit, a lens or mirror to collimate the light sent through the slit into a parallel beam, a transmission or reflection grating to disperse the light, a lens or mirror to focus the light into spectrum lines (which are monochromatic images of the slit in the light of each wavelength passing through it), and an eyepiece for viewing the

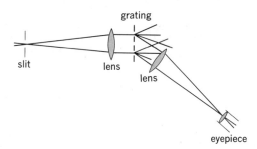

Fig. 1. Transmission grating spectroscope.

spectrum (**Fig. 1**). If a camera is substituted for the telescope, the instrument becomes a grating spectrograph. If a photoelectric cell, a thermocouple, or other radiation-detecting device is used instead of a camera or telescope, the device becomes a grating spectrometer. For some important applications of the latter device SEE INFRARED SPECTROSCOPY.

Echelette grating. This has coarse groove spacing and is designed for the infrared region, the grooves being so shaped and of such size that most of the radiation is concentrated by reflection into a small angular coverage. Radiation of any given wavelength is thus concentrated largely into one order by shaping the point of the ruling diamond to give grooves with comparatively flat sides and by choosing a groove separation which minimizes diffraction effects.

Echelle grating. This is designed for use in high orders and at angles of illumination greater than 45° to obtain high dispersion and resolving power by the use of high orders of interference. Echelles have properties lying midway between those of plane gratings of the ordinary type and interferometers of the reflection echelon type, orders of interference ranging from 100 to 1000 being used. Overlapping orders are separated by using crossed dispersion.

Concave grating. This is a widely used form of reflection grating with which a spectrograph can be formed that has no auxiliary optical parts except a slit and a camera. Being ruled on a concave mirror, this type of grating can both collimate and focus the light that falls upon it. It is made by spacing straight grooves equally along the chord (rather than the arc) of a spherical or paraboloidal mirror surface. Light which passes through a slit and falls on such a grating is dispersed by it into spectra which are in focus on the Rowland circle, a circle drawn tangent to the face of the grating at its midpoint, having a diameter equal to the radius of curvature of the grating surface.

A great advantage of the concave grating is that it provides a dispersing and focusing system free of refracting material, so that it can be used with ultraviolet, visible, or infrared radiation interchangeably so long as its grooves diffract radiation and its surface has adequate reflecting power. A disadvantage is its astigmatism at high angles of incidence or reflection, which can, however, be diminished with various optical devices. A plane reflection grating used with two concave mirrors avoids this difficulty.

Grating mountings. The slit, grating, and camera of a concave grating spectrograph can be placed anywhere on the Rowland circle so that any desired wavelength range can be photographed in the desired order (**Fig. 2**).

The various possible combinations of fixed and moving parts give rise to a number of different grating mountings. In the Rowland mounting, camera and grating are connected by a bar forming a diameter of the Rowland circle, the two running on tracks placed at right angles with the slit fixed at their junction. A spectrum of limited extent having uniform dispersion is then produced at the camera, and camera and grating can be moved on the tracks to shift wavelength coverage. In the Paschen-Runge mounting, slit and grating are fixed, and photographic plates can be clamped to a fixed track almost anywhere on the Rowland circle. In the Eagle mounting, most suited to long, narrow housing, the grating can be rotated and moved toward or away from the slit. The slit is placed close to the camera, which is arranged to rotate so that it can be kept on the Rowland circle.

All these mountings suffer from astigmatism arising from using the grating off-axis. Although this does not markedly reduce the sharpness of the spectrum lines, it may result in a great

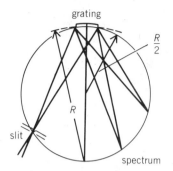

Fig. 2. Rowland circle.

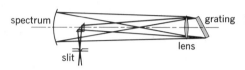

Fig. 3. Littrow mounting of a plane grating.

loss of light intensity when the grating is used at high angles, and it makes difficult the sharp focusing of step filters, sector disks, and interferometer patterns that are placed at the slit.

Astigmatism is greatly reduced in the Wadsworth mounting, in which the slit is placed at the principal focus of a concave mirror, so that the light falling on the grating is in a parallel beam. The grating can be illuminated at any desired angle up to about 40°, and light is taken off along the grating normal, the spectrum being focused on a photographic plate at half the usual distance. The usual dispersion of the grating is halved, but the speed of the spectrograph is increased fourfold, and at high angles the speed is increased much more because of reduction of astigmatism.

Most modern commercial grating spectrographs, because of the need for portability, are based either on the Eagle mounting, with which a rather limited portion of the spectrum can be photographed at one time, or the Wadsworth mounting, which can give greater spectral coverage without resetting but is bulkier and cannot be used at such high values of $m\lambda$. In order to obtain more complete spectrum coverage in a single exposure, an echelle with crossed dispersion or a grating with some other device for separating the orders may be used.

Plane reflection gratings are ordinarily used in the Littrow mounting (**Fig. 3**), in which a single lens serves for collimating and focusing, or in the Ebert mounting in which there are two concave mirrors.

Bibliography. R. Chang, *Basic Principles of Spectroscopy*, 1978; M. C. Hutley, *Diffraction Gratings*, 1982; F. A. Jenkins and H. E. White, *Fundamentals of Optics*, 4th ed., 1976; S. B. Kessler (ed.), *Modern Spectrum Analysis*, vol. 2, 1986; D. C. O'Shea, *Elements of Modern Optical Design*, 1985; D. L. Pavia et al., *Introduction to Spectroscopy*, 1979; D. A. Ramsay, *Spectroscopy*, 1976.

INTERFEROMETRY
James C. Wyant

The design and use of optical interferometers. Optical interferometers based on both two-beam interference and multiple-beam interference of light are extremely powerful tools for metrology and spectroscopy. A wide variety of measurements can be performed, ranging from determining the shape of a surface to an accuracy of less than a millionth of an inch (25 nanometers) to determining the separation, by millions of miles, of binary stars. In spectroscopy, interferometry can be used to determine the hyperfine structure of spectrum lines. By using lasers in classical interferometers as well as holographic interferometers and speckle interferometers, it is possible to perform deformation, vibration, and contour measurements of diffuse objects that could not previously be performed.

Basic classes of interferometers. There are two basic classes of interferometers: division of wavefront and division of amplitude. **Figure 1** shows two arrangements for obtaining

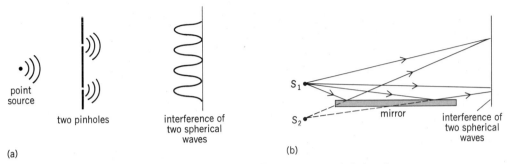

Fig. 1. Interference produced by division of wavefront. (*a*) Young's two-pinhole interferometer. (*b*) Lloyd's mirror.

division of wavefront. For the Young's double pinhole interferometer (Fig. 1*a*), the light from a point source illuminates two pinholes. The light diffracted by these pinholes gives the interference of two point sources. For the Lloyd's mirror experiment (Fig. 1*b*), a mirror is used to provide a second image S_2 of the point source S_1, and in the region of overlap of the two beams the interference of two spherical beams can be observed. There are many other ways of obtaining division of wavefront; however, in each case the light leaving the source is spatially split, and then by use of diffraction, mirrors, prisms, or lenses the two spatially separated beams are superimposed.

Figure 2 shows one technique for obtaining division of amplitude. For division-of-amplitude interferometers a beam splitter of some type is used to pick off a portion of the amplitude of the radiation which is then combined with a second portion of the amplitude. The visibility of the resulting interference fringes is a maximum when the amplitudes of the two interfering beams are equal.

Michelson interferometer. The Michelson interferometer (**Fig. 3**) is based on division of amplitude. Light from an extended source S is incident on a partially reflecting plate (beam splitter) P_1. The light transmitted through P_1 reflects off mirror M_1 back to plate P_1. The light which is reflected proceeds to M_2 which reflects it back to P_1. At P_1, the two waves are again partially reflected and partially transmitted, and a portion of each wave proceeds to the receiver R, which may be a screen, a photocell, or a human eye. Depending on the difference between the distances from the beam splitter to the mirrors M_1 and M_2, the two beams will interfere constructively or destructively. Plate P_2 compensates for the thickness of P_1. Often when a quasimonochromatic light source is used with the interferometer, compensating plate P_2 is omitted.

The function of the beam splitter is to superimpose (image) one mirror onto the other. When the mirrors' images are completely parallel, the interference fringes appear circular. If the mirrors

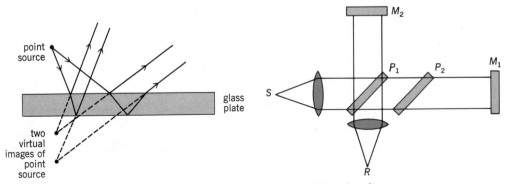

Fig. 2. Division of amplitude.

Fig. 3. Michelson interferometer.

are slightly inclined about a vertical axis, vertical fringes are formed across the field of view. These fringes can be formed in white light if the path difference in part of the field of view is made zero. Just as in other interference experiments, only a few fringes will appear in white light, because the difference in path will be different for wavelengths of different colors. Accordingly, the fringes will appear colored close to zero path difference, and will disappear at larger path differences where the fringe maxima and minima for the different wavelengths overlap. If light reflected off the beam splitter experiences a one-half-cycle relative phase shift, the fringe of zero path difference is black, and can be easily distinguished from the neighboring fringes. This makes use of the instrument relatively easy.

The Michelson interferometer can be used as a spectroscope. Consider first the case of two close spectrum lines as a light source for the instrument. As the mirror M_1 is shifted, fringes from each spectral line will cross the field. At certain path differences between M_1 and M_2, the fringes for the two spectral lines will be out of phase and will essentially disappear; at other points they will be in phase and will be reinforced. By measuring the distance between successive maxima in fringe contrast, it is possible to determine the wavelength difference between the lines.

This is a simple illustration of a very broad use for any two-beam interferometer. As the path length L is changed, the variation in intensity $I(L)$ of the light coming from an interferometer gives information on the basis of which the spectrum of the input light can be derived. The equation for the intensity of the emergent energy can be written as Eq. (1), where β is a constant,

$$I(L) = \int_0^\infty I(\lambda) \cos^2\left(\frac{\beta L}{\lambda}\right) d\lambda \tag{1}$$

and $I(\lambda)$ is the intensity of the incident light at different wavelengths λ. This equation applies when the mirror M_1 is moved linearly with time from the position where the path difference with M_2 is zero, to a position which depends on the longest wavelength in the spectrum to be examined. From Eq. (1), it is possible mathematically to recover the spectrum $I(\lambda)$. In certain situations, such as in the infrared beyond the wavelength region of 1.5 micrometers, this technique offers a large advantage over conventional spectroscopy in that its utilization of light is extremely efficient. SEE INFRARED SPECTROSCOPY.

Twyman-Green interferometer. If the Michelson interferometer is used with a point source instead of an extended source, it is called a Twyman-Green interferometer. The use of the laser as the light source for the Twyman-Green interferometer has made it an extremely useful instrument for testing optical components. The great advantage of a laser source is that it makes it possible to obtain bright, good-contrast, interference fringes even if the path lengths for the two arms of the interferometer are quite different.

Figure 4 shows a Twyman-Green interferometer for testing a flat mirror. The laser beam is expanded to match the size of the sample being tested. Part of the laser light is transmitted to the reference surface, and part is reflected by the beam splitter to the flat surface being tested. Both beams are reflected back to the beam splitter, where they are combined to form interference fringes. An imaging lens projects the surface under test onto the observation plane.

Fringes (**Fig. 5**) show defects in the surface being tested. If the surface is perfectly flat, then straight, equally spaced fringes are obtained. Departure from the straight, equally spaced condition shows directly how the surface differs from being perfectly flat. For a given fringe, the difference in optical path between light going from laser to reference surface to observation plane and the light going from laser to test surface to observation plane is a constant. (The optical path is equal to the product of the geometrical path times the refractive index.) Between adjacent fringes (Fig. 5), the optical path difference changes by one wavelength, which for a helium-neon laser corresponds to 633 nm. The number of straight, equally spaced fringes and their orientation depend upon the tip-tilt of the reference mirror. That is, by tipping or tilting the reference mirror the difference in optical path can be made to vary linearly with distance across the laser beam.

Deviations from flatness of the test mirror also cause optical path variations. A height change of half a wavelength will cause an optical path change of one wavelength and a deviation from fringe straightness of one fringe. Thus, the fringes give surface height information, just as a topographical map gives height or contour information.

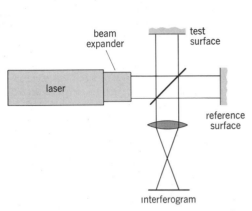

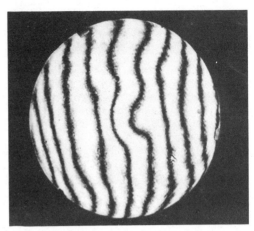

Fig. 4. Twyman-Green interferometer for testing flat surfaces.

Fig. 5. Interferogram obtained with the use of a Twyman-Green interferometer to test a flat surface.

The existence of the essentially straight fringes provides a means of measuring surface contours relative to a tilted plane. This tilt is generally introduced to indicate the sign of the surface error, that is, whether the errors correspond to a hill or a valley. One way to get this sign information is to push in on the piece being tested when it is in the interferometer. If the fringes move toward the right when the test piece is pushed toward the beam splitter, then fringe deviations from straightness toward the right correspond to high points (hills) on the test surface and deviations to the left correspond to low points (valleys).

The basic Twyman-Green interferometer (Fig. 4) can be modified (**Fig. 6**) to test concave-spherical mirrors. In the interferometer, the center of curvature of the surface under test is placed at the focus of a high-quality diverger lens so that the wavefront is reflected back onto itself. After this retroflected wavefront passes through the diverger lens, it will be essentially a plane wave, which, when it interferes with the plane reference wave, will give interference fringes similar to those shown in Fig. 5 for testing flat surfaces. In this case it indicates how the concave-spherical mirror differs from the desired shape. Likewise, a convex-spherical mirror can be tested. Also, if a high-quality spherical mirror is used, the high-quality diverger lens can be replaced with the lens to be tested.

Fizeau interferometer. One of the most commonly used interferometers in optical metrology is the Fizeau interferometer, which can be thought of as a folded Twyman-Green interoferometer. In the Fizeau, the two surfaces being compared, which can be flat, spherical, or aspherical, are placed in close contact. The light reflected off these two surfaces produces interference fringes. For each fringe, the separation between the two surfaces is a constant. If the

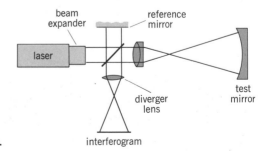

Fig. 6. Twyman-Green interferometer for testing spherical mirrors or lenses.

two surfaces match, straight, equally spaced fringes result. Surface height variations between the two surfaces cause the fringes to deviate from straightness or equal separations, where one fringe deviation from straightness corresponds to a variation in separation between the two surfaces by an amount equal to one-half of the wavelength of the light source used in the interferometer. The wavelength of a helium source, which is often used in a Fizeau interferometer, is 587.56 nm; hence one fringe corresponds to a height variation of approximately 0.3 μm.

Mach-Zehnder interferometer. The Mach-Zehnder interferometer (**Fig. 7**) is a variation of the Michelson interferometer and, like the Michelson interferometer, depends on a amplitude splitting of the wavefront. Light enters the instrument and is reflected and transmitted by the semitransparent mirror M_1. The reflected portion proceeds to M_3, where it is reflected through the cell C_2 to the semitransparent mirror M_4. Here it combines with the light transmitted by M_1 to produce interference. The light transmitted by M_1 passes through a cell C_1, which is similar to C_2 and is used to compensate for the windows of C_1.

The major application of this instrument is in studying airflow around models of aircraft, missiles, or projectiles. The object and associated airstream are placed in one arm of the interferometer. Because the air pressure varies as it flows over the model, the index of refraction varies, and thus the effective path length of the light in this beam is a function of position. When the variation is an odd number of half-waves, the light will interfere destructively and a dark fringe will appear in the field of view. From a photograph of the fringes, the flow pattern can be mathematically derived.

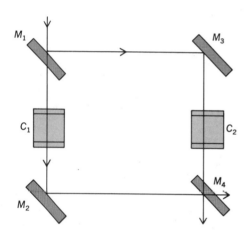

Fig. 7. Mach-Zehnder interferometer.

A major difference between the Mach-Zehnder and the Michelson interferometer is that in the Mach-Zehnder the light goes through each path in the instrument only once, whereas in the Michelson the light traverses each path twice. This double traversal makes the Michelson interferometer extremely difficult to use in applications where spatial location of index variations is desired. The incoming and outgoing beams tend to travel over slightly different paths, and this lowers the resolution because of the index gradient across the field.

Shearing interferometers. In a lateral-shear interferometer, an example of which is shown in **Fig. 8**, a wavefront is interfered with a shifted version of itself. A bright fringe is obtained at the points where the slope of the wavefront times the shift between the two wavefronts is equal to an integer number of wavelengths. That is, for a given fringe the slope or derivative of the wavefront is a constant. For this reason a lateral-shear interferometer is often called a differential interferometer.

Another type of shearing interferometer is a radial-shear interferometer. Here, a wavefront is interfered with an expanded version of itself. This interferometer is sensitive to radial slopes.

The advantages of shearing interferometers are that they are relatively simple and inexpen-

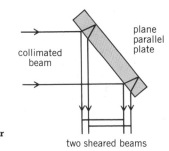

Fig. 8. Lateral shear interferometer.

sive, and since the reference wavefront is self-generated, an external wavefront is not needed. Since an external reference beam is not required, the source requirements are reduced from those of an interferometer such as a Twyman-Green. For this reason, shearing interferometers, in particular lateral-shear interferometers, are finding much use in applications such as adaptive optics systems for correction of atmospheric turbulence where the light source has to be a star, or planet, or perhaps just reflected sunlight.

Michelson stellar interferometer. A Michelson stellar interferometer can be used to measure the diameter of stars which are as small as 0.01 second of arc. This task is impossible with a ground-based optical telescope since the atmosphere limits the resolution of the largest telescope to not much better than 1 second of arc.

The Michelson stellar interferometer is a simple adaptation of Young's two-slit experiment. In its first form, two slits were placed over the aperture of a telescope. If the object being observed were a true point source, the image would be crossed with a set of interference bands. A second point source separated by a small angle from the first would produce a second set of fringes. At certain values of this angle, the bright fringes in one set will coincide with the dark fringes in the second set. The smallest angle α at which the coincidence occurs will be that angle subtended at the slits by the separation of the peak of the central bright fringe from the nearest dark fringe. This angle is given by Eq. (2), where d is the separation of the slits, λ the dominant wavelength

$$\frac{\lambda}{2d} = \alpha \qquad (2)$$

of the two sources, and α their angular separation. The measurement of the separation of the sources is performed by adjusting the separation d between the slits until the fringes vanish.

Consider now a single source in the shape of a slit of finite width. If the slit subtends an angle at the telescope aperture which is larger than α, the interference fringes will be reduced in contrast. For various line elements at one side of the slit, there will be elements of angle α away which will cancel the fringes from the first element. By induction, it is clear that for a separation d' such that the slit source subtends an angle as given by Eq. (3) the fringes from a single slit

$$\alpha' = \frac{\lambda}{d'} \qquad (3)$$

will vanish completely. For additional information on the Michelson stellar interferometer.

Fabry-Perot interferometer. All the interferometers discussed above are two-beam interferometers. The Fabry-Perot interferometer (**Fig. 9**) is a multiple-beam interferometer since the

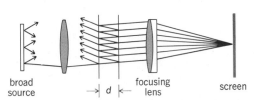

Fig. 9. Fabry-Perot interferometer.

two glass plates are partially silvered on the inner surfaces, and the incoming wave is multiply reflected between the two surfaces. The position of the fringe maxima is the same for multiple-beam interference as two-beam interference; however, as the reflectivity of the two surfaces increases and the number of interfering beams increases, the fringes become sharper.

A quantity of particular interest in a Fabry-Perot is the ratio of the separation of adjacent maxima to the half-width of the fringes. It can be shown that this ratio, known as the finesse, is given by Eq. (4), where R is the reflectivity of the silvered surfaces.

$$\mathcal{F} = \frac{\pi\sqrt{R}}{1-R} \quad (4)$$

The multiple-beam Fabry-Perot interferometer is of considerable importance in modern optics for spectroscopy. All the light rays incident on the Fabry-Perot at a given angle will result in a single circular fringe of uniform irradiance. With a broad diffuse source, the interference fringes will be narrow concentric rings, corresponding to the multiple-beam transmission pattern. The position of the fringes depends upon the wavelength. This is, each wavelength gives a separate fringe pattern. The minimum resolvable wavelength difference is determined by the ability to resolve close fringes. The ratio of the wavelength λ to the least resolvable wavelength difference $\Delta\lambda$ is known as the chromatic resolving power $\mathcal{R}$. At nearly normal incidence it is given by Eq. (5), where n is the refractive index between the two mirrors separated a distance d. For a wave-

$$\mathcal{R} = \frac{\lambda}{(\Delta\lambda)_{min}} = \mathcal{F}\frac{2nd}{\lambda} \quad (5)$$

length of 500 nm, $nd = 10$ mm, and $R = 90\%$, the resolving power is well over 10^6.

When Fabry-Perot interferometers are used with lasers, they are generally used in the central spot scanning mode. The interferometer is illuminated with a collimated laser beam and all the light transmitted through the Fabry-Perot is focused onto a detector, whose output is displayed on an oscilloscope. Often one of the mirrors is on a piezoelectric mirror mount. As the voltage to the piezoelectric crystal is varied, the mirror separation is varied. The light output as a function of mirror separation gives the spectral frequency content of the laser source.

Holographic interferometry. A wave recorded in a hologram is effectively stored for future reconstruction and use. Holographic interferometry is concerned with the formation and interpretation of the fringe pattern which appears when a wave, generated at some earlier time and stored in a hologram, is later reconstructed and caused to interfere with a comparison wave. It is the storage or time-delay aspect which gives the holographic method a unique advantage over conventional optical interferometry.

A hologram can be made of an arbitrarily shaped, rough scattering surface, and after suitable processing, if the hologram is illuminated with the same reference wavefront used in recording the hologram, the hologram will produce the original object wavefront. If the hologram is placed back into its original position, a person looking through the hologram will see both the original object and the image of the object stored in the hologram. If the object is now slightly deformed, interference fringes will be produced which tell how much the surface is deformed. Between adjacent fringes the optical path between the source and viewer has changed by one wavelength. While the actual shape of the object is not determined, the change in the shape of the object is measured to within a small fraction of a wavelength, even though the object's surface is rough compared to the wavelength of light.

Double-exposure. Double-exposure holographic interferometry (**Fig. 10**) is similar to real-time holographic interferometry described above, except now two exposures are made before processing: one exposure with the object in the undeformed state and a second exposure after deformation. When the hologram reconstruction is viewed, interference fringes will be seen which show how much the object was deformed between exposures.

The advantage of the double-exposure technique over the real-time technique is that there is no critical replacement of the hologram after processing. The disadvantage is that continuous comparison of surface displacement relative to an initial state cannot be made, but rather only the difference between two states is determined.

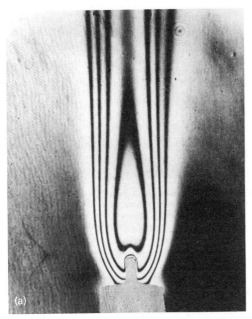 

Fig. 10. Double-exposure holographic interferograms. (a) Interferogram of candle flame. (b) Interferogram of debanded region of honeycomb construction panel. (*From C. M. Vest, Holographic Interferometry, John Wiley and Sons, 1978*)

Time-average. In time-average holographic interferometry (**Fig. 11**) a time-average hologram of a vibrating surface is recorded. If the maximum amplitude of the vibration is limited to some tens of light wavelengths, illumination of the hologram yields an image of the surface on which is superimposed several interference fringes which are contour lines of equal displacement of the surface. Time-average holography enables the vibrational amplitudes of diffusely reflecting surfaces to be measured with interferometric precision.

Speckle interferometry. A random intensity distribution, called a speckle pattern, is generated when light from a highly coherent source, such as a laser, is scattered by a rough surface. The use of speckle patterns in the study of object displacements, vibration, and distortion is becoming of more importance in the nondestructive testing of mechanical components. For example, time-averaged speckle photographs can be used to analyze the vibrations of an object in its plane. In physical terms the speckles in the image are drawn out into a line as the surface vibrates, instead of being double as in the double-exposure technique. The diffraction pattern of this smeared-out speckle-pattern recording is related to the relative time spent by the speckle at each point of its trajectory (**Fig. 12**).

Speckle interferometry can be used to perform astronomical measurements similar to those performed by the Michelson stellar interferometer. Stellar speckle interferometry is a technique for obtaining diffraction-limited resolution of stellar objects despite the presence of the turbulent atmosphere that limits the resolution of ground-based telescopes to approximately 1 second of arc. For example, the diffraction limit of the 200-in.-diameter (5-m) Palomar Mountain telescope is approximately 0.02 second of arc, 1/50 the resolution limit set by the atmosphere.

The first step of the process is to take a large number, perhaps 100, of short exposures of the object, where each photo is taken for a different realization of the atmosphere. Next the optical diffraction pattern, that is, the squared modulus of the Fourier transform of all the short-exposure photographs, is added. By taking a further Fourier transform of each ensemble average diffraction pattern, the ensemble average of the spatial autocorrelation of the diffraction-limited images of each object is obtained.

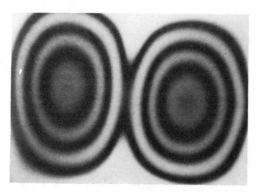

Fig. 11. Photograph of time-average holographic interferogram. (*From C. M. Vest, Holographic Interferometry, John Wiley and Sons, 1978*)

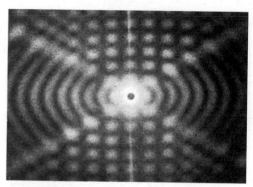

Fig. 12. Diffraction patterns from time-averaged speckle interferogram of a surface vibrating in its own plane with a figure-sight motion. (*From J. C. Dainty, Laser Speckle and Related Phenomena, Springer-Verlag, 1975*)

Phase-shifting interferometry. Electronic phase-measurement techniques can be used in interferometers such as the Twyman-Green, where the phase distribution across the interferogram is being measured. Phase-shifting interferometry is often used for these measurements since it provides for rapid precise measurement of the phase distribution. In phase-shifting interferometry, the phase of the reference beam in the interferometer is made to vary in a known manner. This can be achieved, for example, by mounting the reference mirror on a piezoelectric transducer. By varying the voltage on the transducer, the reference mirror is moved a known amount to change the phase of the reference beam a known amount. A solid-state detector array is used to detect the intensity distribution across the interference pattern. This intensity distribution is read into computer memory three or more times, and between each intensity measurement the phase of the reference beam is changed a known amount. From these three or more intensity measurements, the phase across the interference pattern can be precisely determined to within a fraction of a degree.

SPECTROHELIOSCOPE

ROBERT R. McMATH AND JOHN W. EVANS

An instrument for the monochromatic visual observation of the Sun. A telescope projects an image of the Sun on the first slit of a powerful spectroscope (**Fig. 1**). The resulting spectrum is imaged in the plane of a second slit which permits only a single line element of the spectrum to emerge from the instrument. The emergent line element is a monochromatic image of that part of the Sun that falls on the first slit. The widths of the slits are generally chosen to isolate a spectral interval 0.05 nanometer (0.5 angstrom) or less in width. When the two slits are vibrated synchronously at

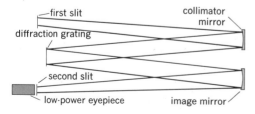

Fig. 1. Hale spectrohelioscope.

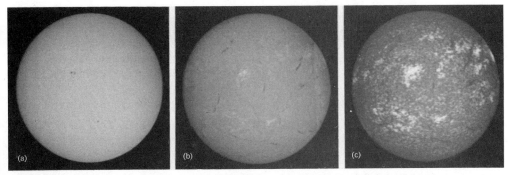

Fig. 2. Spectroheliograms of the Sun (1958 June $20^d 12^h 10^m$ UT) formed of line elements 0.03 nm wide. (a) At 660.0 nm. (b) At 656.3 nm (Hα). (c) At 393.3 nm (K). The slits were moved at a slow rate.

high frequency, persistence of vision permits monochromatic observation of an area of the solar surface. The slits may also be moved at a slow rate and the image recorded photographically (**Fig. 2**). This modification of the spectrohelioscope is a simple form of the spectroheliograph.

SPECTROGRAPHY
WALTER CLARK

The use of photography to record the electromagnetic spectrum displayed in a spectroscope. The technique is used mainly in atomic and molecular physics, in analysis of the chemical composition of materials, and in astronomical photography. The sources of radiation for spectrography are incandescent or electrically excited, and the spectrum contains lines characteristic of the elements and gives definite evidence of their presence. Continuous spectra are also emitted by incandescent sources, but are little used except in absorption measurements. Band spectra are characteristic of molecules. The position and intensity of the lines and bands in the spectrum is a measure of the nature and amount of the elements present.

A great variety of photographic plates and films is made, having a range of speeds, contrast, resolving power, and spectral sensitivity, permitting spectrography from very short-wavelength ultraviolet to the infrared at about 1300 nanometers. *See* SPECTROCHEMICAL ANALYSIS.

GAMMA-RAY DETECTORS
J. W. OLNESS

Devices for detecting gamma rays. Most detectors are capable of detecting and registering the passage of individual gamma rays, or photons. A detector which simultaneously determines the energy of a gamma ray may be properly termed a spectrometer. The term detector is frequently used as a generic name for devices that are also spectrometers. *See* GAMMA RAYS.

Detector-spectrometer devices. The two most popular devices in this category are the (thallium-activated) sodium-iodide [NaI(Tl)] detector and the (lithium-drifted) germanium [Ge(Li)] detector, which were developed in the 1950s and 1960s, respectively. These detectors are constructed from large single crystals of Ge or NaI; the activator element or dopant critical to operation of the device is written in parentheses, for example, (Tl).

Since gamma rays have no intrinsic charge or mass, both detectors depend on the fact that the three processes whereby gamma rays interact with matter (photoemission, Compton effect, and pair production) result in the production of energetic electrons whose total energy is exactly proportional to the original gamma ray energy. The electrons, in turn, lose energy by the ionization of the material (Ge or NaI), producing thereby a number of ions pairs exactly proportional to the electron energy. In both the NaI(Tl) and Ge(Li) detectors, the ionization energy is ultimately converted to a voltage-current pulse proportional to the total ionization produced by the gamma ray interaction. The output pulses from the detector (after amplification) are then analyzed by a pulse-height analyzer to provide a quantitative measure of the pulse-height spectrum.

In the ideal spectrometer, a uniform gamma ray energy should produce a uniform voltage pulse which results in a peak in the pulse-height spectrum at a pulse height proportional to the gamma ray energy. Of the three interactions, however, only the photoemission process results in a direct total conversion of gamma energy to ion energy. In the Compton process, the scattering of a gamma ray of initial energy E_γ results in a photon of lesser energy E'_γ and a broad distribution of electron energies (pulse heights), ranging from 0 to $E_\gamma/(1 + m_0c^2/2E_\gamma)$, is obtained; here m_0c^2 is the rest energy (511 keV) of the scattered electron. In the pair-production process, the gamma energy is converted into a positron and an electron pair of total kinetic energy $E_\gamma - 2m_0c^2$. The positron subsequently annihilates by pairing itself with another electron to produce two gamma rays of energy 511 keV. The lower-energy photons resulting from the Compton and pair-production processes may themselves subsequently interact by means of the first two processes. Thus, if the detector volume is large enough, the increased probability for multiple processes ensures that there also is a reasonable probability for complete conversion of gamma energy to ionization energy.

A major aim of detector development has therefore been the production of large-volume detectors which retain the essential property of high energy-resolving power. These detectors are characterized by relatively large, narrow, full-energy peaks. Because of incomplete conversion, a broad Compton distribution will remain, together with weaker subsidiary peaks (at $E_\gamma = 511$ keV and $E_\gamma = 1022$ keV) resulting from incomplete energy conversion in the pair-production process due to escape from the detector of the 511-keV annihilation radiations.

Ge(Li) semiconductor detectors. Illustration *a* is a schematic illustration of a "coaxial" Ge(Li) detector, showing the structure in terms of the *n*-type (*n*), intrinsic (*i*), and *p*-type (*p*) volumes. This is a single-crystal semiconductor device in which a large voltage applied in the reverse direction between the coaxial *n* and *p* surfaces produces a correspondingly large radial electric field, but a very low radial current.

Two of the three gamma ray interaction processes are shown schematically in illus. *a*, in which the production of ionization with the active (intrinsic) volume by the interacting electron is illustrated. The resultant distribution of pulse heights from these two processes in indicated in illus. *b*. The negative ionization products (electrons) are rapidly swept by the radial electric field to the *n* surface, whereas the positive "holes" are swept to the *p* surface, resulting in a current pulse exactly proportional to the total ionization produced by the interaction electron, and thus also to the initial electron energy. Assuming the detector is large enough so that the gamma ray energy is totally transferred (by means of multiple interactions) to ionizing electrons, the resultant

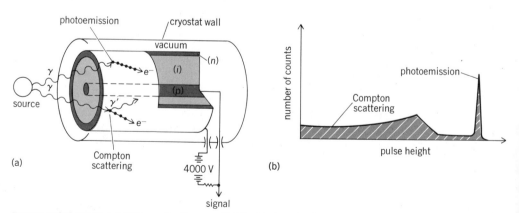

Coaxial Ge(Li) detector. (*a*) Schematic diagram. (*b*) Distribution of pulse heights from photoemission and Compton process.

current pulse is also exactly proportional to the gamma ray energy, and the necessary condition for a useful and efficient spectrometer is realized.

The typical large detector is approximately 6 in.^{3} (100 cm^3) in volume, roughly 2 in. (5 cm) in diameter by 2 in. (5 cm) in length. The resolution, expressed as the full width at half maximum (FWHM) of the full energy peak, is approximately 2.0 keV for 1332-keV gamma rays. The peak efficiency is about 20–30% that of a 3-in. by 3-in. (7.6-cm by 7.6-cm) NaI(Tl) detector; the ratio of peak-to-Compton counts is approximately 30:1.

Ge(Li) detectors are prepared from single crystals of germanium by coating the n surface with lithium and then diffusing the lithium ions radially inward to form the intrinsic volume; within this volume they serve the purpose of compensating for impurity atoms within the crystalline lattice. The crystal is contained in a vacuum cryostat and operated at liquid-nitrogen temperatures. In addition to the coaxial design shown in illustration a, "planar" and "trapezoidal" configurations are also used.

Two other semiconductor detectors, the lithium-drifted silicon [Si(Li)] detector and the intrinsic germanium [Ge(HP)] detector, have been widely used. Both are operated in configurations similar to that described for the Ge(Li) detector. The Si(Li) detector offers slightly improved resolution for gamma rays of energy less than 400 keV, but the detection efficiency for higher-energy gamma rays is relatively poor. Detectors using ultrapure intrinsic germanium without lithium compensation have also been developed. The advantages, in terms of stability and serviceable lifetime, are outweighed in most cases by the smaller size (1.2 in.^{3} or 20 cm^3, as opposed to 6 in.^{3} or 100 cm^3) and relatively high cost of these detectors.

NaI(Tl) detector. This is a member of the family of detectors designated scintillation detectors. Other frequently used scintillators are CsI(Tl) and various organic phosphors, both liquid and solid, of the general form $(CH_2)^n$.

In the NaI(Tl) detector the ionization energy produced by the gamma-ray interaction is converted into a light pulse generated by the subsequent recombination of these ions. The many low-energy photons composing this light pulse are reflected onto the photocathode of a photomultiplier tube, optically coupled to the crystal, to produce finally a current pulse which is then amplified by the photomultiplier to a useful current range. The purpose of the thallium is to "shift" the resultant light spectrum into a region appropriate to the spectral response of the photocathode.

The most popular crystal configurations are cylindrical, ranging in size (diameter by height) from 2 by 2 in. to 5 by 6 in. (5 by 5 cm to 12.5 by 15 cm). Crystals as large as 12 by 12 in. (30 by 30 cm) have been employed, and a variety of smaller detectors of differing configurations have been utilized for specific purposes. The photomultiplier is optically coupled to one of the plane faces, and provides a current gain of 10^5–10^{10}; the remaining surfaces of the crystal are surrounded by a light-reflective shield. The entire assembly must be airtight, because sodium iodide is extremely hydroscopic.

Although the NaI(Tl) detector is more efficient than the Ge(Li) detector, its peak resolution is considerably poorer; for example, for 1332-keV gamma rays the resolution (FWHM) is approximately 80 keV, and the peak-Compton ratio is only about 6. It is very widely employed, however, because of its durability, relatively low cost, and high absolute peak efficiency (30–90%).

Other detectors. Gas-filled ionization chambers have also been used for gamma ray detection. The efficiencies are generally poor compared to those of NaI(Tl) or Ge(Li) detectors, and the energy-resolving capabilities are either inferior or nonexistent. But due to relatively low cost, they are still used in specialized applications, particularly for radiation monitoring for personnel safety.

Other spectrometers. Specific investigation in basic and industrial research have employed specialized spectrometers for gamma ray analysis. Among these, the crystal diffraction spectrometer has been frequently used for measurements of gamma rays up to 1 MeV in energy. The Compton spectrometer and the pair spectrometer, both of which employ magnetic analysis of interaction electrons ejected from thin metallic foils, have also been utilized for analysis of higher-energy gamma rays.

Bibliography. D. A. Bromley, *Detectors in Nuclear Science, Nuclear Instruments and Methods*, vol. 162, 1979; K. Debertin and W. B. Mann, *Gamma and X-ray Spectrometry Techniques and Applications*, 1983; P. J. Ouseph, *Introduction to Nuclear Radiation Detectors*, 1975; W. J. Price, *Nuclear Radiation Detection*, 2d ed., 1964; S. Raman, *Capture Gamma-Ray Spectroscopy and Related Topics 1984: International Symposium*, 1985; E. Segré, *Nuclei and Particles*, 2d ed., 1977.

MATRIX ISOLATION
LESTER ANDREWS

A technique for providing a means of maintaining molecules at low temperature for spectroscopic study. This method is particularly well suited for preserving reactive species in a solid, inert environment. Elusive molecular fragments, such as free radicals that may be postulated as important controlling intermediates for chemical transformations used in industrial reactions, high-temperature molecules that are in equilibrium with solids at very high temperatures, and molecular ions that are produced in plasma discharges or by high-energy radiation all can be examined by using absorption (infrared, visible, and ultraviolet), electron-spin resonance, and laser-excitation spectroscopes.

Experimental apparatus. The experimental apparatus for matrix isolation experiments is designed with the method of generating the molecular transient and performing the spectroscopy in mind. **Figure 1** shows the cross section of a vacuum vessel used for absorption spectroscopic measurements. The optical windows must be transparent to the examining radiation. The rotatable cold window is cooled to 10–20 K (−423 to −441°F) by using closed-cycle refrigeration or 4 K (452°F) by using liquid helium. The matrix sample is introduced through the spray-on line at rates of 1–5 millimoles per hour; argon is the most widely used matrix gas, although neon, krypton, xenon, and nitrogen are also used. The reactive species can be generated in a number of ways: mercury-arc photolysis of a trapped precursor molecule through the quartz window, evaporation from a Knudsen cell in the heater, chemical reaction of atoms evaporated from the Knudsen cell with molecules deposited through the spray-on line, and vacuum-ultraviolet photolysis of molecules deposited from the spray-on line by radiation from discharge-excited atoms flowing through the tube. For laser excitation studies, the sample is deposited on a tilted copper wedge which is grazed by the laser beam, and light emitted or scattered at approximately 90° is examined by a spectrograph. In electron-spin resonance studies, the sample is condensed on a sapphire rod that can be lowered into the necessary waveguide and magnet. The chemical versatility of the matrix technique can perhaps best be described by considering a number of important examples.

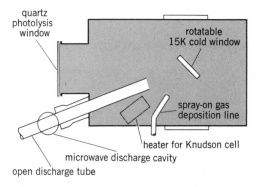

Fig. 1. Vacuum-vessel base cross section for matrix photoionization experiments. 15 K = −433°F.

Applications. The first free radical stabilized in sufficient concentration for matrix infrared detection was formyl HCO. Hydrogen iodide was deposited in a CO matrix and photolyzed with a mercury arc; hydrogen atoms produced by dissociation of HI reacted with CO in the cold solid to produce HCO. The infrared spectrum of HCO provided vibrational fundamentals and information about the chemical bonding in this reactive species.

The first molecular ionic species characterized in matrices, $Li^+O_2^-$, was formed by the cocondensation reaction of lithium atoms and oxygen molecules at high dilution in argon. The infrared spectrum exhibited a weak $(O \leftrightarrow O)^-$ stretching vibration and two strong $Li^+ \leftrightarrow O_2^-$

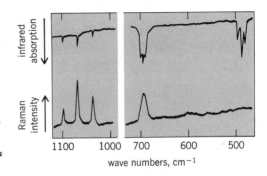

Fig. 2. Infrared and Raman spectra of lithium superoxide, $Li^+O_2^-$, using lithium-7 and 30% $^{18}O_2$, 50% $^{16}O^{18}O$, and 20% $^{16}O_2$ at concentrations of $Ar/O_2 = 100$. The Raman spectrum was recorded using 200 mW of 488-nm excitation and a long-wavelength-pass dielectric filter in the 1000-cm^{-1} region.

stretching vibrations, as shown at the top of **Fig. 2** for the ^7Li and $^{16,18}O_2$ isotopic reaction. The 1-2-1 relative-intensity oxygen isotopic triplets in this experiment showed that the oxygen atomic positions in the molecule are equivalent and indicated an isosceles triangular structure. The ionic model for the bonding in $Li^+O_2^-$ was confirmed by contrasting intensities between the laser-Raman and infrared-absorption spectra shown in Fig. 2.

High-temperature molecules like LiF can be trapped in matrices by evaporating the molecule from the crystalline solid in a Knudsen cell at high temperature or by reacting lithium atoms with fluorine molecules during condensation. The latter method has been used to synthesize the CaO molecule from the calcium atom–ozone reaction for infrared observation of the ground electronic state.

A large number of free radicals have been synthesized and trapped using the lithium atom abstraction reaction. For example, the Li + CCl_4 reaction produced CCl_3 and LiCl, whereas Li + CH_2Cl_2 gave CH_2Cl and LiCl. The C-Cl stretching fundamentals in these chlorocarbon radicals are suggestive of pi bonding between chlorine and the half-filled carbon orbital.

The radical ion Cl_2^- has been formed as the $M^+Cl_2^-$ ion pair and examined by resonance-Raman and optical-absorption techniques. The matrix moderates decomposition of the $M^+Cl_2^-$ species so that laser excitation in the wing of an absorption band can give resonance-enhanced Raman spectra.

Interest in chlorine-oxygen chemistry has been renewed by the possible role of chlorine atom reactions with ozone in the upper atmosphere to give the chlorine oxide free radical. This important free radical was identified in the Raman spectrum of Cl_2O in solid argon following 488-nm laser photolysis, as illustrated in the lower spectrum in **Fig. 3**. Also shown is an interesting isomerization process initiated by the laser, the asymmetric CL-Cl-O species, formed upon rearrangement of the symmetric Cl_2O precursor in the solid argon cage. The infrared spectrum of argon/dichlorine monoxide samples after mercury-arc photolysis, shown at the top of Fig. 3, reveals the same band positions for the Cl-Cl-O photoisomer with different relative intensities; the weak infrared–absorbing ClO free radical was not detected in the infrared spectrum.

The important isolated molecular ion C_2^- was produced by the vacuum ultraviolet photolysis of C_2H_2 for observation of its visible absorption spectrum at 520.7 and 472.5 nm. The C_2 molecular fragment, produced by photolysis of acetylene, captured an electron from the photoionization of C_2H_2 to give C_2^-. Laser excitations of these samples gives rise to a very intense fluorescence spectrum of this stable molecular anion.

Continuous exposure of a condensing sample to argon-resonance radiation during sample condensation has been used to produce molecular cations for spectroscopic study. In the case of CHF_3, photolysis produced the CF_3 radical which may be photoionized by a second 11.6-eV photon to give CF_3^+. The infrared spectrum of CF_3^+ revealed a very high C-F vibrational fundamental which indicates substantial pi bonding in this planar carbocation. Similar studies with CCl_4 produce a very strong blue absorption band which has been identified as arising from CCl_4^+, the parent molecular ion. The blue absorption was destroyed by visible-light photolysis, showing that the carrier is an extremely unstable species. Matrix photoionization experiments with CF_3Cl

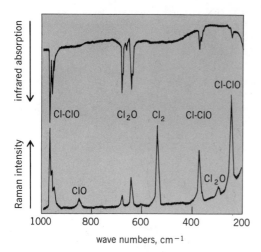

Fig. 3. Infrared and Raman spectra of Cl_2O and its photolysis products. $Ar/Cl_2O = 100$. The infrared spectrum was recorded after 10 min of ultraviolet photolysis. The Raman spectrum was recorded with 300 mW of 488-nm excitation.

yielded the daughter cations CF_2Cl^+ and CF_3^+ and the parent cation CF_3Cl^+. The latter cation photolyzed readily with near-ultraviolet radiation, which helped identify the parent cation.

The matrix isolation technique enables spectroscopic data to be obtained for reactive molecular fragments, many of which cannot be studied in the gas phase. SEE INFRARED SPECTROSCOPY; LASER SPECTROSCOPY; RAMAN EFFECT.

Bibliography. L. Andrews, Laser excitation matrix isolation spectroscopy, *Appl. Spectrosc. Rev.*, 11:125–161, 1976; L. Andrews, Spectroscopy of molecular ions in noble gas matrices, *Annu. Rev. Phys. Chem.*, 30:79–101, 1979; E. Barnes (ed.), *Matrix Isolation Spectroscopy*, 1981; H. E. Hallam, *Vibrational Spectroscopy of Trapped Species*, 1973; M. E. Jacox, The stabilization and spectra of free radicals and molecular ions in rare gas matrices, *Rev. Chem. Intermed.*, 2:1–36, 1978.

ATOMIC AND MOLECULAR SPECTROSCOPY

Atomic spectroscopy	114
Photoacoustic spectroscopy	116
Beam-foil spectroscopy	120
Laser spectroscopy	123
Doppler effect	128
Resonance-ionization spectroscopy	131
Molecular beams	134
Infrared spectroscopy	137
Raman effect	145
Quasielastic light scattering	151
X-ray spectrometry	154
X-ray fluorescence analysis	160
Extended x-ray absorption fine structure (EXAFS)	169
Electron spectroscopy	169

ATOMIC SPECTROSCOPY

Claude Veillon and Rodney K. Skogerboe

R. K. Skogerboe wrote the section Atomic Emission.

Analytical atomic spectroscopy utilizing the emission, absorption, or fluorescence of light at discrete wavelengths by atoms in a vaporized sample for the determination of the elemental composition of the sample. The sample, usually an aqueous solution, is introduced into a high-temperature device, such as an electrical discharge, flame, or furnace, where it is dissociated into its component atoms. In atomic emission spectroscopy, the sample atoms absorb energy by collisions with high-kinetic-energy atoms and molecules in the hot gas, promoting electrons to higher energy levels within the atom (excitation). A fraction of these excited atoms then return to their normal or "ground" state by emission of electromagnetic radiation at discrete energies (and, thus, discrete wavelengths) corresponding to the energy differences between the electronic energy levels involved. Just as excited atoms can emit radiation, ground-state atoms can absorb radiation of the appropriate wavelength at these same discrete energies, and this reverse of the emission process forms the basis of atomic absorption spectroscopy. A third technique, atomic fluorescence spectroscopy, is basically a combination of the first two: the sample atoms first absorb radiation from an external source, becoming excited atoms, and the reradiation (fluorescence) is measured.

Atomic absorption. In its usual configuration, an atomic absorption spectrometer consists of four basic components: (1) a light source, emitting the spectrum of the desired analyte element; (2) a sample atomization cell, such as a flame or graphite tube furnace; (3) a monochromator, to isolate the desired source emission line; and (4) a detector/readout system, to allow measurement of the change in source line intensity by sample atom absorption.

The light source is usually a hollow cathode discharge lamp, prepared with the desired element. Radiation from the source passes through the atomization cell, where it is absorbed by the same atoms (as the source) produced when the sample is introduced. Calibration is achieved with standard solutions containing known concentrations of the element being determined.

In the case of flame atomization, fuel/oxidant mixtures like acetylene/air are used for most easily atomized elements. Many other elements exist mostly as monoxides (that is, molecules, and thus not available for atomic absorption) at the relatively low temperature of an acetylene/air flame. For these, the analytical sensitivity can be greatly enhanced by going to a higher-temperature flame, such as acetylene/nitrous oxide, shifting the exothermic equilibrium $M + O \rightleftarrows MO$ toward the metal (M).

A significant development was the nonflame atomization systems, particularly the graphite tube furnaces. These devices consist basically of a graphite tube, into which a small amount of sample is placed, which is then heated electrically to a high temperature, vaporizing and atomizing the sample. These devices improve the analytical sensitivity for many elements by about 100-fold, compared to chemical flames. This improvement is due to several factors, such as an oxygen-free inert-gas atmosphere sheathing the tube, the reducing power of the hot carbon, and a relatively long residence time of sample atoms in the small-volume optical path. These devices are particularly useful when the amount of sample available is limited, since sample volumes are usually in the microliter range. Not without limitations, these devices often suffer from matrix effects and exhibit relatively poor precision because of the small sample size.

As an analytical technique for trace elemental determinations, atomic absorption spectrometry has important advantages: it has high sensitivity for many elements, with perhaps 60 or so elements measurable in the parts-per-million range (~ 0.1 to 10 micrograms/ml, or $\sim 10^{-7}$ to 10^{-5} g on an absolute basis) with flame atomization, and perhaps half of these measurable at 10–1000 times lower concentrations with nonflame atomization; it is simple, rapid, relatively low-cost, and highly specific (few interferences).

The principal limitations of the technique are: a separate line source is needed for each element to be determined (for many applications, where only one or a few elements are to be determined in a sample, this will not be an important limitation; the technique does not lend itself well to simultaneous multielement determinations); and it has limited dynamic range (for a given determination, all samples must be within about a one-decade concentration interval).

Atomic fluorescence. Atomic fluorescence spectrometry utilizes the same basic components as atomic absorption in a different optical arrangement. The light source(s) is placed at right angles to the optical axis. The fluorescence intensity is directly proportional to the source intensity (over the atom's absorption line width), and so intense sources such as simple electrodeless discharge lamps are used, rather than the hollow cathode lamps used in atomic absorption. The source is not on the optical axis, and so it need not produce narrow lines. In fact, a single, broadband continuum source can be used quite successfully, eliminating the need to change the source for each element determined. Tunable dye lasers show considerable promise as sources because of their extremely high intensity.

Atomic fluorescence spectrometry combines some of the advantages of emission and absorption and has a few of its own. It has the simplicity, speed, cost, and specificity advantages of atomic absorption, the wide dynamic range and adaptability to multielement determinations of atomic emission, and higher demonstrated and potential sensitivity for many elements. The nonflame atomizers are equally advantageous in this technique. Its principal limitation is scattering of source radiation (common to fluorescence techniques), and so caution must be used in the design of atomization systems to eliminate particulate matter from the optical path.

Atomic emission. Atomic emission spectroscopy relies on the use of an energetic medium to promote the electronic excitation of atoms in the gas phase and the subsequent emission of light at wavelengths characteristic of the atoms involved. At typical system consists of the following: the energy medium for vaporization, atomization, and excitation of the material to be analyzed; a dispersive unit which isolates the various wavelengths of light emitted; and a measurement or detection device. Three-component systems of this type are widely used to determine what elements are present in a variety of materials (qualitative analysis) and how much of each element is present (quantitative analysis). The latter relies on the fact that the intensity of light emitted at a particular wavelength is proportional to the concentration of atoms which are capable of emitting at the wavelength.

Energy media which may be used for vaporization, atomization, and excitation include electrical discharges, combustion flames, plasmas, and laser beams. Electrical discharges are produced between two conductive electrodes because it is inexpensive and easy to prepare. Common discharges used include direct-current (dc) arcs, alternating-current (ac) arcs, high-voltage ac sparks, and various combinations of these. Although electrical discharges are most often used for the analysis of solid materials, they may also be utilized for the analysis of gases and liquids. Combustion flames are used almost exclusively for the analysis of liquids. Plasmas produced by the interaction of electrical fields alternating in the radio or microwave frequency range with inert gases such as argon or helium are becoming very popular for the excitation of elements contained in solution samples. High-energy laser beams capable of vaporizing and exciting components of solid samples are often used by focusing the beam through a microscope to permit analysis of very small samples or highly localized sections of larger samples. The choice of the excitation method to be used must be based on the type of material to be analyzed, the elements to be determined, and other requirements imposed by the purpose of the particular analysis.

The dispersion system used to isolate the wavelengths of light emitted by the sample almost exclusively relies on the use of diffraction gratings. If the dispersion system is designed to permit observation of only a single wavelength at any one time by scanning the wavelength region of interest, it is referred to as a monochromator. If it permits the simultaneous observation of several wavelengths without scanning, it may be classed as a polychromator. The monochromator is used to determine different elements in a sequential mode of operation, while the polychromator permits simultaneous measurements. The higher cost of a polychromator may consequently be offset by the reduction in time required to determine several elements.

Photomultiplier tubes or photographic films and plates are widely used to observe the dispersed radiation and measure its intensity at any wavelength(s) characteristic of the element(s) to be determined. Photomultipliers instantaneously provide a photocurrent proportional to the light intensity incident on them. Sensitive ammeters measure the photocurrent, which may be related to the concentrations to be determined. Such readout systems in combination with a wavelength dispersion unit are referred to as spectrophotometers if their wavelength resolution capability is low, or spectrometers if it is high. Direct-reading spectrometers have 10 to 60 photomultiplier tubes, each dedicated to measuring the light intensity emitted by a particular element at a characteristic wavelength. Such units are quite expensive and complex to operate but provide a means of rapidly obtaining analysis results. Dispersion units which rely on photographic detection are known as spectrographs. Such detection takes advantage of the well-known effect of light on photographic emulsions deposited on film or glass plates. All wavelengths covered by the dispersion system are permanently recorded in the form of a negative print; the concentration of an element may be determined by measuring the degree of blackness (optical density) of the negative at the appropriate wavelength with a densitometer. Advantages include the ability to determine a wider range of elements simultaneously (up to 70–80) and lower equipment acquisition costs. Longer times are required, however, to measure the photo image densities and convert these to concentrations for the elements of interest. Both the direct and photographic readout systems must be calibrated by analysis of standard materials for which the concentrations of the elements to be determined are known.

Atomic emission spectroscopy has proven to be a very useful means of analysis; its scope of application is diverse. The approach is widely used in industry, government, and university research laboratories for the identification and determination of many elements as means of solving very complex problems.

Bibliography. P. W. Boumans et al. (eds.), *Profile of Current Developments in Atomic Spectroscopy*, 1985; C. L. Chakrabarti (ed.), *Progress in Analytical Atomic Spectroscopy*, vols. 1–6, 1978–1985; L. Ebdon, *Introduction to Atomic Absorption Spectroscopy: A Self Teaching Approach*, 1981; J. M. Ottaway and A. M. Ure, *Practical Atomic Absorption Spectrometry*, 1987; M. Slavin, *Emission Spectrochemical Analysis*, 1971; J. D. Winefordner (ed.), *Trace Analysis*, Chemical Analysis Series, vol. 46, pp. 123–181, 1976.

PHOTOACOUSTIC SPECTROSCOPY
C. K. N. PATEL

A technique for measuring small absorption coefficients in gaseous and condensed media, involving the sensing of optical absorption by detection of sound. It is frequently called optoacoustic spectroscopy. Although the technique dates back to 1880 when A. G. Bell used chopped sunlight as the source of radiation, it remained dormant for many years, primarily because of the lack of suitable powerful sources of tunable radiation. However, the usefulness of optoacoustic detection for spectroscopic applications was recognized early in its development, and pollution monitoring instruments (called spectrophones) dedicated for detection of specific gaseous constituents have been used intermittently since Bell's work.

Methods of measuring absorption. During the transmission of optical radiation through a sample (gas, liquid, or solid), the absorption of radiation by the sample can be measured by at least three techniques. The first one is the straight-forward detection technique which requires a measurement of the optical radiation level with and without the sample in the optical path. The transmitted power P_{out} and the incident power P_{in} are related through Eq. (1), where α

$$P_{out} = P_{in}e^{-\alpha l} \tag{1}$$

is the absorption coefficient and l is the length of the absorber. With this technique, the minimum measurable αl is of the order of 10^{-4} unless special precautions have been taken to stabilize the source of radiation.

The second of the techniques is the derivative absorption technique where the frequency of the input radiation is modulated at a low radio frequency or audio frequency, ω_m. The transmitted radiation then contains a time-varying component at ω_m, if the optical path contains absorption which has a frequency-dependent structure. (For structureless absorption, modulated absorption spectroscopy does not provide a signal that can characterize the amount of absorption.) For situations where the absorption has well-defined structure, the modulation absorption spectroscopy can be used to measure αl as small as about 10^{-8} for sufficiently high input powers. The ability to measure the small absorption effects is independent of the input and output power levels for the straightforward measurement technique as long as the noise contributed by the detector is not a factor in determining the signal-to-noise ratio. For the derivative absorption technique, the smallest αl that can be measured varies as $(P_{in})^{-1}$ until the shot noise of the detector begins to be appreciable.

The third technique, optoacoustic detection, is a calorimetric method where no direct detection of optical radiation is carried out but, instead, a measurement is made of the energy, with power P_{abs}, absorbed by the medium from the incident radiation, Eq. (2). Thus the optoacoustic

$$P_{abs} = P_{in}(1 - e^{-\alpha l}) \qquad (2)$$

signal, V_{oa}, is given by Eq. (3), where K is the constant describing the conversion factor for

$$V_{oa} = K\{P_{in}(1 - e^{-\alpha l})\} \approx KP_{in}\alpha l \qquad \text{(for } \alpha l \ll 1\text{)} \qquad (3)$$

transforming the absorbed energy into an electrical signal using an appropriate transducer. It has been tacitly assumed that the absorbed energy is lost by nonradiative means rather than by reradiation. The optoacoustic detection scheme implies that the absorbed energy will be converted into acoustic energy for eventual detection.

From Eq. (3), the optoacoustic signal is proportional to the incident power and the absorption-length product αl. Thus, for given sources of noise from the detection transducers, the signal-to-noise ratio improves as the incident energy is increased. Put differently, the smallest amount of absorption that can be measured using the optoacoustic technique varies as $(P_{in})^{-1}$ with no limitation on the level to which P_{in} can be increased for detecting small absorptions. Values of αl as small as 10^{-10} can be measured in the gas phase. The techniques generally used for gases and those used for condensed-phase optoacoustic spectroscopy differ in detail somewhat.

Gases. If optical radiation is amplitude-modulated at an audio frequency, the absorption of such radiation by a gaseous medium that has been confined in a cell with appropriate optical windows for the entrance and exit of the radiation and the subsequent nonradiative relaxation of the medium, will cause a periodic variation in the temperature of the column of the irradiated gas (**Fig. 1**). Such a periodic rise and fall in temperature gives rise to a corresponding periodic variation in the gas pressure at the audio frequency. The audio-frequency pressure fluctuations (that is, sound) are efficiently detected using a sensitive gas-phase microphone. The intrinsic noise limitation to the optoacoustic detection scheme arises from the Brownian motion of gas atoms/molecules, and Kreuzer showed that the minimum detectable absorbed power is $P_{min} \approx 3.6 \times 10^{-11}$ W for a 11.9-cm-long (4.7-in.) cell. Substituting P_{min} for P_{abs} in Eq. (2), and noting, as in Eq. (3), that $(1 - e^{-\alpha l}) \cong \alpha l$ for $\alpha l \ll 1$, it follows that α_{min} varies as (P_{min}/P_{in}) as indicated above. The

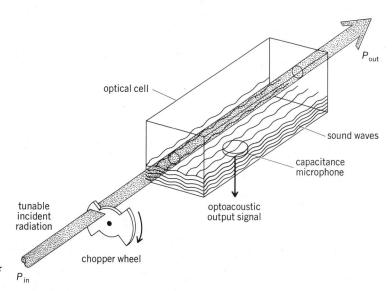

Fig. 1. Optoacoustic cell for gaseous spectroscopy.

usefulness of the optoacoustic detection for measurement of small absorption coefficients became evident with the development of a variety of tunable high-power laser sources which could take advantage of the $(P_{in})^{-1}$ dependence. Using a spin-flip Raman laser tunable in the 5.0- to 5.8-micrometer range, with a power output of approximately 0.1 W, C. K. N. Patel and R. J. Kerl were able to detect α_{min} of approximately 10^{-10} cm^{-1} for a cell length of 10 cm (4 in.). These studies used a miniature optoacoustic cell (**Fig. 2**) with a total gas volume of approximately 3 cm^3 (0.18 in.^{3}). The absorber used was nitric oxide diluted in nitrogen. It is estimated that for a signal-to-noise ratio of approximately 1, and a time constant of 1 s, it is possible to detect a nitric oxide (NO) concentration of approximately 10^7 molecules cm^{-3}, corresponding to a volumetric mixing ratio of approximately $1:10^{12}$ at atmospheric pressure.

Fig. 2. Sensitive optoacoustic gaseous spectroscopy cell.

The capability of measuring extremely small absorption coefficients and correspondingly small concentrations of the absorption gases has many applications, including high-resolution spectroscopy of isotopically substituted gases, excited states of molecules and forbidden transitions, and pollution detection. In the last application, both continuously tunable lasers, such as the spin-flip Raman laser and dye lasers, and step tunable infrared lasers, such as the carbon dioxide (CO$_2$) and carbon monoxide (CO) lasers, have been used as sources of high power radiation. The pollution measurements have demonstrated that the optoacoustic spectroscopy technique in conjunction with tunable lasers can be routinely used for on-line real-time in-place detection of undesirable gaseous constituents at sub-parts-per-billion levels. Specific examples include the measurement of nitric oxide on the ground and in the stratosphere (where nitric oxide plays an important role as a catalytic agent in the stratospheric ozone balance) and measurements of hydrocyanic acid (HCN) in the catalytic reduction of CO + N$_2$ + H$_2$ + $\cdots$ over platinum catalysts. These studies point toward expanding use in the future of optoacoustic spectroscopy in pollution detection. *See* LASER SPECTROSCOPY.

Condensed-phase spectroscopy. A straightforward application of the gas-phase optoacoustic spectroscopy technique to the study of condensed phase (liquid or solid) spectra involves enclosing the condensed phase material within the gas-phase optoacoustic cell (**Fig. 3**). The

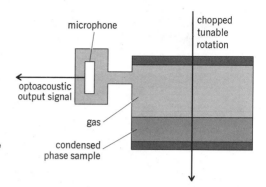

Fig. 3. Arrangement for condensed-phase photoacoustic spectroscopy.

"photoacoustic" signals generated in the sample due to the absorption of optical radiation are communicated to the gas-phase microphone via coupling through the gas filling the chamber. The inefficiency of such a system is high because of the very poor acoustical match (coupling efficiency approximately 10^{-5}) between the condensed-phase sample and the gas. In reality, because of the large acoustical mismatch, the detection scheme is really "photothermal" rather than "photoacoustic," and this scheme provides a capability of measuring fractional absorption at a level of approximately 10^{-4} when a continuous-wave laser power of approximately 10 W is used. A more severe drawback of the scheme, however, lies in the difficulty of interpretation of the data because of the intimate dependence of the observed optoacoustic signal from the microphone on the chopping frequency, absorption depth, and heat diffusion depth. However, in spite of its shortcomings, the gas-phase microphone technique for condensed-phase optoacoustic spectroscopy has found applications.

A very sensitive calorimetric spectroscopic technique has been developed for the study of weak absorption in liquids and solids. This technique uses a pulsed tunable laser for excitation and a submerged piezoelectric transducer, in the case of a liquid, or a contacted piezoelectric transducer, in the case of a solid, for the detection of the ultrasonic signal generated due to the absorption of the radiation and its subsequent conversion into a transient ultrasonic signal (**Fig. 4**). The major distinction between the above condensed-phase "photoacoustic" spectroscopy technique and the pulsed-source, submerged or contacted piezoelectric transducer technique, is the high coupling efficiency of approximately 0.2 for the ultrasonic signal in the liquid to the submerged transducer, or an efficiency of approximately 0.9 for coupling the ultrasonic wave in a solid to a bonded transducer. Because of this high efficiency, the pulsed-laser, submerged or bonded optoacoustic spectroscopy technique has been shown to be useful for measuring fractional absorptions (that is, values of αl) as small as 10^{-7} when using a laser source with pulse energy of approximately 1 millijoule, pulse duration of approximately 1 microsecond, and a pulse repetition frequency of 10 Hz. There is room for improvement by increasing the laser pulse energy. There is a possibility of the electrostriction effect giving rise to an unwanted background signal, but this signal is not dependent on the light wavelength, and can be minimized by proper choice of experimental parameters.

The pulsed-laser, submerged or bonded piezoelectric transducer technique has a further advantage that time-gating of the ultrasonic signal output can be utilized for the rejection of spurious signals since the sound velocity in condensed media is known and hence the exact arrival time of the real optoacoustic pulse can be calculated. This technique has been used for measurement of very weak overtone spectra of a variety of organic liquids, optical absorption coefficients of water and heavy water in the visible, two-photon absorption spectra of liquids, Raman gain spectra in liquids, absorption of thin liquid films, spectra of solids and powders, and weak overtone spectra of condensed gases at low temperatures. Because of the capability of measuring very small fractional absorptions, the technique is clearly applicable to the area of monitoring water pollution, impurity detection in thin semiconductor wafers, transmission studies of ultrapure glasses (used in optical fibers for optical communications), and so forth. Further, even

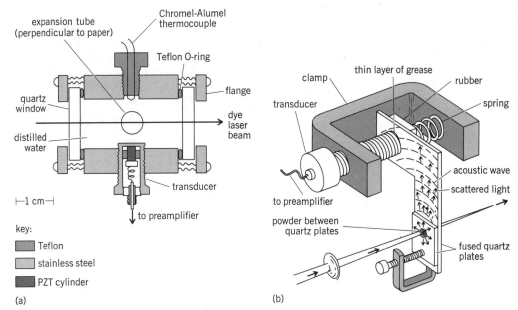

Fig. 4. Arrangement for pulsed-laser (a) immersed and (b) contacted piezoelectric transducer optoacoustic spectroscopy.

though in all of the present studies use is made of only the optical radiation, there is no reason to restrict "optoacoustic" spectroscopy to the optical region. By using pulsed x-ray sources, such as the synchrotron light source or pulsed electron beams, the principle described above for a pulsed-light-source, submerged or bonded piezoelectric transducer, gated-detection technique can be extended to x-ray acoustic spectroscopy and electron-loss acoustic spectroscopy. These extensions are likely to have major impact on materials and semiconductor fabrication technology.

Bibliography. A. G. Bell, On the production and reproduction of sound by light, *Proc. Amer. Ass. Adv. Sci.*, 29:115–129, 1880; L. B. Kreuzer, The physics of signal generation and detection, in Y. M. Pao (ed.), *Optoacoustic Spectroscopy and Detection*, 1977; V. S. Letokhov and V. P. Zharov, *Laser Optoacoustic Spectroscopy*, 1986; E. Luescher et al., *Photoacoustic Effect: Principles and Applications*, 1984; C. K. N. Patel, Spectroscopic measurements of stratosphere using tunable infrared lasers, *Opt. Quantum Electr.*, 8:145–154, 1976; C. K. N. Patel and R. J. Kerl, A new optoacoustic cell with improved performance, *Appl. Phys. Lett.*, 30:578–579, 1977; C. K. N. Patel and E. D. Shaw, Tunable simulated Raman scattering from conduction electrons in InSb, *Phys. Rev. Lett.*, 24:451–454, 1970; C. K. N. Patel and A. C. Tam, Pulsed optoacoustic spectroscopy of condensed matter, *Rev. Mod. Phys.*, 53:517–550, 1981; M. B. Robin and W. R. Harshbarger, The optoacoustic effect: Revival of an old technique for molecular spectroscopy, *Acc. Chem. Res.*, 6:329, 1973; A. Rosencwaig, *Photoacoustics and Photoacoustic Spectroscopy*, 1980.

BEAM-FOIL SPECTROSCOPY
STANLEY BASHKIN

A method of determining the energies and mean lives of excited electronic levels in monatomic ions and in atoms. Beams of particles (ions) of any element are energized in a particle accelerator and then sent in vacuum through a thin foil, usually of carbon. The particles emerge from the foil in various stages of ionization. Numerous energy levels are excited in each of those stages of ionization. The spontaneous loss of the energy of excitation takes place as the beam moves downstream from the foil. That loss is detected by means of the electromagnetic radiation or the electrons which the ions emit. Only experiments dealing with the light emission will be considered, since they represent the great majority of beam-foil researches.

Optical spectra. For optical radiations (4 to 700 nanometers), the wavelength is determined by a spectrometer employing a diffraction grating. More energetic forms of light (x-rays) are studied with energy-sensitive silicon crystals to which some lithium has been added. In both techniques, the radiation is transformed into an electrical pulse which is then amplified and counted, and the number of counts is plotted against the wavelength. Such a graph (a spectrum) contains peaks of various intensity occurring at wavelengths characteristic of the electronic structure of the excited ion. In the example shown in **Fig. 1**, three spectral lines are identified; the quantum-mechanical notation above each peak describes the origin of the lines.

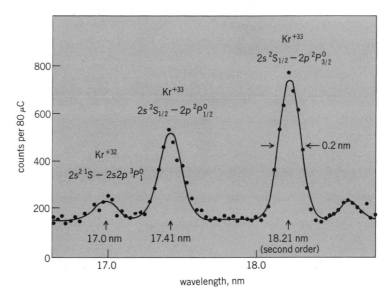

Fig. 1. Spectrum of krypton, excited by sending 714-MeV ions through a carbon foil with a thickness of 600 μg/cm².

Mean lives. The mean life of a level provides sensitive information about the quantum nature of the level itself and the mechanism whereby it connects, by either the absorption or emission of light, to some other level. In beam-foil spectroscopy, this mean life is found from observations on the intensity of a particular spectral line as a function of the separation of the emitter and the exciter foil. An example of the technique is show in **Fig. 2**, in which the excitation mechanism is the same as in Fig. 1. The symbol τ stands for the mean life, the subscript identifying which electronic level is decaying. The 9.11-nm line in Fig. 2 is in first order, while the corresponding second-order line at 18.21 nm appears in Fig. 1.

Degree of ionization. The degree of ionization which can be achieved depends on the atomic number of the ion and its energy. Early work was restricted to a few elements at the low end of the periodic table and to energies of a few million electronvolts. Subsequently, more than 60 of the 94 naturally occurring elements were used, at energies extending to nearly 900 MeV. In the 900-MeV work, ions of rhodium, element 45, were accelerated in the heavy-ion linear accelerator (HILAC) at the Lawrence Berkeley Laboratory of the University of California. Figures 1 and 2 are taken from a similar experiment on krypton, element 36, at 714 MeV. In both cases, the incident particles lost so many electrons that only three-electron (lithiumlike) or four-electron (berylliumlike) ions remained. Studies were made of the spectra (Fig. 1) and mean lives (Fig. 2)

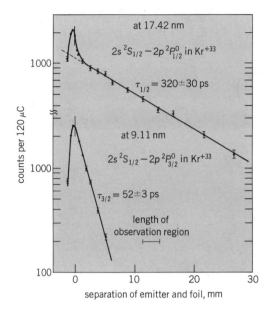

Fig. 2. Lifetime data for two levels in three-electron krypton ions, excited by the same mechanism as in Fig. 1.

reflecting transitions of one of these remaining electrons. To reach the same degree of ionization in a hot plasma would require a temperature of 2.5×10^{7}°C.

Relativistic effects. The Coulomb force acting on the electrons of highly ionized atoms is so large that they move with speeds that are appreciable fractions of the speed of light; this means that the electronic motions must be treated relativistically. Since such theories are complicated, there is some uncertainty as to which kind of calculation is the best representation of the data; the beam-foil spectroscopy work on spectra and lifetimes is done partly to clarify the theory. Thus the krypton experiment and earlier work on iron, element 26, showed that the so-called dipole-length calculations are to be preferred over the dipole-velocity approach to the level lifetimes. An interesting relativistic effect is that electronic transitions which are forbidden under ordinary circumstances can become dominant over the usual allowed transitions. Such effects were seen in the work at HILAC.

Lamb shift. A major advance in atomic theory was made when W. E. Lamb, Jr., showed that the Dirac treatment of the hydrogen atom had to be revised. The revision consisted of a small change in the energy levels of a one-electron system, the Lamb shift. Among other things, it is strongly dependent on the atomic number of the ion, and beam-foil spectroscopy experiments have been done to see if current calculations are applicable to one-electron ions up to chlorine and argon, elements 17 and 18. Preliminary results suggest that the theory, quantum electrodynamics, is satisfactory.

The Lamb shift also occurs in two-electron ions, and that has been investigated up to oxygen, element 8.

Beam-foil interaction. The fundamental process which gives rise to the excitation phenomenon is still poorly understood. Efforts have been made to develop a semiempirical theory of beam-foil spectroscopy, the basic data being the relative populations of the excited levels as a function of particle energy and element. *See* ATOMIC STRUCTURE AND SPECTRA.

Bibliography. D. D. Dietrich et al., Oscillator strengths of the $2s\ ^{2}S_{1/2}-2p\ ^{2}P^{0}_{1/2,\ 3/2}$ transitions in Fe XXIV and the $2s\ ^{1}S_{0}-2s2p\ ^{3}P^{0}_{1}$ transition in Fe XXIII, *Phys. Rev. A*, 18:208–211, 1978; N. A. Jelley et al., Lamb shift and fine structure in $1s3s^{3}S-1s\ 3p\ ^{3}P$ transitions in helium-like oxygen, *J. Phys. B: Atom, Molec. Phys.*, 12:2605–2611, 1979; Proceedings of the 5th International Conference on Beam-Foil Spectroscopy, *J. Phys.*, C1:1–367, 1979.

LASER SPECTROSCOPY
THEO W. HANSCH

Spectroscopy with laser light or, more generally, studies of the interaction between laser radiation and matter. Lasers have led to a rejuvenescence of classical spectroscopy, because laser light can far surpass the light from other sources in brightness, spectral purity, and directionality, and if required, laser light can be produced in extremely intense and short pulses. The use of lasers can greatly increase the resolution and sensitivity of conventional spectroscopic techniques, such as absorption spectroscopy, fluorescence spectroscopy, or Raman spectroscopy. Moreover, interesting phenomena have become observable in the resonant interaction of intense coherent laser light with matter. Some of these effects have become the basis for powerful spectroscopic methods, which offer unprecedented spectral resolution, or which permit the investigation of properties of matter that could not be observed previously. Laser spectroscopy has become a wide and diverse field, with applications in numerous areas of physics, chemistry, and biology.

Tunable sources. Early lasers, such as ruby lasers or helium-neon lasers, worked only at a few discrete wavelengths, determined by narrow spectral lines of the active medium. The vigorous advances of laser spectroscopy since about 1970 have been largely due to the development of laser sources which are highly monochromatic, and in which the wavelength can be tuned continuously over a wide spectral range.

Tunable coherent sources are available for any wavelength region from the far infrared into the near-vacuum ultraviolet. Tunable infrared lasers include high-pressure molecular gas lasers, semiconductor diode lasers, spin-flip Raman lasers, and color-center lasers. In the visible region, organic dye lasers have proved particularly versatile and powerful spectroscopic sources. Both pulsed and continuous-wave dye lasers cover the entire visible spectrum, including the bordering near-infrared and near-ultraviolet regions. Intense shorter-wave ultraviolet radiation can be generated with excimer lasers over limited regions.

The generation of harmonic frequencies or sum frequencies in nonlinear optical crystals or gases provides a valuable method to produce tunable coherent ultraviolet radiation from laser light of longer wavelengths. Optical parametric oscillators and difference-frequency crystal mixers offer corresponding alternatives for the generation of coherent infrared radiation. Stimulated Raman scattering can be used to shift the frequency of a tunable laser by some integer multiple of a molecular vibration frequency and thus to extend the tuning range. These nonlinear frequency-mixing and shifting techniques work best with the intense radiation from pulsed lasers.

Although pulsed tunable lasers can be highly monochromatic by conventional standards, their line width can never be narrower than the inverse pulse duration. The highest resolution is obtained with continuous-wave lasers. Considerable engineering efforts have been devoted to the active frequency stabilization of such lasers. By electronically locking the frequency of a continuous-wave dye laser to some reference interferometer, for instance, a line width of less than 1 MHz or a resolution on the order of 1 part in 10^9 is achieved routinely, and line widths down to a few hundred hertz have been produced with more sophisticated feedback controls.

Much engineering effort has been devoted to the development of highly accurate wavelength and frequency meters, which can do justice to the narrow line width of tunable lasers.

Absorption spectroscopy. Lasers can replace conventional light sources and spectrographs or monochromators for absorption spectroscopy. The spectral purity of laser light can eliminate instrumental resolution limits, the high intensity helps to overcome detector noise, and the good directionality permits long or folded absorption paths. All these factors contribute to improved sensitivity.

Rather than by measuring the attenuation of a laser beam in the sample, the absorption can often be monitored indirectly with still higher sensitivity. For instance, absorption of a modulated laser beam will produce a sound wave in the sample, because some of the light is converted to heat. This sound wave can be picked up by a microphone (optoacoustic detection). Resonant absorption of laser light by atoms or molecules in a gas discharge can change the ionization probability. The resulting changes in discharge current or voltage across the tube are easily measured (optogalvanic detection). The highest sensitivities at visible and ultraviolet wavelengths have been obtained by monitoring the fluorescence or photoionization of laser-excited atoms or molecules in a gas or molecular beam. Single atoms have been selectively detected in this way.

Intracavity absorption. Very high sensitivity has also been obtained by placing an absorbing sample inside the resonator of a broad-band dye laser without any optical tuning elements. Any absorption lines can be detected as dips in the laser emission spectrum, when analyzed by a spectrograph. This intracavity spectroscopy can surpass the sensitivity of a conventional single-pass absorption measurement by a large factor (on the order of 10^5) because the sample is effectively traversed by the light many times, and the competition between many simultaneously oscillating modes strongly disfavors modes with slightly increased losses.

Multiphoton absorption. It has long been known that atoms can be excited to a higher quantum state by simultaneously absorbing two or more photons which together provide the necessary energy. The probability for an N-photon transition grows initially with the Nth power of the intensity. Only the high intensity of laser light has made it possible to observe two- and multiphoton absorption in the optical region. Two-photon spectroscopy permits the study of states with the same parity as the absorbing level, which are not normally reached by single-photon transitions, and it requires photons of less energy, which are sometimes more readily produced.

Fluorescence spectroscopy. Intense laser light is a very effective means to pump a large fraction of an absorbing species to some excited quantum level. Hence, lasers can greatly increase the sensitivity of such classical spectroscopic methods as fluorescence spectroscopy, optical pumping, level-crossing spectroscopy, or double-resonance spectroscopy. Moreover, lasers make is possible to apply these techniques to atomic and molecular transitions at wavelengths where intense spectral lamps are not available, and stepwise excitation permits studies even of highly excited states, including autoionizing levels and very high Rydberg states of atoms.

Studies of the line shape of fluorescent emission at high intensities have permitted interesting tests of the predictions of quantum-electrodynamic theory.

Raman spectroscopy. Lasers have revolutionized Raman spectroscopy, that is, the observation of scattered light at wavelengths other than that of the exciting light. The high intensity of laser light has greatly increased the sensitivity of this form of two-photon spectroscopy, where the energy difference between incident photon and scattered photon corresponds to a resonant transition between states of equal parity. Tunable lasers have enhanced the sensitivity, resolution, and versatility of Raman spectroscopy even more, by making it possible to excite close to some intermediate resonant transition (resonance Raman spectroscopy), or to observe stimulated Raman scattering with the help of a second, tunable probe laser beam. Polarization anisotropies associated with Raman transitions provide additional information and can further increase the detection sensitivity. *See Raman effect.*

Frequency mixing. The high intensity of laser light has also made possible a new class of spectroscopic methods which rely on nonlinear frequency mixing in the sample. For instance, if a sample is irradiated simultaneously by two strong laser beams of frequencies ω_1 and ω_2, the nonlinear response of the driven dipoles leads to the generation of coherent light at a new frequency, $2\omega_1 - \omega_2$. This frequency mixing can be described in terms of a third-order nonlinear susceptibility of the sample.

If the frequency ω_2 is tuned to the (redshifted) Stokes line of a Raman transition, excited by ω_1, a resonance is observed in the intensity of the new wave at the (blueshifted) anti-Stokes frequency. This type of frequency mixing is sometimes referred to as coherent anti-Stokes Raman spectroscopy (CARS). It can be used to study Raman transitions even in the presence of strong incoherent background radiation, such as in the observation of flames.

High-resolution laser spectroscopy. Lasers have led to particularly noteworthy progress in the field of very high-resolution spectroscopy. They have become powerful tools to investigate the structure of atoms, molecules, and ions. They can be used to study fine and hyperfine splittings, Zeeman and Stark splittings, light shifts, collision broadening, collision shifts, and other attributes of spectral lines. Moreover, lasers make it possible to measure the wavelengths of spectral lines with unprecedented accuracy. The laser wavelength can be locked to an atomic or molecular transition, providing an accurate standard length or frequency. Such lasers have become important tools for precision metrology, making possible accurate measurements of fundamental constants and stringent tests of basic physics laws.

Doppler-free spectroscopy. The sharpest spectral lines are generally those in atoms or molecules which are relatively free and undisturbed. Such molecules, however, are almost inevitably moving with high thermal velocity. Molecules which are moving toward an observer appear to emit or absorb light at higher frequencies than molecules at rest, and molecules moving away appear to absorb at lower frequencies. In a gas, with molecules moving at random in all directions, the lines appear blurred, with typical Doppler widths on the order of 1 part in 1,000,000. To take advantage of the very narrow instrumental line width of laser sources, it is necessary to overcome this Doppler broadening of spectral lines. SEE DOPPLER EFFECT.

The oldest method of Doppler-free spectroscopy is the transverse observation of a well-collimated molecular beam, so that the range of velocities along the lines of sight is much restricted. However, molecular-beam equipment tends to be expensive and cumbersome, and it is difficult to observe rare species or molecules in short-living excited states in this way. Fortunately, a number of schemes have been devised which permit Doppler-free spectroscopy of simple gas samples. These include, in particular, techniques of saturation spectroscopy which use the laser light itself to label molecules of slow velocity. SEE MOLECULAR BEAMS.

Saturation spectroscopy. Laser light can easily be intense enough to partly saturate the absorption of a spectral line. That is, those molecules which have absorbed a light quantum are temporarily removed from the initial state. The absorption from that level is reduced or saturated, at least until some relaxation process can replenish the supply of absorbing molecules. In a gas, a monochromatic laser beam will resonantly interact only with those molecules which have the right axial velocity to be Doppler-shifted into resonance. The resulting velocity-selective saturation can be used for Doppler-free spectroscopy.

The first step was the realization by Willis E. Lamb, Jr., in 1963 that the two waves traveling in opposite directions inside a gas laser could work together to saturate the emission (rather than absorption) of those atoms which happen to have a zero component of velocity along the laser axis. Thus the power output would decrease when the laser length was adjusted to produce the light wavelength that would interact with those stationary atom. This Lamb dip was soon observed and was used for high-resolution spectroscopy. However, for some time it was limited to studying the laser transitions themselves, or those few molecular lines which happened to coincide with gas laser wavelengths.

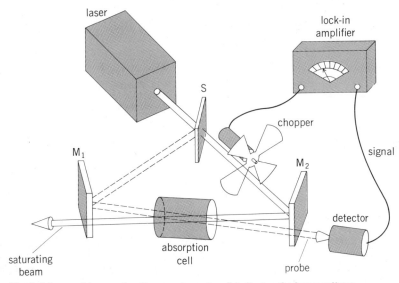

Fig. 1. Scheme of laser saturation spectrometer; S indicates the beam splitter and M_1 and M_2 are mirrors.

In 1970 C. Borde and T. W. Hansch introduced independently a now commonly used form of saturation spectroscopy which achieves good sensitivity with samples outside the laser resonator and which is particularly well suited for use with broadly tunable laser sources. As illustrated in **Fig. 1,** the output of a tunable laser is divided by a beam splitter into a stronger saturating beam and a weaker probe beam that are traversing an absorbing gas sample along the same path but in nearly opposite directions. When the saturating beam is on, it bleaches a path through the cell; that is, it depletes those molecules which are Doppler-shifted into resonance, and a stronger probe signal is received at the detector. As the saturating beam is alternately stopped and transmitted by a chopper, the probe signal is modulated. However, that happens only when both beams interact with the same molecules, and those can only be molecules which are standing still or at most moving transversely. Thus the method picks out those molecules which have near zero component of velocity along the laser beams and ignores others. **Figure 2** illustrates the power of the technique by comparing a Doppler-broadened absorption profile of the well-known Balmer-alpha line of atomic hydrogen with one of the first saturation spectra, recorded with a pulsed dye laser in a glow discharge.

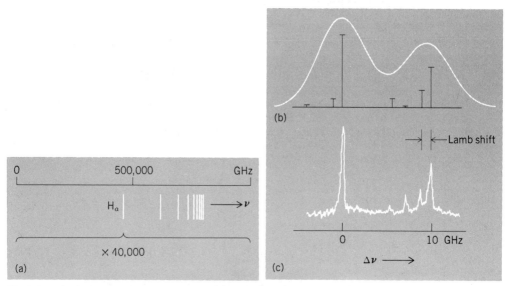

Fig. 2. Laser spectroscopy of atomic hydrogen. (a) Balmer series. (b) Doppler-broadened absorption profile (300 K) of the Balmer-alpha line with theoretical fine structure. (c) Saturation spectrum of Balmer-alpha line with resolved 2S-2P Lamb shift.

The described methods work well only if the sample has noticeable absorption and if the laser is strong enough to excite a substantial fraction of the resonant molecules. Higher sensitivity can sometimes be achieved by observing the absorption of laser light indirectly, via the emitted fluorescence or via acoustooptic or optogalvanic detection. In this case, it is advantageous to monitor the nonlinear interaction of two counterpropagating laser beams via intermodulation; that is, the two laser beams are chopped at two different frequencies f_1 and f_2, and the spectrum is recorded as a modulation in the signal at the sum or difference frequency $f_1 + f_2$ or $f_1 - f_2$.

Polarization spectroscopy. There are also purely optical methods, which can be considerably more sensitive than the older saturated absorption method by suppressing the fluctuating background of probe light on which the signal has to be detected. These include, in particular, the technique of polarization spectroscopy introduced in 1976 by C. Wieman and Hansch. A polarization spectrometer looks similar to a saturation spectrometer but takes advantage of the fact that small changes in light polarization can be detected more easily than changes in light intensity. The probe beam "sees" the sample placed between nearly crossed linear polarizers so that only very little light arrives at the photodetector. The saturating beam is made circularly polarized by a birefringent plate. Alternatively, a linearly polarized beam is used with its polarization axis rotated at 45°.

Normally, in a gas, molecules have their rotation axes distributed at random in all directions. But the probability for absorbing polarized light depends on the molecular orientation. Thus the saturating beam depletes preferentially molecules with a particular orientation, leaving the remaining ones polarized. These can then be detected with high sensitivity because they can change the polarization of the probe beam. The probe acquires a component that can pass through the crossed polarizer into the detector, but again this happens only near the center of a Doppler-broadened line where both beams are interacting with the same molecules.

Polarization spectroscopy makes it possible to observe fewer molecules with lower light intensity so that external causes of line broadening and shifts are more easily avoided. A promising related technique, saturated interference spectroscopy, can work even in spectral regions where good polarizers are not available.

Doppler-free two-photon spectroscopy. There is a completely different approach to Doppler-free laser spectroscopy which does not rely on velocity selection, but which works only for two- or multiphoton spectroscopy, where the frequencies and directions of the exciting photons can be chosen so that the momenta of the absorbed photons add to zero.

Doppler broadening in two-photon excitation can be eliminated to first order simply be reflecting the output of an intense monochromatic laser back onto itself and placing a gas sample in the resulting standing wave field, so that the molecules can absorb two photons of equal energy coming from opposite directions (**Fig. 3**). From a moving molecule, one beam will appear Doppler-shifted toward the blue, and the other will appear shifted toward the red by an equal amount. The sum frequency is hence constant, independent of the molecular velocity. If the number of excited molecules is observed during a laser scan, a sharp resonance appears on a low, Doppler-broadened background, produced by each traveling wave separately. The Doppler-free signal is strongly enhanced because all molecules, regardless of their velocity, contribute.

Resolution limits. The resolution of any of the described methods is ultimately limited by the natural line width of the observed transition. However, in practice, a large number of additional causes of line broadening demand attention, including collision effects, power broadening and light shifts, relativistic "transverse" Doppler shifts, and imperfections in the laser wave fronts. Another important limitation is transit time broadening, in which the number of field oscillations a molecule can see, and hence the resolution, is limited by the finite time during which a molecule traverses the laser beam. Transit time broadening can sometimes be alleviated with the help of two or more spatially separate laser fields. In this case, narrow-band interference fringes appear in the spectrum, similar to the Ramsey fringes commonly observed in molecular-beam spectroscopy with separate radio-frequency fields.

Other approaches are being explored which promise to overcome some of the resolution limits. The resonant radiation pressure of intense laser light can be used to cool gases rapidly to

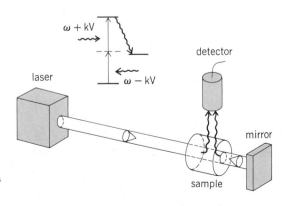

Fig. 3. Setup for Doppler-free two-photon spectroscopy.

very low temperature, far below the equilibrium condensation point. Such radiation cooling has been demonstrated for ions in a radio-frequency quadrupole trap or a Penning trap. It has also been suggested as a means to trap slow atoms with laser light fields, taking advantage of dielectric forces.

Hole-burning spectroscopy. Although most solid samples show only rather broad spectral features, some extremely narrow natural line widths have been observed in certain ions and molecules in host crystals near liquid helium temperature. Generally, such lines appear broadened because crystal-site-dependent statistical field variations cause varying line shifts. The true natural line width can be observed by a method closely related to saturation spectroscopy in gases. A monochromatic laser interacts resonantly only with a group of absorbing ions or molecules at selected crystal sites. By temporarily removing these from their absorbing level, it produces a narrow-band "hole-burning" in the absorption profile, which can be observed with a second tunable probe laser beam.

Time-resolved laser spectroscopy. Lasers permit the generation of extremely short and intense light pulses, down to subnanosecond or subpicosecond duration, which can be powerful tools for studies of transient phenomena in the interaction of light with matter.

Short laser pulses permit, in particular, measurements of excited-state lifetimes and studies of relaxation processes in atoms, molecules, and ions in the gaseous, liquid, and solid phase with unprecedented temporal resolution.

Pulsed lasers can be used to excite atoms or molecules to a superposition of two or more closely spaced levels. Interference effects lead to a modulation of the subsequent spontaneous emission. Such quantum beats provide information about the detailed level structure.

Lasers have also made it possible to observe interesting coherent transient phenomena in the interaction of light with resonant transitions. Such effects include free induction decay, optical nutations, photon echoes, and self-induced transparency. They often have analogs in phenomena previously observed in the microwave region in studies of nuclear magnetic resonance, and they are useful not only for measurements of phase relaxation processes, but also for understanding the intricacies in the interaction of light with matter. SEE NUCLEAR MAGNETIC RESONANCE (NMR).

Bibliography. W. M. Demtroder, *Laser Spectroscopy: Basic Concepts and Instrumentation*, 1981; B. A. Garetz and John R. Lombardi (eds.), *Advances in Laser Spectroscopy*, vols. 1–3, 1982–1986; M. Levinson, *Introduction to Nonlinear Laser Spectroscopy*, 1982; S. Stenholm, *Foundations of Laser Spectroscopy*, 1984; H. P. Weber and W. Lüthy (eds.), *Laser Spectroscopy 6*, Springer Series in Optical Sciences, vol. 40, 1983; W. M. Yen and P. M. Selzer (eds.), *Laser Spectroscopy of Solids*, 2d ed., 1986.

DOPPLER EFFECT
ERNST W. OTTEN

The change in the frequency of a wave observed at a receiver whenever the source or the receiver is moving relative to each other or to the carrier of the wave (the medium). The effect was predicted in 1842 by C. Doppler, and first verified for sound waves by C. H. D. Buys-Ballot in 1845 from experiments conducted on a moving train.

Spectral lines from atomic or molecular gases show a Doppler broadening, due to the statistical Maxwell distribution of their velocities. Hence the spectral profile, that is, the intensity I as a function of frequency ν, is a gaussian centered around ν_0, given by Eq. (1), with full halfwidth given by Eq. (2), where k is the Boltzmann constant, T is the thermodynamic temperature in

$$I(\nu) = I(\nu_0) \exp - \left(\frac{\nu - \nu_0}{\delta\nu}\right)^2 \quad (1) \qquad \delta_{1/2}(\nu) = (2 \ln 2)\, \delta\nu = (2 \ln 2)\frac{\nu_0}{c_0}\sqrt{\frac{2kT}{m}} \quad (2)$$

kelvins, and m is the mass of a molecule. The Doppler width in Eq. (2) is of order $10^{-6}\nu_0$ and exceeds the natural linewidth, given by Eq. (3), by orders of magnitude. Equation (3) is the ulti-

$$\delta\nu_{\text{nat}} = \frac{1}{2\pi\tau} \quad (3)$$

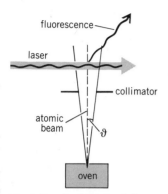

Fig. 1. Doppler-free atomic-beam spectroscopy on thermal atomic beams.

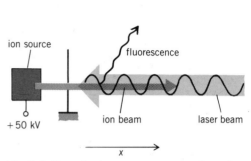

Fig. 2. Collinear laser spectroscopy on accelerated beams.

mate limit of resolution, set by the decay time τ of the quantum states involved in the emission of the line. Thus, precision spectroscopy is seriously hindered by the Doppler width. *See* LINEWIDTH.

Around 1970, tunable, monochromatic, and powerful lasers came into use in spectroscopy and resulted in several different methods for completely overcoming the Doppler-width problem, collectively known as Doppler-free spectroscopy. Lasers also improved the spectroscopic sensitivity by many orders of magnitude, ultimately enabling the spectroscopy of single atoms, and both improvements revolutionized spectroscopy and its application in all fields of science. The most common Doppler-free methods are discussed below.

1. Atomic beams. An atomic beam, emerging from an oven, well collimated and intersecting a laser beam at right angles, absorbs resonant laser light with the Doppler width suppressed by a factor of sin ϑ, where ϑ is the collimation angle (**Fig. 1**). This straightforward method was used before the advent of lasers with collimated light sources. *See* MOLECULAR BEAMS.

2. Collinear laser spectroscopy on fast beams. Another way to reduce the velocity spread in one direction is by acceleration. Consider an ion emerging from a source with a thermal energy spread δE, of the order of the quantity kT, or approximately 0.1 eV, and accelerated by a potential of, say, 50 kV along the x direction (**Fig. 2**). The kinetic energy spread in this direction remains constant, however, and may be expanded as in Eq. (4), where the overbars represent average

$$\delta E \approx kT \approx \delta\left(\frac{m}{2}v_x^2\right) \approx m\bar{v}_x \cdot \delta v_x \approx \frac{mc_0^2}{v_0^2} \cdot \overline{\Delta\nu} \cdot \delta\nu = \text{constant} \qquad (4)$$

values. Since the average velocity v_x has been increased by a factor of 700 in the example given, its spread δv_x, and hence the Doppler width with respect to a collinear laser beam, is reduced by the same factor. Consequently, the product of the Doppler shift $\overline{\Delta\nu}$ of the fast beam and its Doppler

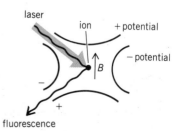

Fig. 3. Laser cooling of a trapped ion.

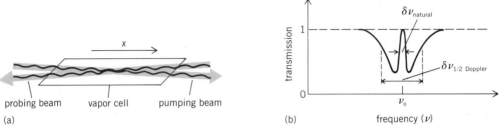

Fig. 4. Saturation spectroscopy. (a) Experimental configuration. (b) Transmission of probing beam.

width $\delta\nu$ is a constant of the motion. This method is very versatile and extremely sensitive.

3. Laser cooling of a trapped ion. Consider an ion trapped in a vacuum by electromagnetic fields (**Fig. 3**). The radial motion is stabilized by a magnetic field B, causing small cyclotron orbits of the ion, whereas axial motion is stabilized by applying a repelling electric potential to the electrodes on the axis. A laser excites the ion at the low-frequency side of the Doppler profile of a resonance line, thus selecting excitations at which the ion counterpropagates to the light. Therefore the ion is slowed down, since it absorbs the photon momentum $h\nu/c$ (where h is Planck's constant) which is opposite to its own. After the ion emits fluorescent radiation, the process is repeated until the ion comes almost to rest (with a thermodynamic temperature on the order of 1 kelvin) in the center of the trap.

4. Saturation spectroscopy. Whereas the velocity distribution of the atoms is manipulated in the above examples, in the following ones this distribution is left untouched, but Doppler broadening is suppressed through the nonlinear optical interaction of the photons with the atoms. The field of nonlinear optics can be roughly characterized by the condition that the atomic system interact with more than one light quantum during its characteristic decay time τ. In saturation spectroscopy, two almost counterpropagating laser beams of the same frequency ν are used, a strong pumping beam and a weak one (**Fig. 4**). The pumping beam is strong enough to saturate the excitation of those atoms in the vapor which match the resonance condition, that is, which compensate the frequency offset from the center $\nu - \nu_0$ by the proper velocity component, given by Eq. (5). Saturation means that the excitation is fast as compared to the decay time, leading to

$$v_x = c(\nu - \nu_0) \quad (5)$$

an equal population of the lower and the upper atomic states. Such a system is completely transparent. The probing beam feeds on atoms with exactly the opposite velocity component given by Eq. (6). It is too weak to saturate, and is absorbed in the vapor except at the center frequency

$$-v_x = c(\nu - \nu_0) \quad (6)$$

$\nu = \nu_0$, where it matches with the saturated part of the velocity spectrum. Resolutions of $\delta\nu/\nu_0 < 10^{-10}$ obtained in this way make possible improved values of fundamental constants and of the standards of length and time.

5. Two-photon spectroscopy. In this method, two counterpropagating laser beams are operated at half the transition frequency ν_0 of the atomic system. Then Bohr's quantum condition,

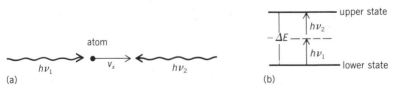

Fig. 5. Two-photon spectroscopy. (a) Relation of atom's motion to photons from counterpropagating laser beams. (b) Energy-level diagram.

that the transition energy ΔE equal $h\nu_0$, can be fulfilled by simultaneous absorption of two photons, say, one from each beam (**Fig. 5**). Including the Doppler effect, the energy balance is then given by Eq. (7). Thus, the Doppler effect cancels completely, and the two-photon resonance occurs

$$h\nu_1 + h\nu_2 = h\nu\left(1 + \frac{v_x}{c}\right) + h\nu\left(1 - \frac{v_x}{c}\right) = 2h\nu = h\nu_0 \quad (7)$$

sharply at half the transition frequency. *See* ATOMIC STRUCTURE AND SPECTRA; LASER SPECTROSCOPY.

RESONANCE-IONIZATION SPECTROSCOPY
G. S. HURST

A form of atomic and molecular spectroscopy in which wavelength tunable light sources are used to remove electrons from (that is, ionize) a given kind of atom or molecule. Laser-based resonance ionization spectroscopy (RIS) techniques have been developed and used with ionization detectors such as the proportional counter to show that single atoms can be detected. Both RIS and one-atom detectors find a wide range of applications in physics, chemistry, and oceanography, and in the environmental sciences.

Theory. When an atom (or molecule) is subjected to a light source that provides photons of an angular frequency ω, these photons can be absorbed by the atom if the photon energy $\hbar\omega$ ($\hbar$ is Planck's constant divided by 2π) is almost exactly the difference in energy between an atom in its normal or ground state and some excited state. Suppose, as in scheme 1 of **Fig. 1**, that a light source is tuned to a frequency which excites a given kind of atom, A. If the light source is a tunable pulsed laser of very narrow bandwidth, it is highly unlikely that any other kind of atom will be excited. But the atoms which are in an excited state can be further excited to the ionization continuum where electrons are set free, provided that the ionization potential of the atoms is less than $2\hbar\omega_1$. While the final ionization step can occur with photons of any energy above a threshold, the entire process is a resonance process. In sharp contrast with other ionization

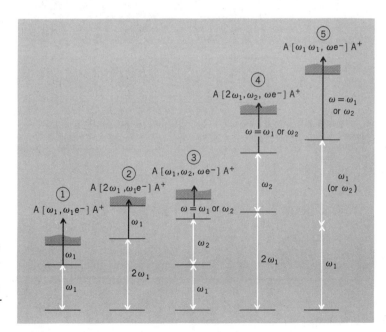

Fig. 1. Various laser schemes used in resonance ionization spectroscopy. (*After Office of Health and Environmental Research, U.S. Department of Energy under contract W-7405-eng-26 with the Union Carbide Corp.*)

means—for example, x-rays or radioactive sources—resonance ionization spectroscopy is a selective process in which only those atoms that are in resonance with the light source are ionized. Modern pulsed lasers are excellent tunable sources for resonance ionization spectroscopy; furthermore, they provide enough light in a single pulse to remove one electron from each atom of the selected type. A laser that provides 100 millijoules of photons in a single pulse of 10^{-6} s duration can be tuned to ionize nearly all of the atoms of a given type that may happen to be gas contained in a virtual test tube whose diameter is 1 cm and which can be very long (even meters), consistent with the divergence of the laser beam. SEE ATOMIC STRUCTURE AND SPECTRA; LASER SPECTROSCOPY.

Laser schemes. Many laser schemes can be used, as shown in Fig. 1. The notation $A[\omega_1, \omega_1 e^-]A^+$, taken from the standard notation for describing nuclear reactions, is used for the two-step process described above. On the other hand, the frequency of a laser can be doubled to $2\omega_1$, so that scheme 2 requires only one laser, while schemes 3, 4, and 5 involve two lasers to generate photons at frequencies of ω_1 and ω_2. With these five schemes only, it is possible to selectively ionize every known element in the periodic table except two of the noble gases, helium and neon (**Fig. 2**).

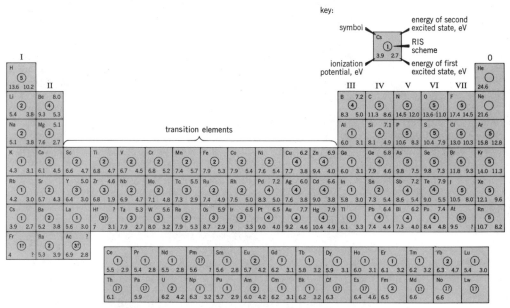

Fig. 2. Resonance ionization spectroscopy schemes for the elements of the periodic table. Numbers identifying schemes are those given in Fig. 1. (*After Office of Health and Environmental Research, U.S. Department of Energy under contract W-7405-eng-26 with the Union Carbide Corp.*)

One-atom detection. Both the high selectivity and the extraordinary sensitivity of resonance ionization spectroscopy were demonstrated by pulsing a laser directly through a proportional counter (**Fig. 3**). It was shown by S. C. Curran, J. Angus, and A. L. Cockcroft in 1949 that an improved version of the 1908 Rutherford-Geiger electrical counter (now known as a proportional counter) can be used to count single electrons at thermal energy. Therefore, if lasers are used to remove one electron from all of the atoms of a selected type, one-atom detection is possible. Proportional counters are normally filled with gases like argon (90%) and methane (10%). A pulsed laser tuned, for example, to 455.5 nanometers can detect even one atom of cesium without producing background ionization of the counting gas. In the original demonstration of one-atom de-

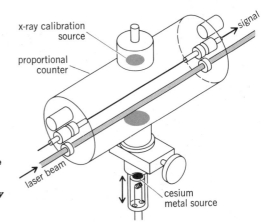

Fig. 3. Apparatus for experiment conducted to prove that resonance ionization spectroscopy can be used to detect a single atom. (*After Office of Health and Environmental Research, U.S. Department of Energy under contract W-7405-eng-26 with the Union Carbide Corp.*)

tection, it was proved that one atom of cesium could be selected out of 10^{19} atoms of the counting gas (argon and methane).

Another important form of one-atom detection involves the time-resolved detection of a single daughter atom in flight following the decay of a parent atom. Thus, it was shown that an atom of cesium could be detected from the fission decay of an individual atom of the isotope ^{252}Cf. The energy released in the fission process generated a signal that triggered the laser used to accomplish the resonance ionization spectroscopy process $Cs[\omega_1, \omega_1 e^-]Cs^+$. The success of that experiment proved that daughter atoms can be detected in coincidence with the decay of parent atoms. Such techniques could eventually work for most of the daughter atoms associated with radioactive decay, and could possibly be used to greatly reduce backgrounds in low-level counting facilities.

Applications. Resonance ionization spectroscopy and one-atom detection find a variety of interesting applications.

Classical chemical physics. The capability for detecting a population of just a few atoms has made it possible to investigate some problems in classical chemical physics which previously were difficult or impossible. Precision measurements of the diffusion of free atoms among other atoms and molecules have been made in sufficient detail to test the basic diffusion equation in both time and space domains. The determination of rates of reactions of extremely reactive substances (such as alkali atoms) with other atoms or molecules is now possible. Since only a few of these reactive atoms need to be produced, several problems concerning corrosion of the apparatus and the production of complicated chemical by-products are avoided. Study of a population of a few atoms (for example, 10) to observe their statistical behavior has been made.

Modern physics. One-atom detection makes it feasible to detect extremely rare events. Measurements of neutrinos from the Sun are crucial to testing both solar models and neutrino physics. After prolonged exposure to the Sun, the neutrinos may produce on the order of 100 atoms of a particular type in a very large tank. Previously, targets have been rich in ^{37}Cl so that neutrino capture would produce ^{37}Ar, a radioactive atom which can be counted by the standard methods of radioactivity. Resonance ionization spectroscopy and one-atom detection are making possible a much wider variety of neutrino targets. For example, in a lithium-rich target neutrinos produce ^{7}Be, which can be detected by observing daughter (lithium) atoms in time coincidence with the decay of the parent atoms.

Another experiment involves bromine-rich targets, where the neutrino capture produces ^{81}Kr. Radioactive ^{81}Kr can be counted directly (before it decays) by another technique made possible by resonance ionization spectroscopy. Other problems in weak interaction physics are also amenable to the one-atom detection techniques. Some meson interactions with nuclei have extremely low cross sections and thus produce only a few product atoms which, however, can be detected with resonance ionization spectroscopy techniques.

Environmental. Oceanographers have considered the use of ^{39}Ar as a tracer for ocean water circulation. Measurements of ^{81}Kr in the natural environment to obtain the ages of polar ice caps and old groundwater deposits have been suggested. Several techniques made possible by the development of resonance ionization spectroscopy and one-atom detection have been developed for these applications.

Bibliography. S. C. Curran, J. Angus, and A. L. Cockroft, Investigation of soft radiations by proportional counters—I, *Phil. Mag.*, 40:36–52, 1949; R. Davis, Jr., D. S. Harmer, and K. C. Hoffman, Search for neutrinos from the Sun, *Phys. Rev. Lett.*, 20:1205–1209, 1968; G. S. Hurst et al., Resonance ionization spectroscopy and one-atom detection, *Rev. Mod. Phys.*, 51:767–819, 1979.

MOLECULAR BEAMS
James E. Bayfield

Utilization of well-directed streams of atoms or molecules in vacuum. This is a cornerstone technique in the investigation of molecular structure and interactions. Molecular beams are usually formed at sufficiently low particle density for the interaction of one beam molecule with another to be negligible. This ensemble of truly isolated molecules is available for the spectroscopic study of molecular energy levels using photon probes from the radio-frequency to optical portions of the electromagnetic spectrum. Some of the best-determined fundamental knowledge of physics comes from spectroscopic molecular-beam experiments. Beyond this, beams can be applied as probes of the multifaceted nature of gases, plasmas, surfaces, and even the structure of solids. An application intermediate in complexity is the study of molecular interactions determining the properties of plasma and electric discharge devices, the nature of the upper atmosphere, and some aspects of the cooler astrophysical regions.

Production and detection. One simple means of forming a beam is to permit gas from an enclosed chamber to escape through a small orifice into a second chamber maintained at high vacuum by means of large pumps. **Figure 1a** shows such effusion into a collimating chamber from an oven chamber, generally heated to control the vapor pressure of the gas. The molecules coming from the orifice are distributed in angle according to a cosine law, illustrated by the circle downstream of the orifice which represents the envelope of relative beam flux vectors. A useful number of molecules passes forward along the horizontal axis of the apparatus. A well-collimated beam is then formed by requiring that those molecules entering the test chamber where an experiment is to be performed pass not only through the orifice but also through a second small hole separating the collimating and test chambers. A property of beams formed this way is that the velocities of the individual molecules have a large thermodynamic spread in values centered on a mean value of order 10^3 m/s (3.3×10^3 ft/s) determined by the oven temperature.

Charge-exchange system. If higher velocities are desired, then a charge-exchange beam system can be used. In this scheme, ions are produced by some ionizing process such as electron impact on atoms within a gas discharge. Since the ions are electrically charged, they can be accelerated to the desired velocity and focused into a beam using electric or magnetic fields. The

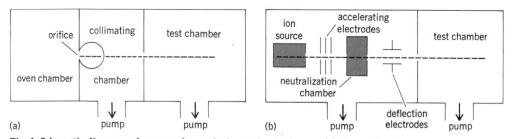

Fig. 1. Schematic diagrams of systems for producing molecular beams. (*a*) Conventional oven-beam system. (*b*) Charge-exchange beam system.

last step in neutrally charged beam formation is to pass the ions through a neutralizing gas where electrons from the gas molecules are transferred to the beam ions in charge-exchange molecular collisions. If the acceleration voltage is relatively high, then the ion-beam velocity will be 10^5 m/s (3.3×10^5 ft/s) or greater upon entering the neutralizer. In many cases, such energetic charge-exchange collisions produce beam atoms or molecules in internally excited energy levels rather than just in the ground-state level of lowest energy occupied by atoms in low-temperature gases. On the other hand, useful beams of internally excited molecules can be formed from ground-state molecules which have been excited by energetic electron- or ion-beam impact or by photon absorption from a laser or other light beam.

Secondary electron detection. The faster types of neutral molecular-beam particles are easy to detect by secondary electron ejection from a solid. A collision of a beam molecule with a surface is sufficiently violent for one or more electrons to be ejected. For intense beams, the rate of electron production is so high that the electrons can be collected and measured electronically as an electric current. At lower rates, the effect of each electron can be multiplied by means of an alternating sequence of electron acceleration and surface ejection steps, to produce a burst of 10^6 or more electrons from a single beam molecule. This pulse of current is adequate for electronic pulse counting, leading to a very sensitive overall beam-detection technique useful at beam intensities as low as five molecules per minute.

Other detection techniques. For molecular beams effusing from an oven, a variety of detection techniques have been devised. Alkali atoms have such small ionization potentials that their valence electrons can be transferred to a heated metal surface having a large work function, such as tungsten. The resultant ions can then be detected by current measurement or particle multiplication techniques, depending again upon the beam intensity. A second special technique useful for some beams of reactive or excited molecules is electron ejection from a surface, powered by the internal energy of these special beam molecules. A universal detector for any kind of slow molecule employs initial conversion into ions by electron- or light-beam impact; the ions are then accelerated into a fast beam for easy detection.

Special beams. Occasionally a beam containing atoms or molecules in a specific quantum-mechanical state is needed. Energy-resonant transitions between the ground state and the specific excited state can be utilized, often induced by single-frequency laser radiation. Beams of slow molecules possessing magnetic (or electric) dipole moments can be selected according to the direction of orientation of their moments; spatial separation into component beams is achieved through the orientation-dependent interaction between the dipole moment and an externally applied strong nonuniform magnetic (or electric) field. Some precision spectroscopic molecular-beam experiments use such beams, and form the basis for atomic-clock precision time standards.

Molecular-beam spectroscopy. Much of molecular spectroscopy involves the absorption or emission of light by molecules in a gas sample. The frequency of the light photon is proportional to the separation of molecular energy levels involved in the spectroscopic transition. However, the molecule density in typical gas samples is so high that the energy levels are slightly altered by collisions between molecules, with the transition frequency no longer characteristic of the free molecule. The use of low-density molecular beams with their sensitive detection techniques can reduce this collision alteration problem, with the result that atomic properties can be measured to accuracies of parts per million or even better. If the very simplest atoms or molecules are employed, then the basic electromagnetic interactions holding the component electrons and nuclei together can be precisely studied. This is of great importance to fundamental physics, since theoretical understanding of electromagnetic interactions through quantum electrodynamics represents the most successful application of quantum field theory to elementary particle physics problems.

Properties measured. Among the basic quantities of physics measured by molecular beams are: the fine-structure constant $\alpha = e^2/\hbar c = 1/137.0361$ (e is the electron charge in electrostatic units, $\hbar$ is Planck's constant divided by 2π, and c is the speed of light), obtained from microwave spectroscopy studies of the fine and hyperfine energy-level splittings in one- and two-electron atoms; the value 1.5210326×10^{-3} of the magnetic moment of the proton in Bohr magnetons, obtained with the hydrogen maser; the purely quantum-electrodynamic shifts of atomic energy levels called Lamb shifts; the nuclear magnetic dipole moments of several hundred isotopes; the equality in magnitude of the unit electron and nuclear charges to an accuracy of better than parts in 10^{18}; and the absence of intrinsic electric dipole moments for the electron and proton, which is a test of parity and time reversal as symmetry properties obeyed by the electromagnetic interaction. The isolated, unperturbed nature of atoms in a beam is needed for the device used as the length standard, based upon the 605.7-nanometer wavelength of krypton atoms being reproducible to parts in 10^9. The time standard is the cesium-beam atomic clock operating on the ground-state hyperfine splitting microwave transition with a reproducibility over hours of parts in 10^{12}.

Apparatus. The molecular-beam magnetic resonance apparatus used in a number of the spectroscopy experiments is shown in **Fig. 2**, along with a diagram showing typical molecular trajectories. On the left end of Fig. 2a is an oven beam source, here of alkali atoms; on the right end is a hot-wire surface ionization detector of these atoms. Between the ends are three separate regions where external fields are applied to the beam. The A region is a state selector, and the B region a state analyzer; that is, some property of the beam atoms is well defined in the A region, and then this situation is checked in the B region. In the example of Fig. 2 the selected property is the direction of the atoms' magnetic moment. The center of the apparatus is the transition

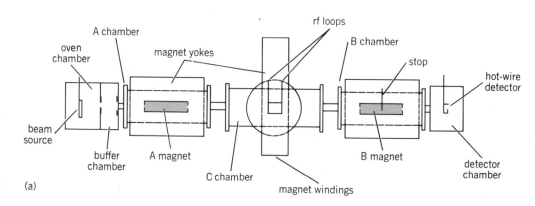

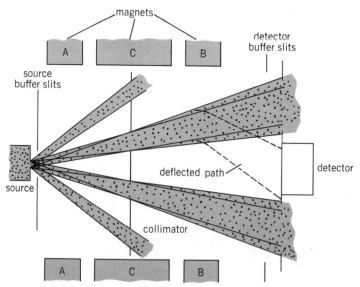

Fig. 2. Molecular-beam magnetic resonance spectroscopy experiment. (a) Apparatus used in the experiment. (b) Spatially resolved multiple-beam configuration showing typical molecular trajectories.

region marked C, where a resonant radio-frequency transition can be induced between atomic energy levels associated with different directions for the magnetic moment. This alteration of the atoms entering the analyzing region results in a change in atom trajectories, with a corresponding change in the intensity of the beam striking the hot-wire detector. For example, in Fig. 2b, molecules undergoing a transition within the C region follow the deflected path and are detected. When the frequency of the radio waves in the C region is not resonant, such a change does not occur. Determining the frequency for maximum effect results in a measure of the atomic energy level splitting, which for the example of Fig. 2 is directly related by theory to the atom's nuclear magnetic moment.

The power of molecular-beam techniques derives, in large measure, from the ability, in principle, of the A and B regions to contain any kind of state-selective device utilizing combinations of static or time-varying electromagnetic fields. Also, the interaction region C can involve resonant interactions with light or other photons or even nonresonant-state destructive interactions such as collisions with atoms in an introduced gas. If the A and B regions do not control beam atom trajectories, but instead some other property such as the state of internal energy, then correspondingly the nature of the beam detector might need to be sensitive to the controlled property.

Use of fast beams. The fast molecular beams produced by charge exchange increasingly have been employed in spectroscopy experiments and offer some advantages in addition to easy detection. Among these is the fractionally well-defined and controllable molecule velocity determined by the ion acceleration voltage, a feature useful in the study of time-dependent quantum-mechanical interference effects on transition rates. The ability to transport rapidly decaying excited atoms through an apparatus is also enhanced using fast beams. A highly accurate atomic fine-structure measurement employed a fast hydrogen atom beam, as did an experiment on multiphoton microwave transitions between highly excited atomic states.

Laser excitation. The development of tunable, strong laser sources of single-frequency light beams added another dimension to molecular-beam experiments. With laser radiation resonantly tuned to excite a molecule from its normal ground state to one of its infinite number of vibrationally, rotationally, and electronically excited states, the number of possible studies and applications of excited molecular beams becomes enormous. Of basic importance is the fact that excited molecules can be either highly reactive or good carriers of stored potential energy. SEE MOLECULAR STRUCTURE AND SPECTRA.

Bibliography. Faraday Society, *Molecular Beam Scattering*, vol. 55, 1973; P. Kusch and V. W. Hughes, *Atomic and Molecular Beam Spectroscopy*, vol. 37 of *Handbuch der Physik*, 1958; G. zu Putlitz, E. W. Weber, and A. Winnacker, *Atomic Physics 4*, 1975; N. F. Ramsey, *Molecular Beams*, 1956; C. Schlier, *Molecular Beams and Reaction Kinetics*, 1970.

INFRARED SPECTROSCOPY
RICHARD C. LORD

The study of the interaction of material systems with electromagnetic radiation in the infrared region of the spectrum. The infrared region is valuable for the study of the structure of matter because the natural vibrational frequencies of atoms in molecules and crystals fall in the infrared range. Some gaseous molecules also have rotational frequencies in the far-infrared range, and certain frequencies corresponding to the energy levels of electrons in solids and in large molecules lie in the near infrared. For a detailed discussion of molecular vibration and rotation SEE MOLECULAR STRUCTURE AND SPECTRA.

The infrared absorption spectrum of a molecule is highly characteristic, and often has been referred to as a molecular fingerprint. The spectrum can thus be used for molecular identification. Because the absorption of radiation at various infrared frequencies is quantitatively related to the number of absorbing molecules in a system, quantitative analysis is also possible.

The usefulness of an infrared absorption spectrum for identification and chemical analysis was recognized as long ago as 1890. In the early 1900s the American physicist W. W. Coblentz determined the infrared spectra of hundreds of substances and clearly demonstrated the potential value of such spectra. Unfortunately, the instrumentation of that day was cumbersome and necessarily homemade, so that few physicists and chemists were attracted by Coblentz's work. Only after the development of commercial electronic devices for amplification and recording of a continuously scanned spectrum in the 1940s was extensive use made of the technique.

Instrumentation and techniques. The usual arrangement for measurement of an infrared spectrum is shown schematically in **Fig. 1**. A source Q sends a beam of continuous infrared radiation to a spherical condensing mirror C, which passes the beam through S, the sample to be studied. Some of the infrared frequencies in the beam are absorbed strongly, some weakly. The reduced beam passes on and comes to a focus at the entrance slit of the monochromator M. The latter is an infrared spectrometer which disperses the radiation into a spectrum. One frequency at a time appears at the exit slit of M, from which the radiation of that frequency is passed by a suitable optical system to the detector T. The detector (a thermocouple or other device) converts the radiant energy into an electrical signal, which is amplified electronically at A and recorded by a chart recorder CR.

An infrared spectrum is a record of intensity of infrared radiation as a function of frequency or wavelength. To produce such a record, the chart recorder is driven in synchronism with the dispersing system of the monochromator M by some common driving mechanism D. In this way a given position on the chart corresponds directly to a given frequency setting of M, at which setting radiation of that frequency is emerging from M.

For many basic reasons—atmospheric absorption, variation of source intensity with frequency, changing dispersion in the spectrometer, and the like—the electrical output of the detector would not be constant even if the sample S were completely transparent. To correct for these variations it is necessary to determine two spectra, one with the sample S in the beam and one with S removed from the beam. The absorption of S as a function of frequency can then be computed from these two spectra. The individual spectra on which the computation is based are called single-beam spectra.

The computation is laborious, time consuming, and potentially unreliable because of changes in the entire system between the two determinations of spectra. These difficulties are avoided if a second optical path, shown in broken lines in Fig. 1, is introduced. The second optical path, called the reference beam, is made as nearly like the first as possible, except for the absence of the sample. In fact, the reference beam may contain an absorption cell R which differs from S only in the absence of the sample itself. For instance, if the sample S were in solution, R would contain the same amount of solvent as S.

The operation of the double-beam spectrometer, often called a spectrophotometer, consists of a rapid switching of the beam (say 10 times per second) back and forth between S and R by alternately placing plane mirrors 1 and 1' in the optical system. The identical mirrors 2 and 2' are permanently placed. The spectrum is scanned continuously as for single-beam operation, but the beams through S and R are compared 10 times per second and the chart records the energy passing through S relative to that through R. In this way the variations mentioned cancel out.

Typical spectra. Typical mid-infrared spectra, plotted automatically as percent transmission of the sample on a linear frequency scale (wave number in cm^{-1}), are shown in **Fig. 2**. Samples of gases, liquids, and solids can be readily measured. Techniques for high and low tem-

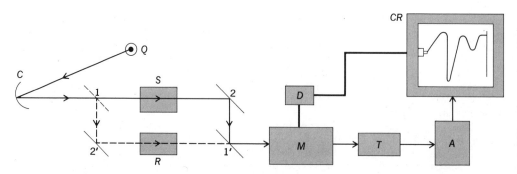

Fig. 1. Recording infrared spectrometer.

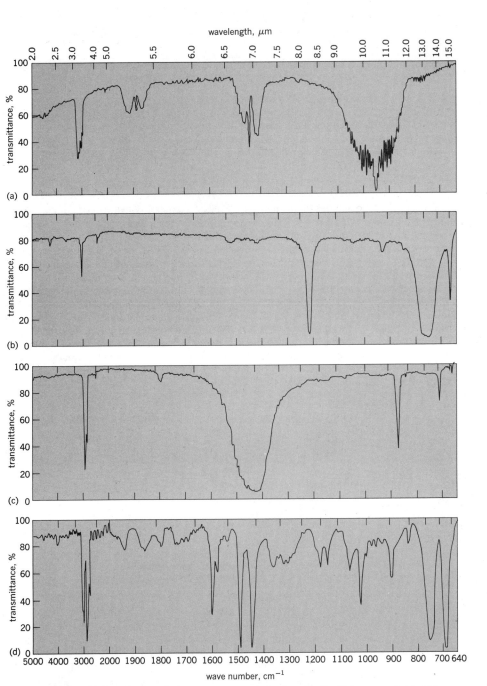

Fig. 2. Typical mid-infrared transmission spectra, recorded automatically. Note compressed scale on left portion of abscissa. (a) Spectrum of ethylene gas. Transmission minima in range 850–1050 cm^{-1} result from modulation of ethylene vibrational frequency at 950 cm^{-1} by molecular rotational frequencies. (b) Spectrum of liquid chloroform. (c) Spectrum of powdered crystalline calcium carbonate. Powder was suspended in mineral oil to obtain spectrum. Transmission minimum at 1430 cm^{-1} is characteristic of carbonate ion, and that at 2900 cm^{-1} is characteristic of CH groups in the oil. (d) Spectrum of an amorphous high polymer (polystyrene). Detail here shows why infrared spectra are sometimes called molecular fingerprints by workers in this discipline of science.

perature of sample and for small samples (down to about 1 mg or less in special cases) are in common use.

Percent transmission T, the quantity usually plotted by commercial instruments, is defined in Eq. (1). Here $I_{0,\nu}$ is the intensity of infrared radiation of frequency ν entering the sample and I_ν

$$T_\nu = \frac{100\, I_\nu}{I_{0,\nu}} \qquad (1)$$

is the intensity of the same radiation after passing through the sample. The percent transmission T_ν at frequency ν is different in principle at different values of ν. A quantity of fundamental importance, the absorbance A_ν, is defined in Eq. (2). The absorbance A_ν is proportional to the

$$A_\nu = \log\left(\frac{I_{0,\nu}}{I_\nu}\right) = -\log\left(\frac{T_\nu}{100}\right) \qquad (2)$$

number of absorbing molecules, and by evaluating the proportionality constant at frequency ν for a given kind of molecule in a particular system, the number of such molecules in other systems of the same kind may be measured quantitatively.

Interferometric methods. Infrared spectra are measured in many applications by the technique of Fourier transform spectroscopy. In dispersive (prism or grating) spectroscopy a spectrum is recorded by continuous scanning of the spectrum at successive frequencies (Fig. 1). In Fourier transform spectroscopy the entire frequency range of interest is passed simultaneously through an interferometer, which produces an output signal containing all these frequencies. The quantitative way in which this signal varies as the condition for interference within the interferometer is varied is called an interferogram (**Fig. 3**). The interferogram can be made to yield the spectrum as a function of frequency by the mathematical procedure known as a Fourier transform. Although this procedure is complicated, small and powerful digital computers are available so that an interferometer and a digital computer can be combined into a single unit which produces the transformed spectrum with negligible delay.

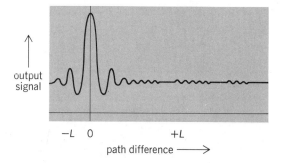

Fig. 3. A representative interferogram.

A block diagram of an interferometric spectrometer and auxiliary components is shown in **Fig. 4**. The source Q sends a beam containing a complete range of infrared frequencies into the interferometer I. The beam is first divided at the semitransparent beam splitter BS, which transmits half to the variable plane mirror M_V and reflects half to the fixed plane mirror M_F. The separate beams are returned by M_V and M_F to the beam splitter, where they are reunited after having traveled distances differing by some amount L which is continuously variable. The reunited beams interfere at BS, where they are partially reflected and sent out of the interferometer to the sample S. After passing through S, the radiation is converted to an electrical signal at the detector det. The output of det as L changes is shown in Fig. 3. This output is processed electronically at A, stored in the computer memory, and then transformed to the spectrum. The spectrum is read out in suitable form, for example, as a plot on the chart recorder CR of percent transmission versus

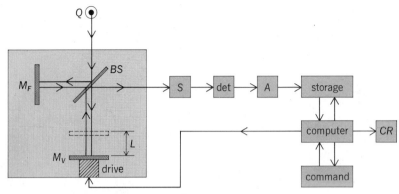

Fig. 4. Block diagram of a Fourier transform system.

frequency. Appropriate instructions to the computer are provided from the command post, which may be a teletypewriter. For double-beam operation the optical system at S may resemble the two-beam arrangement through S and R in Fig. 1.

The virtues of interferometric spectroscopy as compared to dispersive (that is, grating or prism) spectroscopy are:

1. Enormous superiority in the effective use of the limited radiant power in infrared sources. This superiority, which leads to much larger signal-to-noise ratios in the transformed spectra, arises from two fundamental differences between interferometers and spectrometers: first, the interferometer processes all frequencies in the input radiant power simultaneously (the multiplex advantage), whereas the dispersive spectrometer processes them one at a time. Second, the spectrometer needs narrow entrant and exit slits to do this processing, and the solid angle of radiant power accepted from the source is correspondingly restricted. The interferometer does not have this restriction and thus can accept a much larger solid angle of input radiation (the throughput advantage). The combination of these two advantages may lead to a superiority of several orders of magnitude in signal-to-noise ratio of Fourier transform spectroscopy over that of a dispersive spectrometer having the same resolution and scanning rate. Alternatively, Fourier transform spectroscopy will be able to record at the same signal-to-noise ratio and scanning rate with an increase in the resolution of more than an order of magnitude. This latter superiority is illustrated in **Fig. 5**, which shows the absorption spectrum of ethylene gas recorded from 940 to 960 cm^{-1} by Fourier transform spectrometer. The effective scanning rate in cm^{-1}/s and the signal-to-noise ratio are about the same in Fig. 2a (dispersive spectrum of ethylene) and in Fig. 5, but the resolution (as measured by the reciprocal of the spectral bandpass $\Delta \nu$) is 8 times higher in the latter ($\Delta \nu \simeq 2$ cm^{-1} at 950 cm^{-1} in Fig. 2a, $\Delta \nu = 0.25$ cm^{-1} in Fig. 5).

2. The interferometer does not require any filters or other order-sorting devices.

3. The wavelength or wave-number scale of the interferogram is automatically provided by the scanning parameter of the interferometer. This parameter is usually controlled to high precision by an auxiliary laser interferometer.

4. The spectral bandpass $\Delta \nu$ computed from the interferogram is constant throughout the spectrum. It is equal to, or somewhat larger than, $1/(2L_{max})$ where L_{max} is the maximum excursion of the moving mirror, depending on how the data are processed.

5. The computer needed to calculate the Fourier transform can be used in addition for the manipulation of the spectroscopic data at any stage. It can also be programmed to control the mechanical operation of the interferometer, the electronics system, and the read-out device (for example, the recorder).

Use of tunable lasers. The sharpness of the frequency and the high power per unit solid angle and unit spectral bandpass in a laser beam make it attractive for infrared spectroscopy at ultrahigh resolution. The main problem is the tuning of the laser frequency, that is, varying some parameter in the laser system so that its frequency may be varied continuously and accurately.

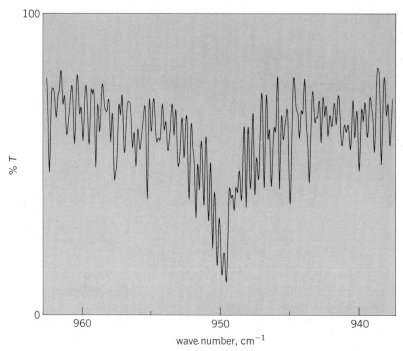

Fig. 5. Fourier transform spectrum of ethylene gas, 940–960 cm^{-1}. Path length of sample, 10 cm; gas pressure, 30 torr (4.0 × 10^3 Pa); recording time, ~ 30 min; spectral bandpass $\Delta \nu$, 0.25 cm^{-1}.

The **table** shows the frequency ranges over which various kinds of infrared lasers can be tuned. It is apparent from the table that their extremely high resolution (small spectral bandpass) makes infrared spectroscopy with tunable lasers a quite different kind of enterprise from dispersive or Fourier transform spectroscopy. An example of this resolution is shown in **Fig. 6**, where the absorption spectrum of ethylene gas at 950 cm^{-1} measured with a tunable semiconductor diode laser is illustrated (compare Figs. 2a and 5). The triangular background with maximum at 949.3 cm^{-1} is the slit function of the dispersive spectrometer used to eliminate unwanted semiconductor diode laser modes.

Tunable infrared lasers have been used mainly to measure the energy levels due to the vibration and rotation of molecules in the gas phase at low pressures. By this means, very accurate values of such molecular parameters as internuclear distances, electric dipole moments, vibrational frequencies, internal force fields, and the like may be measured. Other anticipated uses of laser infrared spectroscopy include the monitoring of the composition of the atmosphere. For example, the amount of carbon monoxide has been monitored in a horizontal path of 2000 ft (600 m) at ground level with a semiconductor diode laser; the sensitivity achieved was of the order of 5 parts per billion.

Applications. An infrared spectrum consists of a plot of T or A as a function of ν (or of wavelength λ). The basic information provided by the spectrum is a set of ν values at which the substance is absorbing strongly, that is, at which T_ν is a minimum (Fig. 2) or A_ν is a maximum. These frequencies of maximum absorption usually correspond to the actual vibrational frequencies of the absorbing molecules or to some arithmetical combination of such vibrational frequencies. If the molecules are in the vapor phase, absorption maxima may also be observed at frequencies which are combinations of frequencies of molecular rotation and vibration. The qualitative usefulness of an infrared spectrum lies in the fact that the set of observed vibrational frequencies characterizes the absorbing molecule.

Frequency ranges of tunable infrared lasers*

Type of laser	Frequency range, cm^{-1}	Minimum spectral bandpass $\Delta\nu$, cm^{-1}
Semiconductor diode laser (SDL)	300–10,000*	3×10^{-6}
Spin-flip Raman laser (SFR)	1600–2000 700–1000	3×10^{-6} 3×10^{-2}
Zeeman-tuned gas laser (ZTG)	1100–3300	3×10^{-3}
High-pressure CO$_2$ laser (HPG)	900–1100	3×10^{-4}
Nonlinear devices		
Optical parametric oscillator (OPO)	900–10,000$^+$	3×10^{-2}
Difference frequency generator (DFG)	1600–3300	5×10^{-4}
Two-photon mixer (TPM)	900–1100	3×10^{-5}
Four-photon mixer (FPM)	400–5000	1×10^{-1}

*Adapted from K. W. Nill, Tunable infrared lasers, *Opt. Eng.*, 13:516–522, 1974.

Qualitative chemical analysis. Infrared spectra can be used for the following purposes.

1. To identify pure chemical compounds by comparison of the spectrum of an unknown with previously recorded spectra of pure compounds. Catalogs of spectra are available, and in addition there are practical methods for encoding the information in the spectra and storing it in a computer memory system or on punched cards. The stored data can then be used for fast identification of an unknown spectrum.

2. To identify the constitutents of mixtures. When a mixture contains only two or three constituents, it is often possible to identify them directly from the spectrum of the mixture. More commonly it is necessary to fractionate the mixture by gas chromatography or other technique and to identify the individual compounds in the mixture by their infrared spectra. For this purpose Fourier transform spectroscopy is advantageous because of its speed and superior signal-to-noise ratio; combined chromatographic and Fourier transform equipment is available.

3. To show the presence of a group of atoms, a so-called functional group, in a molecule of unknown or doubtful structure. It has been known since the 1890s that certain groups of atoms—for example, a methyl group (CH_3), a carbonyl group (CO), a nitrate ion (NO_3^-)—have characteristic absorption frequencies that are relatively independent of the rest of the molecule or crystal in which the group occurs. Literally hundreds of such group frequencies are known.

Quantitative chemical analysis. There is a linear quantitative relationship between the absorbance A_ν, defined in Eq. (2), and the number of absorbing molecules. Thus the quantitative analysis of mixtures by infrared means is feasible. The infrared method is not particularly sensitive, the limit of detection and measurement of minor constituents being in the range 0.1–1.0%, except in favorable circumstances. An example of the latter is the detection of a component of a gaseous mixture. If the component has strong absorption in a spectral range where the rest of the

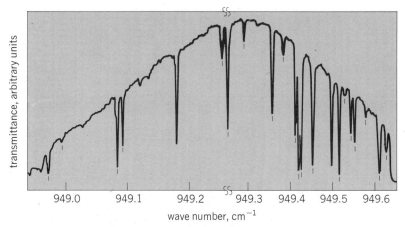

Fig. 6. Absorption spectrum of ethylene gas obtained by tuning a $Pb_xSn_{1-x}Te$ semiconductor diode laser in the range 948.9–949.7 cm^{-1}. Path length, 10 cm; pressure, 0.08 torr (1.1×10^2 Pa); effective spectral bandpass, $\Delta\nu < 0.001$ cm^{-1}. (*After G. P. Montgomery, Jr., and J. C. Hill, High-resolution diode-laser spectroscopy of the 942.9 cm^{-1} band of ethylene, J. Opt. Soc. Amer., 65:579–585, 1975*)

mixture is transparent, much smaller quantities may be detected. In general, the higher the resolution (the narrower the spectral bandpass) of the infrared instrument, the lower the minimum concentration that may be detected, provided the width of the spectral lines of the absorbing constituent is of the same order as the spectral bandpass. The advantage of narrow spectral bandpass disappears if it is much smaller than the spectral line width. Thus the sensitivity of the semiconductor diode laser measurement of atmospheric carbon monoxide (5 parts per billion over a path of 2000 ft or 600 m) would be much lower (that is, much improved) if the spectral line width of CO (~0.2 cm^{-1} at atmospheric pressure) were reduced to that of the semiconductor diode laser used (~0.0001 cm^{-1}).

The precision of quantitative measurement is mainly limited by the signal-to-noise ratio, and Fourier transform spectroscopy therefore offers an advantage for quantitative work. With dispersive spectroscopy, precision of measurement is seldom better than 1% of the quantity being measured and may be considerably worse. Infrared methods are especially useful in the quantitative determination of isomeric substances and in measurement of constituents of a chemical equilibrium.

Determination of molecular structure. Structures of molecules can be determined to varying degrees of refinement from infrared spectra. If only a few independent parameters (interatomic distances and bond angles) are required to specify the structure, as is the case with a small symmetrical molecule, these can be evaluated from the moments of inertia of the molecule, which can in turn be measured from rotational frequencies, usually observed as fine structure in a vibrational absorption. The structural parameters of carbon dioxide, methane, ethylene (Figs. 2a and 5), and ethane, for example, have been evaluated with high precision from their infrared spectra.

If the number of parameters is too large to be determined in this way, it may nevertheless be possible to draw conclusions about the molecule's shape without measuring its size. The number of vibrational frequencies which appear in the infrared spectrum is related to the molecular symmetry, and it is often possible to infer the symmetry from the observed spectrum. Such inferences are more reliable if they are based on combined data from both infrared and Raman spectra. SEE RAMAN EFFECT.

It is still possible to say something about the structure of large molecules of little or no symmetry from their spectra if one is content with a statement about the presence or absence of various functional groups. The organic chemist often finds such statements very valuable. The nature of functional groups in the molecules of high polymers or of natural products such as the steroids can be determined from their infrared spectra, and this permits information about their structure to be obtained.

Study of solids, catalysts, and matrices. The infrared spectra of crystalline solids give information about modes of vibration of crystals, about hydrogen-bond vibrations when such bonds are present in them, and about electronic energy states in semiconductors and superconductors. Fourier transform spectrometers are regularly used by solid-state physicists, particularly for their advantages in the spectral range 1–200 cm^{-1}. At higher frequencies Fourier transform spectrometers are useful in the study of materials absorbed on the surfaces of catalysts. The Fourier transform spectroscopy signal-to-noise advantage is especially important because of the optical heterogeneity of these samples. A similar heterogeneity in samples trapped at low temperatures in condensed rare gases or other types of matrices likewise makes Fourier transform spectroscopy the technique of choice for the investigation of trapped transient species.

Bibliography. L. J. Bellamy, *Infrared Spectra of Complex Molecules*, vol. 1, 3d ed., 1975, vol. 2, 2d ed., 1982; R. J. H. Clark and R. E. Hester, *Advances in Infrared and Raman Spectroscopy*, vols. 1–12, 1975–1985; J. R. Ferraro and J. L. Basile (eds.), *Fourier Transform Infrared Spectroscopy: Applications to Chemical Systems*, vol. 1, 1978, vol. 2, 1979, vol. 3, 1982, vol. 4, 1985; P. R. Griffiths and J. A. DeHaseth, *Fourier Transform Infrared Spectrometry*, 1986; G. R. Harrison, R. C. Lord, and J. R. Loofbourow, *Practical Spectroscopy*, 1948; G. Herzberg, *Infrared and Raman Spectra of Polyatomic Molecules*, 1945.

RAMAN EFFECT
RICHARD C. LORD

A phenomenon observed in the scattering of light as it passes through a material medium, whereby the light suffers a change in frequency and a random alteration in phase. Raman scattering differs in both these respects from Rayleigh and Tyndall scattering, in which the scattered light has the same frequency as the unscattered and bears a definite phase relation to it. The intensity of normal Raman scattering is roughly one-thousandth that of Rayleigh scattering in liquids and smaller still in gases.

Discovery. Because of its low intensity, the Raman effect was not discovered until 1928, although the scattering of light by transparent solids, liquids, and gases had been investigated for many years before. Prompted by A. H. Compton's observation of frequency changes in x-rays scattered by electrons (Compton effect), the Indian physicists C. V. Raman and K. S. Krishnan examined sunlight scattered by a number of liquids. With the help of complementary filters, they found that there were frequencies in the scattered light that were lower than the frequencies in the filtered sunlight. They then showed, by using light of a single frequency from a mercury arc, that the new frequencies in the scattered radiation were characteristic of the scattering medium. Within a few months of Raman and Krishnan's first announcement of their discovery, the Soviet physicists G. Landsberg and L. Mandelstam communicated their independent discovery of the existence of the effect in crystals. In Soviet literature the phenomenon is referred to as combination scattering, and not Raman effect.

The development of the laser has led to a resurgence of interest in the Raman effect and to the discovery of a number of related phenomena. A beam of laser radiation is intense, polarized, and coherent; it can be made monochromatic, small in diameter, and highly collimated. The laser is therefore nearly ideal for the production of the Raman effect, and other kinds of sources are seldom employed. Many different wavelengths in the visible spectrum and adjacent regions are available. The argon-ion and krypton-ion lasers are most commonly used, since they have high continuous-wave power (1 to 10 W), but tunable dye lasers are also often employed in excitation of resonance Raman scattering.

Raman spectroscopy. Raman scattering is analyzed by spectroscopic means. The collection of new frequencies in the spectrum of monochromatic radiation scattered by a substance is characteristic of the substance and is called its Raman spectrum. Although the Raman effect can be made to occur in the scattering of radiation by atoms, it is of greatest interest in the spectroscopy of molecules and crystals.

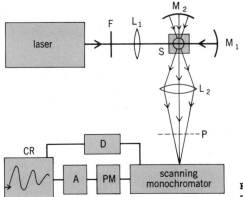

Fig. 1. Laser-Raman spectroscopic system.

Because of the laser beam's small diameter and high collimation, it can easily be used to excite the Raman effect. A typical optical arrangement is shown in **Fig. 1**. Monochromatic radiation from the laser impinges on the sample S in an appropriate transparent cell. It may be desirable to condense or expand the laser beam by means of a lens system L_1 and to remove unwanted radiation from the beam by a narrow-band optical filter F. A concave mirror M_1 can return unscattered radiation for a second passage through the sample.

Raman scattering is approximately uniform in all directions and is usually studied at right angles (Fig. 1). In this way the intense radiation of the laser beam interferes least with the observation of the weak scattered light. This light is collected by a lens system L_2 and focused on the slit of a scanning monochromator, which analyzes it spectroscopically. As the spectrum is scanned, the dispersed radiation from the monochromator is detected by a photomultiplier PM, further amplified and processed electronically at A, and then recorded by a strip-chart recorder CR. The recorder is driven in synchronism with the monochromator by a suitable mechanism D. The concave mirror M_2 may be used to augment the amount of scattered radiation by collecting light scattered at $-90°$ and returning it to the $+90°$ direction. The polarization characteristics of the scattered radiation are frequently of interest, especially since the laser radiation itself is linearly polarized. An analyzing device for evaluating the degree of polarization of the scattered radiation may be inserted at point P.

The appearance of a photoelectrically recorded Raman spectrum of liquid carbon tetrachloride as excited by the red line of the helium-neon laser at 632.8 nm (6328 Å, power incident on the sample of about 50 mW) is shown in **Fig. 2**. Intensity of the scattered light on an arbitrary scale is plotted vertically against the wave number in cm^{-1} measured with respect to the wave number of the exciting line taken as zero. For convenience, it is usual to express the data of Raman spectroscopy in cm^{-1} rather than frequency units (s^{-1}). (Frequency ν in $s^{-1} = c\bar{\nu}$ in cm^{-1}, where c is the velocity of light in vacuum in cm/s.)

A spectrum of the most intense line in Fig. 2 is shown at higher resolution (smaller spectral slit width $\Delta\bar{\nu}$) in **Fig. 3**. The line is seen to consist of several closely spaced components. These result from the presence of the two isotopes of chlorine, ^{35}Cl and ^{37}Cl, which produce five isotopic species $C^{35}Cl_n{}^{37}Cl_{4-n}$, $n = 0, 1, 2, 3, 4$. The line due to the least abundant species, $n = 0$, is not visible, but the other four are readily identified.

Theory. The mechanism of the Raman effect can be envisaged either by the corpuscular picture of light or from the point of view of the wave theory. Both pictures merge in the basic quantum theory of radiation. The corpuscular model of light scattering envisages light quanta or photons as particles which have linear and angular momenta. On passing through a material medium, these particles collide with atoms or molecules. If the collision is elastic, the photons bounce off the molecules with unchanged energy E and momentum, and hence with unchanged frequency ν. Such a process gives rise to Rayleigh scattering. If the collision is inelastic, the

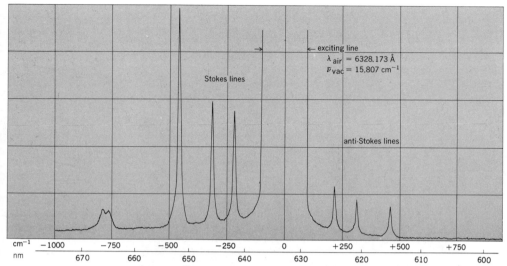

Fig. 2. Photoelectric recording of the Raman spectrum of carbon tetrachloride excited by He-Ne laser line at 632.8 nm. Intensity of the radiation is recorded vertically against the horizontal wave-number scale (cm^{-1}) measured from the exciting line as zero. The Rayleigh-scattered exciting line is three orders of magnitude more intense than the Raman lines, and its maximum is therefore far off-scale. The Stokes lines appear at lower frequencies, and the less intense anti-Stokes lines at higher frequencies, than those of the exciting line. The lower scale shows wavelengths in nanometers.

photons may gain energy from, or lose it to, the molecules. A change ΔE in the photon energy by Planck's relationship, $E = h\nu$, must produce a change in the frequency $\Delta \nu = \Delta E/h$. Such inelastic collisions are rare compared to the elastic ones, and the Raman effect is correspondingly much weaker than Rayleigh scattering.

In the wave picture of the effect, the electromagnetic waves which constitute the incoming monochromatic radiation sweep through the material medium. Since the atoms and molecules composing the medium are made up of negatively charged electrons and positively charged nuclei, the electric field of the light waves sets the electrons to oscillating, chiefly with the frequency of the incoming radiation. The oscillating electrons recreate the alternating electric field of the incoming light, thus passing the light wave along through the medium. This process is analogous to the elastic collisions which are given by the corpuscular picture.

The ability of the electrons and nuclei in a molecule to be displaced by an electric field is called the molecular polarizability α. It is not a simple property of the molecule, but depends in a complicated way on the frequency of the electric field, on the orientation of the molecule, and on the internal motions of the nuclei and electrons. Thus the molecular polarizability α varies periodically with molecular rotation and vibration, and thereby the effect of a light wave on the electrons and nuclei of a molecule can be changed.

When a monochromatic light wave sweeps through a transparent medium containing rotating and vibrating molecules, most of the wave is recreated unchanged by the oscillating electrons, but because of the periodic changes produced in α by rotation and vibration, new frequencies are added to the light wave. The appearance of these new frequencies, whose values are determined by the rotational and vibrational energies of the molecules, is analogous to the result of the inelastic collisions of the corpuscular model. For the wave picture of the Raman effect, the quantity α is the basic quantity. The intensity of the Raman effect depends on the magnitude of the changes produced in α by molecular rotation and vibration, and the number and values of new frequencies (usually expressed as frequency shifts $\Delta \nu$ from the original monochromatic frequency) depend on the variation of α with the frequencies of rotation and vibration.

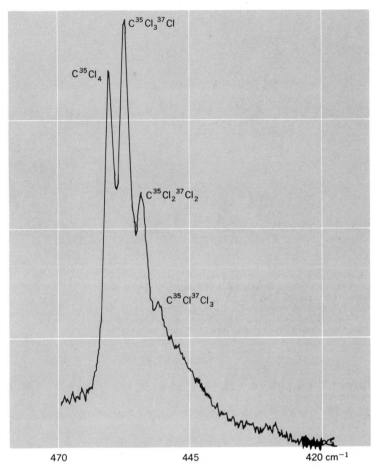

Fig. 3. The Stokes line at -460 cm^{-1} in the Raman spectrum of carbon tetrachloride. This spectrum, taken at about 10 times better resolution than that of Fig. 2, shows lines due to four of five isotopic species resulting from the 3:1 ratio of the chlorine isotopes ^{35}Cl and ^{37}Cl. The instrumental wave-number scale requires a calibration correction to increase the wave numbers by 1.4 cm^{-1}.

The temperature of the scattering molecules is an additional factor which affects the intensity of Raman frequencies higher than the exciting frequency (the anti-Stokes lines of Fig. 2). The anti-Stokes lines, having higher frequencies, correspond to photons which have higher energy than that of the exciting light, and this energy must come from the molecules. If the molecules do not have any available vibrational or rotational energy, that is, if they are at the absolute zero of temperature, there is no possibility of inelastic collisions in which energy is transferred from a molecule to a photon. So, anti-Stokes lines vanish at absolute zero. At nonzero temperatures the intensity ratio of an anti-Stokes line to a Stokes line is approximated by the ratio of the number of molecules which can give up the corresponding energy to the number which can accept it from the light wave.

Special forms. The development of lasers resulted in the discovery of a number of kinds of Raman scattering.

Resonance Raman effect. When the exciting radiation falls within the frequency range of a molecule's absorption band in the visible or ultraviolet spectrum, the radiation may be scattered by two different processes, resonance fluorescence or the resonance Raman effect. Both these processes give much more intense scattering than the normal nonresonant Raman effect. Resonance fluorescence differs from the resonance Raman effect in that the absolute frequencies of the fluorescent spectrum do not shift when the exciting radiation's frequency is changed, so long as the latter does not move outside the absorption band. The absolute frequencies of the resonance Raman effect, on the contrary, shift by exactly the amount of any shift in the exciting frequency, just as do those of the normal Raman effect. Thus the main characteristic of the resonance as compared to the normal Raman effect is its intensity, which may be greater by two or three orders of magnitude. SEE FLUORESCENCE.

The resonance Raman effect was anticipated by G. Placzek in 1934 in his pioneering development of the polarizability theory of Raman scattering. It was actually observed before the discovery of lasers, but tunable lasers are the most effective sources for the study of its various aspects. A typical resonance Raman spectrum is shown in **Fig. 4**, in which oxyhemoglobin is excited by the 568.2-nm (5682 A) wavelength of singly ionized krypton. The top spectrum $I_\parallel$ is taken with the polarizer P of Fig. 1 set to pass the components parallel to the direction of laser polarization; the bottom spectrum $I_\perp$ is taken with P set to pass perpendicular components. Lines in which $I_\perp$ is much greater than $I_\parallel$ are said to have inverse polarization and are seen only in the resonance Raman effect; these include the lines at 1305, 1342, and 1589 cm^{-1} and numerous others.

Hyper-Raman effect. The nature of this effect is most easily described in terms of the corpuscular picture of the Raman effect. With an intense laser source, the number of monochromatic photons impinging on the molecules of a medium per unit volume and unit time may be extremely large. If so, the probability that two photons will collide simultaneously with the same molecule is very much larger than in normal scattering, and there is considerable chance that the two photons will unite and be scattered as a single photon of approximately twice the frequency. The rules governing the scattering in such three-photon processes (two incoming and one outgoing photons) are quite different from those for normal (two-photon) Rayleigh and Raman scattering. For example, in molecules that are centrosymmetric, the collision must be inelastic, that is, the molecule must absorb or give up an amount of energy ΔE during the process. The frequency of the scattered photon will therefore not be exactly twice the frequency of the incident photons but will differ from it by $\Delta \bar{\nu} = \Delta E/hc$. Such scattered radiation is called the hyper-Raman effect. Even in molecules that are not centrosymmetric, the likelihood of elastic collisions is much smaller than in the normal case, so that the intensity of hyper-Rayleigh scattering may be substantially weaker than the hyper-Raman scattering.

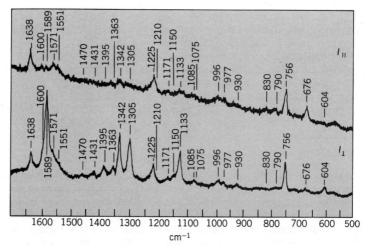

Fig. 4. Resonance Raman spectrum of oxyhemoglobin. (*After T. G. Spiro and T. C. Strekas, Resonance Raman spectra of hemoglobin and cytochrome c, Proc. Nat. Acad. Sci. USA, 69:2622–2626, 1972*)

As implied above, the selection rules for the vibrational and rotational transitions in the hyper-Raman effect are different from those of the normal Raman effect. Thus certain transitions are observable in the hyper-Raman effect that are normally forbidden. This is one virtue of the hyper-Raman effect; the other is that it is observed in a spectral region whose frequency is far removed from that of the incoming radiation (and is, in fact, twice that of the latter). The effect is therefore observable without interference from the normal Rayleigh line.

Stimulated Raman effect. The mechanism of the stimulated Raman effect depends on the coherent pumping of the molecules of the sample into an excited vibrational state by the powerful electric field of the laser beam. In view of the large discrepancy of one or two orders of magnitude between the frequency of the vibration and the frequency of the laser, this can be accomplished only if the field of the light wave has a very high value (the threshold power) and if the mismatch in frequency is compensated by the generation of coherent radiation with a frequency equal to that of the laser minus the vibrational frequency. The coherent radiation so produced is called stimulated Raman scattering. It was first observed by R. Woodbury and A. Ng in 1962. They found the effect in liquid nitrobenzene, which they were using as an electrooptical shutter within a laser system.

In addition to its high intensity and its coherence, there are other new features of stimulated Raman scattering. Since the pumping power of the incident laser beam must exceed a certain threshold for the scattering to take place, when the laser power is used up in exciting one vibrational mode, there is insufficient power available to excite other modes. Therefore the stimulated Raman effect usually contains only one frequency, though in rare cases the power may be divided between two vibrational modes of roughly the same threshold. However, the power in the scattered radiation may itself produce further stimulated Raman emission by a repetition of the intitial process. This results in a new frequency, which is the laser frequency minus exactly twice the frequency of vibrational mode that is being scattered. This fact shows that the mechanism does not involve a double jump in the vibrational levels; such a double jump would give a frequency shift that is not exactly twice that of the vibrational fundamental because of vibrational anharmonicity.

Another striking and unusual effect in stimulated Raman scattering is the excitation of intense anti-Stokes radiation. This radiation may be even stronger than the Stokes radiation in certain circumstances. Moreover, it can be observed at such low temperatures that the initial populations of the excited vibrational levels needed for normal anti-Stokes Raman scattering are zero. It arises from the above-mentioned pumping of molecules from the ground vibrational state into upper excited states by the intitial laser power. These excited molecules can then be pumped by further radiation back into the ground state, with a simultaneous stimulated emission of coherent radiation at a frequency that equals the laser plus the molecular vibrational frequency.

The development of tunable lasers has led to a special technique for stimulated Raman scattering called coherent anti-Stokes Raman spectroscopy (CARS). In this technique, two lasers are used, one of fixed and the other of tunable frequency. The two beams enter the sample at angles differing only by some appropriate small amount (approximately 2°) and simultaneously impinge on the sample molecules. Whenever the frequency difference between the two lasers coincides with the frequency of a Raman-active vibration of the molecules, emission of coherent radiation (both Stokes and anti-Stokes) is stimulated. Thus the total Raman spectrum can be scanned in stimulated emission by varying the frequency of the tunable laser. An advantage of CARS, in addition to the high intensity of the scattering, is that its elevated frequency avoids interference from sample fluorescence, which always has frequencies below that of the exciting radiation.

Applications. Raman spectroscopy is of considerable value in determining molecular structure and in chemical analysis. Molecular rotational and vibrational frequencies can be determined directly, and from these frequencies it is sometimes possible to evaluate the molecular geometry, or at least to find the molecular symmetry. *See Molecular structure and spectra.*

Even when a precise determination of structure is not possible, much can often be said about the arrangement of atoms in a molecule from empirical information about the characteristic Raman frequencies of groups of atoms. This kind of information is closely similar to that provided by infrared spectroscopy; in fact, Raman and infrared spectra often provide complementary data about molecular structure. The complex structures of biologically important molecules, for example, are the subjects of current spectroscopic research. Both normal and resonance Raman spectroscopy are valuable techniques in molecular biology (see Fig. 4). Raman spectra also provide information for solid-state physicists, particularly with respect to lattice dynamics but also concerning the electronic structures of solids. SEE INFRARED SPECTROSCOPY.

Bibliography. N. Bloembergen, *Non-Linear Optics*, 1965; A. J. Clark and R. E. Hester (eds.), *Advances in Infrared and Raman Spectroscopy*, vols. 1–12, 1975–1985; G. L. Eesley, *Coherent Raman Spectroscopy*, 1981; J. G. Grasselli et al., *Chemical Applications of Raman Spectroscopy*, 1981; G. Herzberg, *Infrared and Raman Spectra of Polyatomic Molecules*, vol. 2, 2d ed., 1945; D. A. Long, *Raman Spectroscopy*, 1977; M. C. Tobin, *Laser Raman Spectroscopy*, 1971, reprint 1981; A. T. Tu, *Raman Spectroscopy in Biology: Principles and Applications*, 1982; A. Weber (ed.), *Raman Spectroscopy of Gases and Liquids*, 1979.

QUASIELASTIC LIGHT SCATTERING
ROBERT PECORA

Small frequency shifts or broadening from the frequency of the incident radiation in the light scattered from a liquid, gas, or solid. The term quasielastic arises since the frequency changes are usually so small that, without instrumentation specifically designed for their detection, they would not be observed and the scattering process would appear to occur with no frequency changes at all, that is, elastically. The technique is used by chemists, biologists, and physicists to study the dynamics of molecules in fluids, mainly liquids and liquid solutions.

Several distinct experimental techniques are grouped under the heading of quasielastic light scattering (QLS). Intensity fluctuation spectroscopy (IFS) is the technique most often used to study such systems as macromolecules in solution and critical phenomena where the molecular motions to be studied are rather slow. This technique, also called photon correlation spectroscopy and, less frequently, optical mixing spectroscopy, is used to measure the dynamical constants of processes with relaxation time scales slower than about 10^{-6} s. For faster processes, dynamical constants are obtained by utilizing techniques known as filter methods, which obtain direct measurements of the frequency changes of the scattered light by utilizing a monochromator or filter much as in Raman spectroscopy. SEE RAMAN EFFECT.

Static light scattering. If light is scattered by a collection of scatterers, the scattered intensity at a point far from the scattering volume is the result of interference between the wavelets scattered from each of the scatterers and, consequently, will depend on the relative positions and orientations of the scatterers, the scattering angle θ, and the wavelength λ of the light used. The structure of scatterers in solution whose size is comparable to $(4\pi\lambda) \sin \theta/2 (\equiv q)$ where q is the length of the scattering vector, may be studied by this technique, variously called static light scattering, integrated intensity light scattering, or in the older literature simply light scattering. It was, in fact, developed in the 1940s and 1950s to measure equilibrium properties of polymers both in solution and in bulk. Molecular weights, radii of gyration, solution virial coefficients, molecular optical anisotropies, and sizes and structure of heterogeneities in bulk polymers are routinely obtained from this type of experiment. Static light scattering is a relatively mature field, although continued improvements in instrumentation (mainly the use of lasers and associated techniques) are steadily increasing its reliability and range of application.

Both static and quasielastic light scattering experiments may be performed with the use of polarizers to select the polarizations of both the incident and the scattered beams. The plane containing the incident and scattered beams is called the scattering plane. If an experiment is performed with polarizers selecting both the incident and final polarizations perpendicular to the scattering plane, the scattering is called polarized scattering. If the incident polarization is perpendicular to the scattering plane and the scattered polarization lies in that plane, the scattering is called depolarized scattering. Usually the intensity associated with the polarized scattering is much larger than that associated with the depolarized scattering. The depolarized scattering from relatively small objects is zero unless the scatterer is optically nonspherical.

Intensity fluctuation spectroscopy. The average intensity of light scattered from a system at a given scattering angle depends, as stated above, on the relative positions and orientations of the scatterers. However, molecules are constantly in motion due to thermal forces, and are constantly translating, rotating and, for some molecules, undergoing internal rearrangements. Because of these thermal fluctuations, the scattered light intensity will also fluctuate. The intensity will fluctuate on the same time scale as the molecular motion since they are proportional to each other.

Figure 1 shows a schematic diagram of a typical intensity fluctuation apparatus. Light from a laser source traverses a polarizer to ensure a given polarization. It is then focused on a small volume of the sample cell. Light from the scattering volume at scattering angle θ is passed through an analyzer to select the polarization of the scattered light, and then through pinholes and lenses to the photomultiplier (PM) tube. The output of the photomultiplier is amplified, discriminated, sent to a photon counter, and then to a hard-wired computer called an autocorrelator, which computes the time autocorrelation function of the photocounts. The autocorrelator output is then sent to a computer for further data analysis.

The scattered light intensity as a function of time will resemble a noise signal. In order to facilitate interpretation of experimental data in terms of molecular motions, the time correlation function of the scattered intensity is usually computed by the autocorrelator. The autocorrelation function obtained in one of these experiments is often a single exponential decay, $C(t) = \exp(-t/\tau_r)$, where τ_r is the relaxation time.

The upper limit on decay times τ_r that can be measured by intensity fluctuation spectroscopy is about a microsecond, although with special variations of the technique somewhat faster decay times may be measured. For times faster than this, filter experiments are usually performed by using a Fabry-Perot interferometer.

Fabry-Perot interferometry. Light scattered from scatterers which are moving exhibits Doppler shifts or broadening due to the motion. Thus, an initially monochromatic beam of light from a laser will be frequency-broadened by scattering from a liquid, gas, or solid, and the broadening will be a measure of the speed of the motion. For a dilute gas the spectrum will usually be a gaussian. For a liquid, however, the most common experiment of this type yields a single lorentzian line with its maximum at the laser frequency $I(\omega) = A/\pi[(1/\tau_r)/(\omega^2 + 1/\tau_r^2)]$. **Figure 2** shows a schematic of a typical Fabry-Perot interferometry apparatus. The Fabry-Perot interferometer acts as the monochromator and is placed between the scattering sample and the photomultiplier. Fabry-Perot interferometry measures the (average) scattered intensity as a function of frequency change from the laser frequency. This intensity is the frequency Fourier transform of the time correlation function of the scattered electric field. Intensity fluctuation spectroscopy experiments utilizing an autocorrelator measure the time correlation function of the intensity (which equals the square of the scattered electric field). For scattered fields with gaussian amplitude

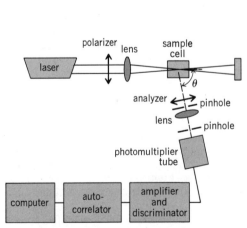

Fig. 1. Schematic diagram of an intensity fluctuation spectroscopy apparatus.

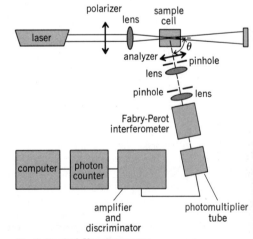

Fig. 2. Typical filter apparatus.

distributions the results of these two types of experiment are easily related. Sometime intensity fluctuation spectroscopy experiments are performed in what is sometimes called a heterodyne mode. In this case, some unscattered laser light is mixed with the scattered light on the surface of the photodetector. Intensity fluctuation spectroscopy experiments in the heterodyne mode measure the frequency Fourier transform of the time correlation function of the scattered electric field.

Translational diffusion coefficients. The most widespread application of quasielastic light scattering is the measurement of translational diffusion coefficients of macromolecules and particles in solution. For particles in solution whose characteristic dimension R is small compared to $q^{-1} = (4\pi/\lambda \sin \theta/2)^{-1}$, that is, $qR < 1$, it may be shown that the time correlation function measured in a polarized intensity fluctuation spectroscopy experiment is a single exponential with relaxation time $1/\tau_r = 2q^2D$, where D is the particle translational diffusion coefficient. For rigid, spherical particles of any size an intensity fluctuation spectroscopy experiment also provides a measure of the translational diffusion coefficient.

Translational diffusion coefficients of spherical particles in dilute solution may be used to obtain the particle radius R through use of the Stokes-Einstein relation [Eq. (1)], where k_B is

$$D = \frac{k_B T}{6\pi\eta R} \qquad (1)$$

Boltzmann's constant, T the absolute temperature, and η the solvent viscosity. If the particles are shaped like ellipsoids of revolution or long rods, relations known, respectively, as the Perrin and Broersma equations may be used to relate the translational diffusion coefficient to particle dimensions. For flexible macromolecules in solution and also for irregularly shaped rigid particles, the Stokes-Einstein relation is often used to define a hydrodynamic radius (R_H).

This technique is routinely used to study such systems as flexible coil macromolecules, proteins, micelles, vesicles, viruses, and latexes. Size changes such as occur, for instance, in protein denaturation may be followed by intensity fluctuation spectroscopy studies of translational diffusion. In addition, the concentration and, in some cases, the ionic strength dependence of D are monitored to yield information on particle interactions and solution structure.

Intensity fluctuation spectroscopy experiments are also used to obtain mutual diffusion coefficients of mixtures of small molecules (for example, benzene–carbon disulfide mixtures) and are also used to measure the behavior of the mutual diffusion coefficient near the critical (consolute) point of a binary liquid mixture. Experiments of this type have proved to be very important in formulating theories of phase transitions.

Rotational diffusion coefficients. Rotational diffusion coefficients are most easily measured by depolarized quasi-elastic light scattering. The instantaneous depolarized intensity for a nonspherical scatterer depends upon the orientation of the scatterer. Rotation of the scatterer will then modulate the depolarized intensity. In a similar way, the frequency distribution of the depolarized scattered light will be broadened by the rotational motion of the molecules. Thus, for example, for dilute solutions of diffusing cylindrically symmetric scatterers, a depolarized intensity fluctuation spectroscopy experiment will give an exponential intensity time correlation function with the decay constant containing a term dependent on the scatterer rotational diffusion coefficient D_R [Eq. (2)]. A depolarized filter experiment on a similar system will give a single lorentzian

$$1/\tau_r = 2(q^2D + 6D_R) \qquad (2)$$

with $1/\tau_r$ equal to one-half that given in Eq. (2). For small molecules (for example, benzene) and relatively small macromolecules (for example, proteins with molecular weight less than 30,000) in solution, filter experiments are used to determine rotational diffusion coefficients. In these cases, the contribution of the translational diffusion to τ_r is negligible. For larger, more slowly rotating macromolecules, depolarized intensity fluctuation spectroscopy experiments are used to determine D_R.

Quasielastic light scattering is the major method of studying the rotation of small molecules in solution. Studies of the concentration dependence, viscosity dependence, and anisotropy of the molecular rotational diffusion times have been performed on a wide variety of molecules in liquids as well as liquid crystals.

Rotational diffusion coefficients of very large (≥ 100 nm) nonspherical particles may also be measured from polarized intensity fluctuation spectroscopy experiments at high values of q.

Other applications. There are many variations on quasielastic light scattering experiments. For instance, polarized filter experiments on liquids also give a doublet symmetrically placed about the laser frequency. Known as the Brillouin doublet, it is separated from the incident laser frequency by $\pm C_s q$, where C_s is the hypersonic sound velocity in the scattering medium. Measurement of the doublet spacing then yields sound velocities. This technique is being extensively utilized in the study of bulk polymer systems as well as of simple liquids.

In a variation of the intensity fluctuation spectroscopy technique, a static electric field is imposed upon the sample. If the sample contains charged particles, the molecule will acquire a drift velocity proportional to the electric field strength $v = \mu E$, where μ is known as the electrophoretic mobility. Light scattered from this system will experience a Doppler shift proportional to v. Thus, in addition to particle diffusion coefficients, quasielectric light scattering can be used to measure electrophoretic mobilites.

Quasielectric light scattering may also be used to study fluid flow and motile systems. Intensity fluctuation spectroscopy, for instance, is a widely used technique to study the motility of microorganisms (such as sperm cells); it is also used to study blood flow.

Bibliography. L. P. Bayvel and A. R. Jones, *Electromagnetic Scattering and its Applications*, 1981; B. J. Berne and R. Pecora, *Dynamic Light Scattering*, 1976; B. Chu, *Laser Light Scattering*, 1974.

X-RAY SPECTROMETRY
L. S. BIRKS

A rapid and economical technique for quantitative analysis of the elemental composition of specimens. It differs from x-ray diffraction, whose purpose is the identification of crystalline compounds. It differs from spectrometry in the visible region of the spectrum in that the x-ray photons have energies of thousands of electronvolts and come from tightly bound inner-shell electrons in the atoms, whereas visible photons come from the outer electrons and have energies of only a few electronvolts.

In x-ray spectrometry the irradiation of a sample by high-energy electrons, protons, or photons ionizes some of the atoms, which then emit characteristic x-rays whose wavelength λ depends on the atomic number Z of the element ($\lambda \propto 1/Z^2$), and whose intensity is related to the concentration of that element. Generally speaking, the characteristic x-ray lines are independent of the physical state (solid or liquid) and of the type of compound (valence) in which an element is present, because the x-ray emission comes from inner, well-shielded electrons in the atom.

Figure 1 illustrates the removal of one of the innermost, K-shell, electrons by a high-energy photon. The photon energy must be greater than the binding energy of the electron; the difference in energy appears as the kinetic energy of the ejected electron. The K-ionized atom is unstable, and one of the L- or M-shell electrons drops into the K-shell vacancy within 10^{-14} to 10^{-6} s. As this transition occurs, a characteristic x-ray photon is emitted with an energy equal to the difference in energy between the K and the L (or M) shell, or an additional electron, called an Auger electron, is ejected from the atom. Either the x-rays or the Auger electrons may be used for

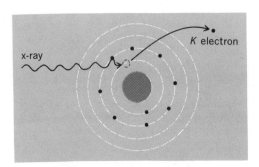

Fig. 1. Removal of a K electron from an atom by a primary x-ray photon. (*After L. S. Birks, X-Ray Spectrochemical Analysis, 2d ed., Wiley-Interscience, 1969*)

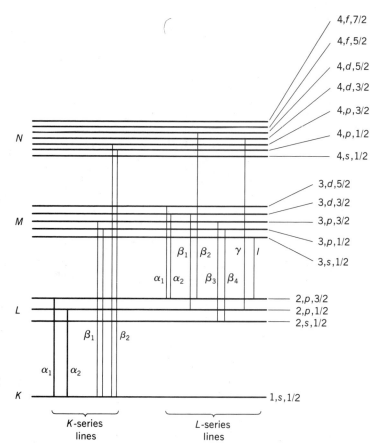

Fig. 2. Partial energy-level diagram showing the transitions leading to the K and L series lines. (After L. S. Birks, *X-Ray Spectrochemical Analysis*, 2d ed., Wiley-Interscience, 1969)

analysis, but in this article the discussion is concerned exclusively with the x-rays. SEE AUGER EFFECT.

Figure 2 shows some of the allowed transitions and the naming of the lines. There is a selection rule in atomic physics which says that only certain ones of the outer electrons are allowed to fill a vacancy in an inner shell. The rule can only be stated in terms of quantum

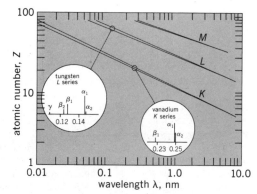

Fig. 3. Atomic number Z versus wavelength λ of the characteristic lines ($\lambda \propto 1/Z^2$). (After L. S. Birks, *X-Ray Spectrochemical Analysis*, 2d ed., Wiley-Interscience, 1969)

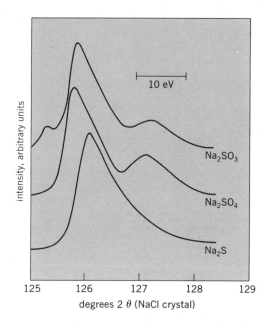

Fig. 4. Variations of the sulfur $K\beta$ line with valence state of sulfur. (After L. S. Birks and J. V. Gilfrich, X-ray fluorescence analysis of the concentration and valence state of sulfur in pollution samples, Spectrochim. Acta, 33B, no. 7:305, 1978)

mechanics: the transition must be from one shell to another, and the second (orbital) quantum number must change by ± 1, that is, $p \rightleftarrows s$, $d \rightleftarrows p$, and so on. **Figure 3** shows the λ versus Z relationship for the strongest K- and L-series lines. Figure 2 shows that some of the transitions may come from the valence shell, in which case there will be slight alteration of wavelength or line shape of characteristic lines with valence. This alteration can be measured for multivalent elements such as sulfur by using spectrometers designed for high resolution (**Fig. 4**).

Spectrum analysis. Photon generation of characteristic spectra is the most common and is called x-ray fluorescence. It is carried out with an x-ray tube as the source of primary radiation. There are two ways of analyzing the spectra: wavelength dispersion and energy dispersion.

Wavelength dispersion. This is shown in **Fig. 5**a. The characteristic emission from the sample is usually excited by a chromium or tungsten target x-ray tube which is operated at 2–3 kW. The emitted radiation is limited to a parallel beam by the blade collimator and is diffracted, one wavelength at a time, by an analyzer crystal. Bragg's law ($n\lambda = 2d \sin \theta$) relates the diffraction angle θ to the wavelength λ for crystal planes with an interatomic-spacing distance d. The term n is the order of the diffraction, 1, 2, 3, etc., as it is in classical optics. As shown in Fig. 3, the characteristic wavelength decreases as the atomic number increases. The $2d$ spacing of the analyzing crystal must be greater than the wavelength being diffracted, but if it is too much greater the spectral lines will be crowded toward small θ. It has become the practice to use a crystal of lithium fluoride for elements of atomic number Z greater than 20, a pentaerythritol crystal for Z between 13 and 20, and a potassium acid phthalate crystal for Z between 8 and 13. X-ray fluoescence analysis is not generally suitable below about $Z = 9$, but with special instruments and techniques it can be extended down to about $Z = 5$.

The detectors used for wavelength dispersion are either gas proportional counters or scintillation counters. Both of these count individual photons, but the gas counters are most suitable for wavelengths longer than about 0.2 nanometer, while the scintillation counters are most suitable for shorter wavelengths. The amplitude of the output pulse for each photon is proportional to the energy of the x-ray photon it represents. However, the statistical variation in the amplitude for each specific photon energy means that characteristic lines from neighboring elements are not resolved by either gas proportional or scintillation counters. **Figure 6** shows the resolution for the gas proportional counter alone and with a crystal spectrometer, as well as for the silicon solid-

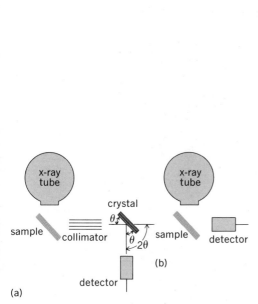

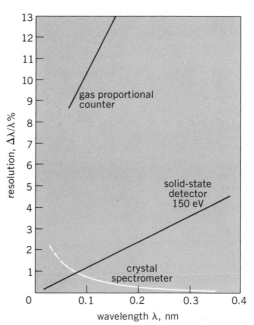

Fig. 5. Wavelength and energy dispersion methods. (a) Wavelength dispersion with a crystal spectrometer. (b) Energy dispersion with a solid-state detector.

Fig. 6. Resolution of a solid-state detector and a gas proportional counter alone or in conjunction with a crystal spectrometer.

state detector used for energy dispersion analysis, which is discussed below; resolution with a scintillation counter is almost a factor-of-three worse than a gas proportional counter. In wavelength dispersion it is the resolution of the crystal spectrometer, however, which determines the separation between neighboring wavelengths, and the crystal resolution is better than any of the detectors.

Energy dispersion. In this method all of the radiation emitted by the sample enters an energy-sensitive detector, usually a silicon solid-state detector (Fig. 5b). Such detectors are operated at liquid-nitrogen temperature to reduce electronic noise and allow an energy resolution of about 150 eV (Fig. 6). This resolution is adequate to distinguish the $K\alpha$ and $K\beta$ lines of a single element, but not adequate to separate the $K\beta$ line of one element from the $K\alpha$ line of the next higher atomic number element (for example, $CrK\beta$ from $MnK\alpha$). **Figure 7** compares the wavelength and energy-dispersion spectra from a sample containing a range of atomic number elements. Resolution is better with the crystal spectrometer for most of the spectral range of interest.

In spite of its relatively poor resolution, energy dispersion has become widely accepted because of two advantages it has over wavelength dispersion. First, all the characteristic lines are recorded simultaneously, which makes the method faster than a scanning crystal spectrometer (but not as fast as the multiple crystal-spectrometer instruments). Second, the solid angle of radiation accepted by the detector is 10–100 times greater than the solid angle accepted by the crystal; this allows the primary-source power to be reduced proportionally. With energy dispersion, it is feasible to use electron excitation at beam currents below 10^{-8} ampere in the scanning electron microscope, compared to beam currents of about 10^{-6} A required in electron probes used with crystal spectrometers. Likewise, with energy dispersion, it is feasible to use proton or alpha-particle beams from Van de Graaff or cyclotron accelerators. For proton beams, the proton energy must be 1–5 MeV, compared to electron energies of 10–50 keV for the same x-ray yield; for alpha particles, the energy should be even higher, 10–50 MeV. Compared to photon excitation, positive-ion excitation results in lower background intensity and improves the limit of detection by as much as a factor of 10. On the other hand, direct electron excitation results in much higher

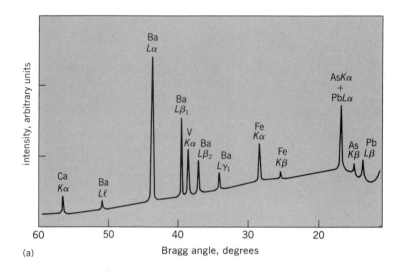

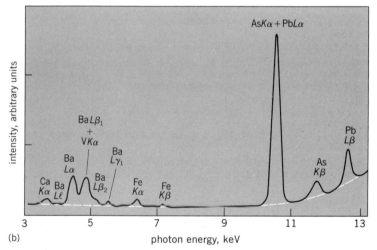

Fig. 7. X-ray spectra of the same sample as measured by (a) wavelength dispersion and (b) energy dispersion. (*After L. S. Birks, Pinpointing airborne pollutants, Environ. Sci. Tech.*, **12**:150, 1978)

background intensity because it is generated in the sample rather than merely scattered by the sample; this degrades the limit of detection by a factor of 10–100.

With energy dispersion, it is even feasible to use some of the radioactive isotope sources as the primary radiation to excite the characteristic x-ray spectra in the sample. There are isotopes such as tritium which undergo beta decay and, when used in a metal matrix, produce continuum and characteristic matrix x-rays; other isotopes such as americium-241 undergo alpha decay and produce gamma rays or x-rays. The advantage of isotope sources is their compact size and the elimination of power supplies and so forth. However, advances in low-power air-cooled x-ray tubes (10–50 W) and compact power supplies have largely eliminated the need for isotope sources.

Types of samples and data interpretation. Two general classes of samples are analyzed routinely by x-ray spectrometry: thin and bulk samples.

Thin samples. Samples smaller than several milligrams per square centimeter such as air-pollution particles collected on a filter, constitute such a thin layer that the primary radiation can penetrate easily and excite emission from each element in proportion to the mass per unit area of that element. Quantitative calibration for each element is accomplished experimentally by determining the sensitivity S in photons per second per microgram per square centimeter for that element. Accuracy (or more properly, precision) depends on the number N of photons counted; the expected standard deviation σ of a measurement is approximated by $\sigma = \sqrt{N}$. Limit of detection C_L depends not only on sensitivity but on the background intensity as well. If the measured intensity of the background is N_B and that at the line peak is N_p, the limit of detection is defined as the amount of material C_L which gives a signal above background of $3\sigma_B$, that is, $N_p - N_B = 3\sqrt{N_B}$ and $C_L = 3\sqrt{N_B}/(S \times t)$, where t is the counting interval in seconds. Thus it is important to minimize the background intensity N_B by careful instrumental design to eliminate scattering from material other than the sample, and also to minimize the mass per square centimeter of the substrate on which the sample is mounted. With photon excitation and optimized design, the limit of detection varies from about 1 nanogram/cm^2 for elements around atomic number 20 to 10–50 nanograms/cm^2 for the extremes of high or low Z. The limit is approximately the same for either wavelength or energy dispersion when photon excitation is used. As was stated in the previous portion of this subject area, the limit of detection can be improved by about a factor of 10 with proton excitation, but the analysis is then generally limited to energy dispersion.

Bulk samples. These consist of solids, powder, or liquids and present quite a different problem from thin samples. X-ray absorption limits the penetration of the primary radiation through the sample matrix and the depth from which characteristic radiation may emerge. In addition, the characteristic radiation from some of the elements in the specimen may excite the characteristic radiation of other elements by secondary fluorescence. Matrix absorption and secondary fluorescence depend on the sample composition and determine the x-ray intensity versus composition relationship for each element. Thus, instead of a single, linear calibration as was described for thin samples, the calibration is generally nonlinear, and a family of curves is needed for each element as the matrix composition changes.

If I_{ai} is the intensity from a pure sample of element i, and I_{bi} is the intensity from an unknown concentration of element i in a matrix of other elements, then the most useful parameter is the relative x-ray intensity $R_i = I_{bi}/I_{ai}$. To a first approximation, R_i can be expressed in terms of the concentration C_i and the absorption term μ which incorporates both the absorption of the incident primary radiation and the emerging characteristic radiation, as shown in Eq. (1), where

$$R_i = AC_i/\mu \tag{1}$$

A is a constant which depends on a number of instrumental parameters.

In x-ray spectrometry the analyst can measure R_i but cannot determine C_i directly, because μ depends on C_i and the concentration of each other element C_j as well. However, it is possible to write an equation containing R_i and C_i in terms of individual absorption and secondary-fluorescence effects of each element on the intensity from each other element. The expression is Eq. (2),

$$C_i/R_i = 1 + \sum \alpha_{ij} C_j \tag{2}$$

where the Σ symbol means the sum of a number of terms, one for each other element in the specimen. The coefficient α_{ij} means the effect on element i by the presence of element j; it is often referred to as an influence coefficient and may be determined experimentally by measuring mixtures of elements i and j. The terms may also be calculated from some of the fundamental properties of atoms and radiation.

There are many variations of the mathematical expressions for determining concentration from x-ray intensity; most of them require computers to evaluate. Whatever the method of data interpretation, a few generalizations may be made about the analysis of bulk specimens. The accuracy (precision) is about 1–2% of the amount of an element present for major constituents, but degrades to 5–10% of the amount present for concentrations to 10–100 ppm. The limit of detection is generally about 1 ppm for middle-range atomic number elements, but varies from less than 0.1 ppm for metals in biological tissue to 10 ppm or more for low-Z elements such as carbon in a middle-Z matrix such as steel.

Improvements and limitations. X-ray spectrometry generally does not require any separation of elements before measuring, because the x-ray lines are easily resolved. However, preconcentration methods are sometimes useful as a means for improving the limit of detection. An example is the precipitation or ion-exchange collection of soluble elements in water. Likewise, dilution is sometimes useful to reduce matrix variability or inhomogeneity. An example is the solution of mineral samples in borax.

One limitation of x-ray spectrometry is the progressive difficulty of measurement below atomic number 11. There are several reasons for this, including the strong absorption of such long-wavelength radiation by the x-ray tube and detector windows and also the reduced intensity due to a lower number of x-ray photons emitted per atom ionized (the fluorescent yield factor). Although the elements from boron (5) to fluorine (9) may be measured with specially designed equipment, such measurement cannot be considered routine. In practice, photoelectron spectroscopy and Auger electron spectroscopy are favored for the lower-atomic-number elements, but electron methods may also require special instrumentation and techniques. SEE ELECTRON SPECTROSCOPY.

Bibliography. E. P. Bertin, *Principles and Practice of X-ray Spectrometric Analysis*, 2d ed., 1975; L. S. Birks, *X-ray Spectrochemical Analysis*, 2d ed., 1969; K. F. Heinrich, *Electron Beam X-ray Microanalysis*, 1980; R. Jenkins, R. W. Gould, and D. Gedke, *Quantitative X-ray Spectrometry*, 1981.

X-RAY FLUORESCENCE ANALYSIS
WILLIAM PARRISH AND MICHAEL MANTLER

A nondestructive physical method used for chemical elemental analysis of solids and liquids. The specimen is irradiated by photons or particles of sufficient energy to cause the elements in it to emit (fluoresce) their characteristic x-ray line spectra. The detection system allows the determination of the energies of the emitted lines and their intensities. Elements in the specimen are identified by their spectral line energies or wavelengths for qualitative analysis, and the intensities are related to their concentrations for quantitative analysis. Computers are widely used in this field both for automated data collection and for reducing the x-ray data to weight- and atomic-percent chemical composition. SEE FLUORESCENCE.

The materials to be analyzed may be solids, powders, liquids, or thin foils and films. The crystalline state and the state of chemical bonding normally have no effect on the analysis. All elements above atomic number 12 can be routinely analyzed in a concentration range from 0.1 to 100 wt %. Special techniques are required for the analysis of elements with lower atomic numbers or of lower concentrations. The counting times required for analysis range from a few seconds to several minutes per element, depending upon specimen characteristics and required accuracy, but may be much longer for trace analysis. The results are in good agreement with wet chemical and other methods of analysis.

Basis of method. The theory of the method has its origin in the classic work by H. G. J. Moseley, who in 1913 measured x-ray wavelengths of a series of elements. He found that each element had a simple x-ray spectrum and characteristic wavelengths and there was a linear relationship between $1/\sqrt{\lambda}$ and Z, where λ is the x-ray wavelength and Z is the atomic number of the element emitting the x-ray. **Figure 1** is a plot of Moseley's law for the K and L x-ray lines. Aside from the discovery of the element hafnium in zirconium ores by G. von Hevesy, only a few practical uses of the relationship were reported until about 1950, when the introduction of modern x-ray equipment made it feasible to use x-rays for routine spectrochemical analysis of a large variety of materials.

An x-ray source is used to irradiate the specimen, and the emitted x-ray fluorescence radiation is analyzed with a crystal spectrometer and scintillation or proportional counter detectors. The fluorescence radiation is diffracted by a crystal at different angles to separate the wavelengths and identify the elements, and the concentrations are determined from the relative intensities. This procedure is widely used and is called the wavelength dispersive method.

Around 1965 lithium-drifted silicon and germanium [Si(Li) and Ge(Li)] solid-state detectors became available for x-ray analysis. These detectors have better resolution, and the average pulse amplitudes are directly proportional to the energies of the x-ray quanta which can be sorted electronically with a multichannel pulse-height analyzer. This eliminates the need for the crystal and is called the energy dispersive method.

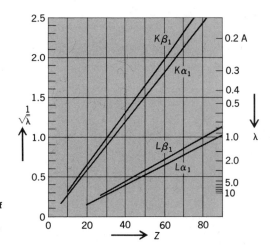

Fig. 1. Plot of Moseley's law, showing dependence of characteristic x-ray-line wavelengths λ on atomic number Z. 1 Å = 0.1 nm. (*Philips Tech. Rev., vol. 17, no. 10, 1956*)

X-ray spectra. The origin of x-ray spectra may be understood from the simple Bohr model of the atom in which the electrons are arranged in orbits within the K, L, M, . . . shells. If a particle or photon with sufficient energy is absorbed by the atom, an electron may be ejected from one of the inner shells and is promptly replaced by an electron from one of the outer shells. This results in the emission of a characteristic x-ray spectral line whose energy is equal to the difference of the binding energies of the two orbits involved in the electron transition. The new vacancy is filled by an additional transition from the outer shells, and this is repeated until the outermost vacancy is filled by a free electron. The sum of energies of all photons emitted during the vacancy-refilling cascade is the ionization energy. The energy of the emitted line from the first transition in the cascade has a slightly lower energy than the ionization energy. For example, the ionization energy for the copper K shell is 8.98 keV, and the observed lines have energies of 8.90 keV (Cu$K\beta$) and 8.04 keV (Cu$K\alpha$); the corresponding wavelengths are 0.138, 0.139, and 0.154 nanometer.

Optical spectra result from electron transitions in the outer ("valence") shells, producing complex spectra with a large number of lines. By contrast, the x-ray lines arise only from a limited number of transitions between the high-energy levels of the inner shells so that the x-ray spectrum of an element consists of relatively few lines. Lines are named after the shell where the corresponding electron transition ends (*K, L, M,* . . . lines). The most probable transition yielding the highest line intensity in this series is named alpha, followed by beta, gamma, and others, and the indices 1, 2, 3, . . . define a specific transition within the subseries. Depending on the number of energy sublevels in each shell, there are usually only a few important lines in the *K* spectrum ($K\beta$, $K\alpha_1$, $K\alpha_2$,) and a dozen or more lines in the *L* spectrum. The *M* lines are rarely used in x-ray analysis.

Auger effect. Occasionally, instead of the emission of the characteristic photon in the course of an electron transition, inner atomic absorption occurs ("internal conversion" or the Auger effect) when the photon ionizes the atom in an additional shell. The ejected Auger electron has a well-defined energy, namely, the energy of the internally absorbed photon minus its ionization energy, and can be used for chemical analysis. The probability that no Auger effect occurs, that is, that the photon is actually emitted from the atom and can be used for analysis, is called fluorescence yield. For low-atomic-number elements, the Auger effect dominates and the fluorescence is low. *See* AUGER EFFECT; ELECTRON SPECTROSCOPY.

X-ray absorption. The type of absorption of the photon or particle leading to the original ionization of the atom is called photoabsorption, to distinguish it from absorption by coherent scattering or Compton scattering. The probability of photoabsorption decreases gradually with increasing photon (or particle) energy, but abruptly increases by an order of magnitude when the photon energy exceeds the ionization energy of a shell. This energy is also called the absorption-edge energy and is shown for the *K* and *L* edges of molybdenum and silver in **Fig. 2**. Thus the x-

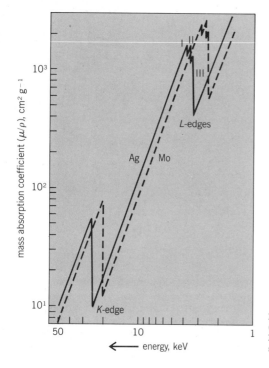

Fig. 2. Mass absorption coefficients of molybdenum (Mo) and silver (Ag) in the 1- to 50-keV region. Roman numerals indicate edges associated with subshells of the L shell.

rays with energies just higher than the absorption-edge energy are most efficient in generating x-ray fluorescence. The efficiency decreases as the photon energy E is further increased from the edge as $1/E^3$ or λ^3. Photons with smaller energies than the absorption edge have no effect in exciting fluorescence.

The absorption of x-rays is usually given as a mass absorption coefficient μ/ρ (usually expressed in $cm^2\ g^{-1}$) and is independent of the physical state of the material. If more than one element is present, the weighted average of the coefficients of the individual elements is used. Tables of mass absorption coefficients have been compiled. The decrease of intensity of x-rays as they traverse the material is given by the linear absorption coefficient μ (usually expressed in cm^{-1}), obtained by multiplying the mass absorption coefficient by the density ρ of the material. The intensity decreases to $e^{-\mu x}$ of its original value when the x-rays pass through a layer x centimeters thick.

Radiation sources. There are two general methods for producing x-ray spectra for fluorescence analysis. The most common method is to expose the specimen to the entire spectrum emitted from a standard x-ray tube. The other, used in electron microscopes and the electron microprobe, uses an electron beam directly on the specimen, and each element generates its own x-ray spectrum, under electron bombardment, as in an x-ray tube. The first method is sometimes modified by using a secondary target material (or monochromator) outside the x-ray tube to excite fluorescence. This has the advantage of selecting the most efficient energy close to the absorption edge of the element to be analyzed and reducing or not exciting other interfering elements, but the intensity is reduced by two or three orders of magnitude.

The wavelengths of the lines are determined by the target element. In addition, the deceleration of the electrons generates a broad continuous spectrum whose intensity and short-wavelength limit are determined primarily by the x-ray-tube voltage or electron beam energy.

X-ray tubes. The primary x-ray-tube targets are usually tungsten, copper, rhodium, molybdenum, silver, and chromium. It is usually necessary to avoid the use of a tube whose target is identical with an element in the specimen because the line spectrum from the target is scattered through the system, adding to the element signal. It is also desirable to select a target whose characteristic line energies lie closely above the absorption edges of the elements to be analyzed; for example, the WL lines and CuK lines are more efficient in exciting fluorescence in the transition elements chromium to copper than are the MoK lines; RhL lines are most useful to excite K lines of elements below sulfur in the periodic table. Tubes for fluorescence analysis usually have a single thin beryllium window placed at the side of the tube.

Equipment is normally operated at x-ray-tube voltages up to 50–60 kV in dc operation at 3 kW or more with water cooling. These voltages generate the K spectra of all the elements up to the rare earths and the L spectra of the higher-atomic-number elements. Since the detector is moved from point to point, it is essential to have a constant primary intensity and to stabilize the voltage and tube current.

Radioactive isotopes. Radioactive isotopes which produce x-rays, such as iron-55 which emits MnK x-rays and americium-241 (NpL x-rays), are used in place of an x-ray tube to excite fluorescence in some applications. These sources are much weaker than x-ray tubes and must be placed close to the specimen. They are often used in field applications where portability and size may be problems. Alpha particles have been occasionally used.

Synchrotron radiation. Synchrotron radiation, although not yet widely used for fluorescence analysis, has many potential advantages. The continuous radiation is several orders of magnitude more intense than x-ray tubes and can be used with a crystal spectrometer. In addition, a tunable crystal monochromator can be placed in the incident beam to select the optimum wavelength for fluorescing each element in the specimen.

Crystal spectrometer. A single-crystal plate is used to separate the various wavelengths emitted by the specimen. Diffraction from the crystal occurs according to Bragg's law, Eq. (1),

$$n\lambda = 2d \sin \theta \qquad (1)$$

where n is a small integer giving the order of reflection, λ the wavelength, d the spacing of the particular set of lattice planes of the crystal that are properly oriented to reflect, and θ the angle between those lattice planes and the incident ray.

Reflection for a particular λ and d occurs only at an angle 2θ with respect to the incident ray, and it is therefore necessary to maintain the correct angular relationship of the crystal planes at one-half the detector angle. This is done by the goniometer which is geared to rotate the crystal at one-half the angular speed of the counter tube, and therefore both are always in the correct position to receive the various wavelengths emitted by the specimen (**Fig. 3**). For a given d, there is only one angle (for each order of reflection) at which each wavelength is reflected, the angle increasing with increasing wavelength.

The angular separation of the lines, or the dispersion, given by Eq. (2), increases with

$$\frac{d\theta}{d\lambda} = \frac{n}{2d \cos \theta} \qquad (2)$$

decreasing d. It is thus easy to increase the dispersion simply by selecting a crystal with a smaller d. Reducing d also limits the maximum wavelength that can be measured since $\lambda = 2d$ at $2\theta = 180°$; the maximum 2θ angle that can be reached in practice with the goniometer is about 150°.

Soller slits. The crystals are usually mosaic, and the reflection is spread over a small angular range. To increase the resolution, that is, decrease the line breadth, it is necessary to limit the angular range over which a wavelength is recorded. Parallel or Soller slits are used for this purpose (Fig. 3). These slits consist of thin (0.002-in. or 0.05-mm) equally spaced flat foils of such materials as nickel and iron, and the angular aperture is determined by the length and spacing. A typical set for fine collimation would have 0.005-in. (0.13-mm) spacings and 4-in. (100-mm) length with angular aperture 0.15° and cross section 0.28 in. (180 mm) square. The absorption of the foils is sufficiently high to prevent rays that are inclined by more than the angular aperture to extend beyond the specimen area and enter the counter tube. Two sets of parallel slits may be used, one set between the specimen and crystal and the other between crystal and detector. This greatly increases the resolution and peak-to-background ratio, and causes a relatively small loss of peak intensity.

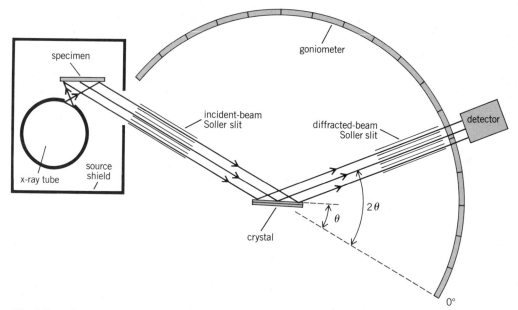

Fig. 3. X-ray fluorescence spectrograph (not to scale). Diffracted-beam Soller slit is optional.

Diffracting crystals. Crystals commonly used in spectrometers are lithium fluoride (LiF) with reflecting plane (200) or (220), silicon (111) and (220), pentaerythritol (001), acid phthalates of potassium and thallium (001), ethylene diamine d-tartrate (020), and thin layers of lead-stearate. It is essential that the crystal be of good quality to obtain sharp, symmetrical reflections. Unless the crystal is homogeneous, the reflection may be distorted and portions of the reflections may occur at slightly different angles. Such effects would decrease the peak intensities of the wavelengths by varying amounts, causing errors in the analysis. Reflection gratings made of multiple very thin film layers have been recently developed for long-wavelength x-rays.

Rapid analysis systems. In certain industrial applications such as the manufacture of cement, steels, glass, and geological exploration, large numbers of specimens containing up to a dozen or more elements must be rapidly analyzed. In some cases the analysis must be done in a few minutes to correct the composition of a furnace which is standing by. Generally the same qualitative compositions have to be routinely analyzed, and a number of prepositioned crystals and detectors, each set for a different element, are used. Instead of sequentially scanning over the wavelength regions, a number (up to 30) of fixed crystals and detectors are positioned around the specimen and allow simultaneous measurements of several elements at peak and background positions. Automated trays load the specimens into the spectrometer.

Detectors. The detectors generally used are scintillation counters with thin beryllium windows and thallium-activated sodium iodide [NaI(Tl)] crystals for higher energies (above 4 keV), and gas flow counters with very low absorbing windows and argon/methane gas for the low-energy region (below 6 keV). A single-channel pulse-amplitude analyzer limits photon counting to a selected energy interval to improve the peak-to-background ratio and to eliminate higher-order reflections. However, no sharp energy separation is possible due to the rather limited energy resolution of these detectors. *See* GAMMA-RAY DETECTORS.

Energy dispersive systems. Solid-state detectors with good energy resolution are used in conjunction with a multichannel pulse-amplitude analyzer. No crystals are required, and the detector and specimen are stationary during the measurement. The method is used with either electron beam excitation in electron microscopes or with x-ray-tube sources. The photons of various energies are registered, and their energies are determined as soon as they enter the detector.

As this occurs statistically for the various fluorescence line energies, the acquisition of the spectral data appears to be simultaneous for all lines.

Solid-state detectors. Lithium-drifted silicon [Si(Li)] detectors are generally used for the lower energies of fluorescence analysis, while lithium-drifted germanium [Ge(Li)] detectors are more often used for nuclear high-energy gamma-ray detection. The energy resolution of good Si(Li) detectors is around 145 eV (full width at one-half maximum) for Mn$K\alpha$ radiation. The lithium-drifted detectors require permanent cooling with liquid nitrogen.

Analyzer. The output signals from the detector are fed into the analyzer, where the photon counts are stored in memory locations (1024 to 8192 channels are generally used) that are related to the energies of these photons. This also allows visual observation on a cathode tube screen of the accumulated spectrum and of the simultaneous counting process. Analyzers are usually provided with cursor markers to easily identify the peaks in the spectrum. Computer memories can be used for storage of the spectral counts, thus providing efficient access to computer routines for further data evaluation.

Use. Energy dispersive x-ray spectrometers are useful to accumulate spectra in short time intervals (for example, 1 min) that often allow a preliminary interpretation of the qualitative and quantitative composition of the specimen. The instruments are comparatively small because they are designed to accept a large aperture of radiation. They require only low-power x-ray tubes that sometimes can be air-cooled.

Limitations. An important limitation of energy dispersive systems is the energy resolution which is about an order of magnitude poorer in the lower energy region that that of crystal spectrometers. For example, the $K\alpha$ lines of the transition elements overlap with the $K\beta$ lines of the element preceding it in atomic number, causing severe analytical difficulties in an important region of the spectrum. The peak-to-background ratio is significantly lower than in crystal spectrometers because of the lower resolution. Another limitation is that the maximum number of photons that can be processed by the electronic circuits is limited to about 15,000 to 50,000 counts per second. This is the total photon count from the entire detected spectral region. Trace elements with low count rates in a matrix of high-count elements are therefore difficult to detect with sufficient statistical accuracy. Various attempts have been made to overcome this drawback by selectively exciting the elements of interest by using selective filters or secondary targets which also greatly reduces the amount of x-ray-tube radiation that is scattered into the detector.

Microanalysis. The electron microprobe is widely used for elemental analysis of small areas. An electron beam of 1 micrometer (or smaller) is used, and the x-ray spectrum is analyzed with a focusing (curved) crystal spectrometer or with an energy dispersive solid-state detector. Usually two or three spectrometers are used to cover different spectral regions. Light elements down to carbon can be detected. An important use of the method is in point-to-point analysis with a few cubic micrometers of spatial resolution. X-Y plots of any element can be made by moving the specimen to determine the elemental distribution. Microanalysis on a much smaller scale can be done with electron microscopes.

Figure 4 illustrates the spectra obtained with three of the most frequently used methods of analysis. The specimen, a high-temperature alloy of the type used in aerospace and other industries, was prepared by the National Bureau of Standards (1208-2) with stated composition in weight percent: molybdenum (Mo) 3.13, niobium (Nb) 4.98, nickel (Ni) 51.5, cobalt (Co) 0.76, iron (Fe) 19.8, chromium (Cr) 17.4, titanium (Ti) 0.85, and aluminum (Al) 0.085, total 99.27%. The spectral lines are identified in the figure legend.

Figure 4a shows the high-resolution spectrum obtained in about an hour with a lithium fluoride [LiF] (200) crystal spectrometer using 50-kV, 12-mA x-ray-tube excitation and scintillation counter. This spectrum also contains the second-order (II) and third-order (III) crystal reflections of molybdenum and niobium whose $K\beta_1$ and $K\beta_3$ components are resolved. The lower resolution of the energy dispersive method is shown in Fig. 4b, recorded in about 10 min using 50-kV, 2-microampere x-ray-tube excitation, Si(Li) detector, and 40 eV per channel (about 400 channels are shown). The spectral range includes the unresolved molybdenum and niobium L lines and titanium. Figure 4c is an energy dispersive spectrum excited by a 25-keV electron beam. The molybdenum and niobium spectra are weakly excited at this low voltage and are not visible on the scale used in the plot. The differences in the relative intensities of the lines in the spectra arise from

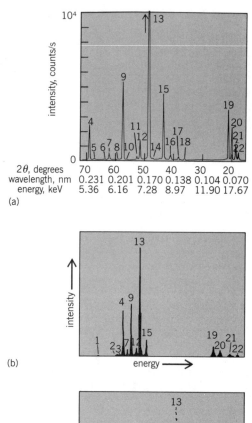

(a)

(b)

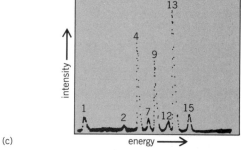

(c)

Fig. 4. Fluorescence spectra of high-temperature alloy obtained with (a) crystal spectrometer, (b) energy dispersive method with x-ray-tube excitation, and (c) energy dispersive method with electron-beam excitation. Spectral lines: 1, Mo + NbLα + Lβ. 2, TiKα. 3, TiKβ. 4, CrKα. 5, NbKα$_{1,2}$III. 6, MoKα$_{1,2}$III. 7, CrKβ. 8, NbKβ$_{1,3}$III. 9, FeKα. 10, MoKβIII. 11, CoKα. 12, FeKβ. 13, NiKα. 14, CoKβ. 15, NiKβ. 16, MoKα$_{1,2}$II. 17, NbKβ$_{1,3}$II. 18, MoKβ$_{1,3}$. 19, NbKα. 20, MoKα. 21, NbKβ$_{1,3}$. 22, MoKβ$_{1,3}$.

differences in the conditions of excitation and detection, and illustrate the necessity of using the proper correction factors for each method of analysis to derive the correct weight percent composition.

Specimen preparation. The specimens may be in the form of powders, briquettes, solids, or liquids. The surface exposed to the primary x-ray beam must be flat, smooth, and representative of the sample as a whole because usually only a thin surface layer contributes to the fluorescent beam in a highly absorbing specimen. The layer may be only a micrometer or less for electron beam excitation and 10 μm or more for x-rays. The degree of surface roughness, which is difficult to measure quantitatively, causes losses in intensity and results in errors in the analysis. Consequently, solid samples are generally polished, and then if necessary are lightly etched or specially cleaned to remove contaminants.

Powders. Powders are processed in one of two ways. The first is to press the ground material into briquettes. The pressure should be several tons per square centimeter (1 ton/cm^2

equals approximately 1.5×10^4 lb/in.^{2} or 100 megapascals), and in most cases organic binders have to be used to improve the mechanical stability. The second way is to use fusion techniques, where the powders (mostly mineralogical or metal oxides) are dissolved at high temperatures in borax or similar chemicals, and glassy pellets are obtained after cooling. The advantage of the second method is a high homogeneity of the specimen and a reduction of interelement effects (discussed below), but the intensities are reduced.

Liquids. Liquids can be analyzed by using small containers with a thin window cover. Examples are sulfur determination in oils during the refining process, lubrication oil additives, the composition of slurries, and the determination of lead, zinc, and other elements in ore processing. Low concentrations of elements in solution can be concentrated with specific ion-exchange resins and collected on filter papers for analysis. Gases containing solid particles can be filtered and the composition of the particles determined as for atmospheric aerosol filters for environmental studies. In certain industrial applications, liquids are continuously analyzed while flowing through a pipe system with a thin window in the x-ray apparatus.

Quantitative analysis. The observed fluorescent intensities must be corrected by various factors to determine the concentrations. These include the spectral distribution of the exciting radiation, absorption, fluorescence yield, and others. Two general methods have been developed to make these corrections, the fundamental parameter method and the empirical parameter method.

Fundamental parameter method. In the fundamental parameter method, a physical model of the excitation is developed and described mathematically. The method derives its name from the fact that the physical constants, like absorption coefficients and atomic transition probabilities, are also called fundamental parameters. Primary and secondary excitation are taken into account; the first is the amount of fluorescent radiation directly excited by the x-ray tube. Secondary excitation is caused by other elements in the same specimen, whose fluorescent radiation has sufficient energy to excite the characteristic radiation of the analyzed element. In practical applications, the count rate must be calibrated for each element by comparing it to the count rate from a standard of accurately predetermined composition. A standard may contain several elements or can be a pure element.

The fundamental parameter method is capable of accuracies around $\pm 1\%$ (absolute weight percentage) for higher concentrations, and between 2 and 10% (relative) for low concentrations. The method has the advantage of allowing the use of pure-element standards. Higher accuracies can be obtained with standard specimens of similar composition to the unknown.

Empirical parameter method. The empirical parameter method is based upon simple mathematical approximation functions, whose coefficients (empirical parameters) are determined from the count rates and concentrations of standards. A widely used set of approximation functions is given by Eq. (3), where c_i is the concentration of the analyzed element i in the unknown

$$\frac{c_i}{r_i} = \frac{1 + \sum_{j \neq i}^{n} \alpha_{ij} c_j}{R_i} \qquad i = 1, \ldots, n \qquad (3)$$

specimen, r_i is the corresponding count rate, R_i is the count rate from a pure-element specimen, i, c_j are the concentrations of the other elements in the unknown specimen, n is the number of elements, and α_{ij} are the empirical parameters (also called alpha coefficients).

A minimum of $n - 1$ standard specimens, each of which contains the full set of n elements (or a correspondingly higher number, if they contain fewer elements), is required to calculate the empirical parameters, α_{ij}, before actual analysis of an unknown is possible. In practical applications, however, at least twice as many standards should be used to obtain good accuracy, thus requiring considerable effort in standard preparations. The empirical parameter method is therefore mainly used in routine applications, where large numbers of similar specimens must be analyzed. The accuracy of the method depends upon the concentration range covered by the standards; around $\pm 0.1\%$ can be obtained if a set of well-analyzed standards with similar compositions to the unknowns are used.

Trace analysis. In trace analysis, where the matrix does not noticeably change with the concentration of the trace element, the relationship between concentration and count rates is practically linear. The minimum detection limit is defined by that concentration, for which the peak is just statistically significant above background level B, usually $3B^{1/2}$. The background arising from scattered continuous radiation from the x-ray tube is a limiting factor in determining the peak-to-background ratio. Since intensity measurements can theoretically be made arbitrarily accurate by using long counting times, the minimum detection limits could be indefinitely low. However, in practice, the limiting factors are the background level and long-term instrument drift. Depending upon excitation conditions, matrix, and counting times, traces in the parts-per-million region may be detected.

Thin-film analysis. In the analysis of very thin films (a few tens of nanometers), the count rates are also a linear function of element concentration and of film thickness. Absorption and interelement effects must be taken into account in the analysis of thicker films and foils. This can be done with special fundamental parameter methods, but requires medium- or large-scale computers for efficient data evaluation.

Limitations on accuracy. In both the fundamental parameter and empirical parameter methods, limitations of the accuracy are mainly due to uncertainties in the composition of the standards and variations in the specimen preparation; intensity fluctuations due to counting statistics, and instrument instabilities may also contribute.

Supplemental methods. As in all analytical methods, it is sometimes necessary to supplement the chemical data from fluorescence analysis with data by other methods to properly characterize the material. Minor trace elements and the first 12 elements in the periodic table (hydrogen through magnesium) cannot be routinely measured in small concentrations, and they may be crucial in the characterization. Examples are carbon in steels, and oxygen in rocks and oxide samples, which may require optical emission, atomic absorption, Auger and electron spectroscopy, or other analytical methods. SEE TRACE ANALYSIS.

An important supplementary method is x-ray polycrystalline diffraction in which the crystalline chemical phases are identified by comparing the pattern of the unknown with standard patterns. Computer methods are widely used to search the 40,000 phases currently contained in the Powder Diffraction File published by the International Center for Diffraction Data, Swarthmore, Pennsylvania. Mixtures of phases can be quantitatively determined, and there are no limitations on the chemistry of the substances. By combining the chemical data from fluorescence with the phase data from diffraction, the relation between the constituents of the sample and its properties can be established.

Applications. X-ray fluorescence analysis has been used in thousands of applications since its introduction about 1950.

The method is widely used for compositional control in large-scale industrial processing of metals and alloys, cements, the petroleum industry, and inorganic chemicals. Among the many other major applications are geological exploration and mineralogical analysis, soils and plants, glasses, corrosion products, the analysis of raw materials, and the measurement of plating coating thickness. It is an important method in materials characterization for research and technology, providing chemical information without destroying the sample. It is the only feasible method for many complex analyses which would require extremely long times by conventional wet chemical methods on materials such as the refractory metals, high-speed cutting steels, and complex alloys.

Besides the large-scale industrial applications, the method has been used in a variety of analyses in the medical field, for environment protection and pollution control, and many research applications. Examples are trace analysis of heavy metals in blood; analysis of airborne particles, historic coins, potteries, lead and barium in Roman skeletons, and various elements in archeological specimens; analysis of pigments to establish authenticity of a painting; quality control of noble metals in alumina-based exhaust catalysts for cars; and analysis of ash and sulfur in coals, slags from furnace products, and surface deposits on bulk metals. The method is also widely used in forensic problems where it is often combined with x-ray powder diffraction.

Bibliography. *Advances in X-ray Analysis*, annually; E. P. Bertin, *Introduction to X-ray Spectrometric Analysis*, 1978; K. F. J. Heinrich, *Electron Beam X-ray Microanalysis*, 1981; K. F. J. Heinrich et al. (eds.), *Energy Dispersive X-ray Spectrometry*, National Bureau of Standards Spec. Publ. 604, 1981; R. Jenkins, *An Introduction to X-ray Spectrometry*, 1974; R. Jenkins, R. W. Gould, and D. Gedke, *Quantitative X-ray Spectrometry*, 1981; R. Tertian and F. Claisse, *Principles of Quantitative X-ray Fluorescence Analysis*, 1982.

EXTENDED X-RAY ABSORPTION FINE STRUCTURE (EXAFS)

C. Denise Caldwell

Oscillations in the total x-ray absorption cross section, observable from several tens to several hundreds of electronvolts above an inner-shell ionization threshold, produced by scattering of the ionized electron off neighboring atoms. This backscattered wave interferes with the direct wave, leading to a maximum if the interference is constructive and a minimum if it is destructive. A single equation, the EXAFS equation, relates the amplitudes and phases of these maxima and minima to the number, type, and positions of the scatterers. The principal contribution to the interference comes from those atoms nearest the one being ionized; thus the oscillations are a reflection of the short-range order in the system.

With the development of electron synchrotrons and storage rings as high flux sources of tunable x-radiation, EXAFS has become a powerful analytic tool. It is element-specific and can be observed in any system which contains more than one atom (from diatomic molecules to highly ordered crystals) and in any physical state (solid, liquid, or vapor). Data are accumulated by measuring the absorption of the x-rays as they pass through the medium or by the detection of a secondary decay process. The resulting spectrum is analyzed numerically with the EXAFS equation as basis, to obtain a set of fit parameters, the most important being the number and average distances of the scatterers from the newly formed ion. This information is the principal result of the experiment and may represent, for example, the bonding of a metal atom in a biological molecule, the coordination of a metal catalyst on a substrate, or the nature of the adhesion of a molecule to a surface. Because of the versatility of the technique, its application has expanded rapidly, and it will become increasingly valuable whenever the nature of the coordination of an element in a particular bond situation is to be investigated.

Bibliography. K. O. Hodgson et al. (eds.), *EXAFS and Near Edge Structure III*, 1984; B. K. Teo and D. C. Joy (eds.), *EXAFS Spectroscopy: Techniques and Applications*, 1981.

ELECTRON SPECTROSCOPY

Kai Siegbahn and Hans Siegbahn

A form of spectroscopy which deals with the emission and recording of the electrons which constitute matter—solids, liquids, or gases. The usual form of spectroscopy concerns the emission or absorption of photons [x-rays, ultraviolet (uv) rays, visible or microwave wavelengths, and so on]. Electron spectra can be excited by x-rays, which is the basis for electron spectroscopy for chemical analysis (ESCA), or by uv photons, or by ions (electrons; see **Fig. 1**). By means of x-ray or uv photons with energy $E_{h\nu}$, photoelectron spectra (PES) are produced when electrons with binding energies E_b are emitted with energy $E_{kinetic}$ from bound molecular states, according to the equation below. E_r is a usually negligible small recoil energy, and φ a small work function correction which

$$E_{kinetic} = E_{h\nu} - E_b - E_r - \varphi$$

can be attributed to contact potentials. For insulating solid materials, precautions are taken for stabilizing the surface potential. For solids, the binding energies are preferably referred to the Fermi level, and for gases, to the vacuum level.

Modes of excitation. By means of ESCA, complete sets of photoelectron lines can be excited from the internal (core) levels as well as from the external (valence) region (**Fig. 2**). Also, complete sequences of the Auger electron lines are automatically obtained in this mode. A convenient source of excitation is the x-radiation from Al at 1487 eV. With the use of spherically bent quartz crystals this radiation can be further monochromatized to an inherent width of 0.2 eV. The commonly used light source for uv excitation of photoelectron spectra within the valence electron region is an He lamp that produces highly monochromatic radiation at 21.2 eV [internal width less than 10 MeV] and (with less intensity) at 40.8 eV. Intermediate in energy for excitation are ultrasoft x-rays (Y Mξ at 132 eV). Another source of excitation in the intermediate region is provided

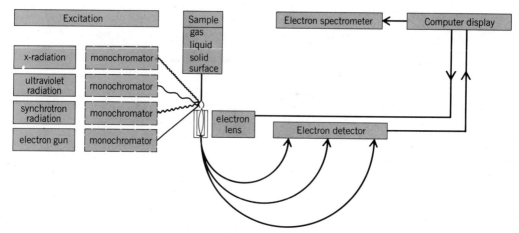

Fig. 1. Excitation of electron spectra recorded with high-resolution instruments. (*After H. Siegbahn and L. Karlsson, Photo-electron spectroscopy, Handbuch der Physik, vol. 31, 1982*)

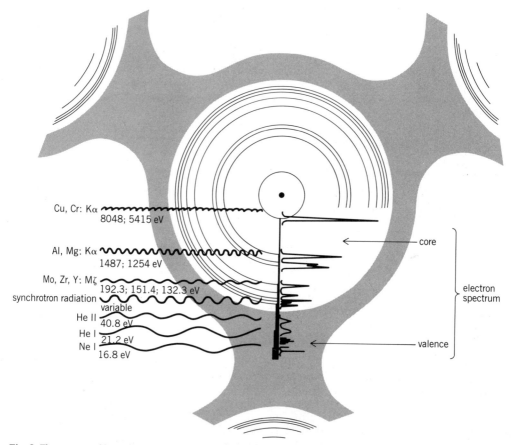

Fig. 2. The system of levels of an atom in a molecule can be divided into a valence electron region and an atomic-core region. Excitation of electron lines from the various regions can be made at different photon energies.

by synchrotron radiation, which can be continually varied by means of a suitable monochromater. Excitation by means of an electron beam is an alternative mode to obtain Auger electron lines that has the advantage of ease of production. Ordinary photoelectron lines are not produced by that radiation. In order to compensate for the low-signal-to-background ratio when Auger electron spectra are excited by an electron beam impinging on a solid surface, the spectral distribution is frequently differentiated. *See Auger effect*.

Applications. In ESCA, only electrons which are expelled from a surface layer of less than 50 nanometers of a solid material contribute to the electron line with the kinetic energy given above. Electrons from the interior of the material are scattered out from the line, and form a low background which does not interfere with the line character of the ESCA spectrum. The electron lines are extremely sharp and well suited for precision measurements. With a high-resolving ESCA spectrometer which has a magnetic or electrostatic focusing dispersive system, the electron lines have widths which are set by the limit caused by the uncertainty principle (the "inherent" widths of atomic levels). With a suitable choice of radiation, electron spectroscopy reproduces directly the electronic level structure from the innermost shells (core electrons) to the atomic surface (valence or conduction band; see **Fig. 3**). Furthermore, all elements from hydrogen to the heaviest ones can be studied even if the element occurs together with several other elements and even if the element represents only a small part of the chemical compound.

When applied to solid materials, ESCA is a typical surface spectroscopy with applications to problems such as chemical surface reactions, for example, corrosion or heterogeneous catalysis. In such cases, ultrahigh vacuum conditions are usually required, with a vacuum less than 10^{-9} torr (10^{-7} pascal) ESCA also reproduces bulk matter properties such as valence electron band

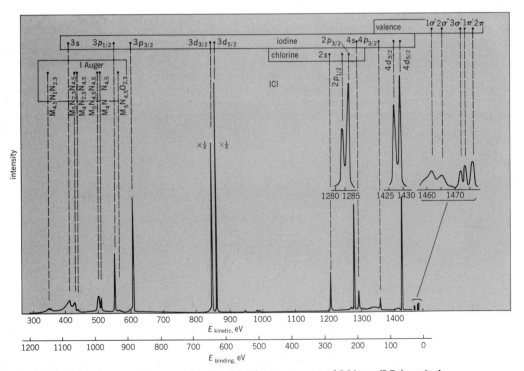

Fig. 3. Electron spectrum of the vapor of iodine chloride at a pressure of 0.04 torr (5 Pa), excited by monochromatic Al $K\alpha$ radiation with the ESCA instrument (scan spectrum). The M- and N-shell levels are excited in iodine, and the L levels in chlorine. At low binding energies (high kinetic energies), the valence spectrum of the molecule is recorded. The *MNN* Auger electron lines of iodine can also be seen.

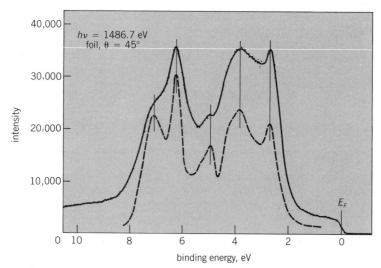

Fig. 4. Valence-band electron spectrum of gold excited by monochromatized Al Kα radiation (solid curve). Comparison is made with a theoretically calculated density of states (broken curve) which has been folded with a 0.25-eV gaussian corresponding to the experimental resolution. The energy scale of the theoretical density-of-states curve has also been expanded by 7.5%, which leads to a near-perfect correspondence between theoretical and experimental band profiles. E_F is the Fermi edge. (*After U. Gelius et al., Experimental spectrum, Nucl. Instr. Meth., B1:85, 1984; and N. E. Christensen and B. O. Seraphin, Theoretical density of states, Phys. Rev., B4:3321, 1971*)

structures (**Fig. 4**). With uv excitation, the resolution is sufficient to resolve vibrational structures in valence electron spectra for gases (**Fig. 5**). In fact, electron spectroscopy can supply a detailed knowledge of the valence orbital structure for all molecules which can be brought into gaseous form with pressures of 10^{-5} torr (10^{-3} Pa) or more. A convenient pressure region is 10^{-2} torr (10 Pa). Liquids and solutions of various compositions can also be studied by ESCA techniques (**Fig. 6**).

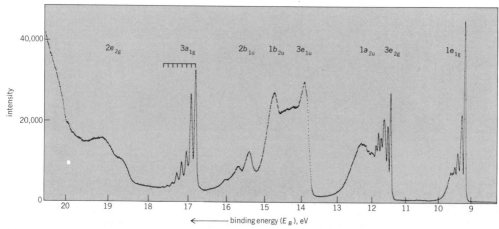

Fig. 5. Benzene valence electron spectrum excited by He-I radiation at 21.22 eV, showing vibrational structures.

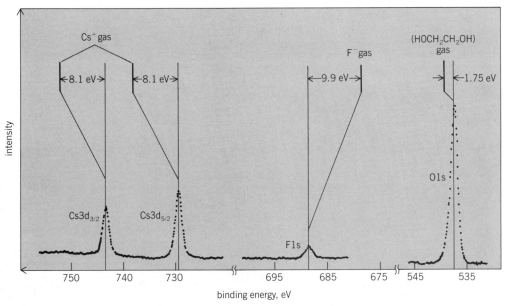

Fig. 6. ESCA spectrum of cesium fluoride (CsF) in glycol (HOCH$_2$CH$_2$OH) solution. Comparison is made with the corresponding electron-binding energies in the gas phase. (*After H. Siegbahn et al., Phys. Scripta, 27:431, 1983*)

ESCA chemical shifts. When atoms are brought close together to form a molecule, the electronic orbitals of each atom are perturbed. Inner orbitals, that is, those with higher binding energies, may still be regarded as atomic and belonging to specified atoms within the molecule, whereas the outer orbitals combine to form the valence-level system of the molecule. These orbitals play a more or less active part in the chemical properties. The chemical bonds affect the charge distribution so that the original atoms can be regarded as charged to various degrees; a neutral molecule has a net charge of zero. The individual atoms in the molecule can be regarded as spheres with different charges set up by the transfer of certain small charges from one atomic sphere to the neighboring atoms taking part in the chemical bond. Inside each charged sphere the atomic potential is constant, in accordance with classical electrostatic theory. The result of this atomic potential is to shift the whole system of inner levels in an atom by a small amount, the same amount for each level. Levels belonging to different atoms in the molecule are generally shifted differently, however, and if the ESCA chemical shifts for individual atoms in the molecule are measured, a mapping can be made of the distribution of charge or potential in the molecule. This, then, is a reflection of the chemical bondings between the atoms, which in turn can be described by the orbitals in the valence-level system.

A unique feature of ESCA is that, if the exact position of the electron lines characteristic of the various elements in the molecule is measured, the area of inspection can be moved from one atomic species to another in the molecular structure. **Figure 7** shows the electron spectrum from the 1s level of the carbon in ethyl trifluoroacetate. All four carbon atoms in this molecule are distinguished in the spectrum. The lines appear in the same order from left to right, as do the corresponding carbon atoms in the structure shown in the figure. If the structure of the molecule is known, the charge distribution can be estimated in a simple way by using, for example, the electronegativity concept and assuming certain resonance structures. More sophisticated quantum-chemical treatments can also be applied. Conversely, if, by means of ESCA, the approximate charge distribution is known, conclusions about the structure of the molecule can be drawn.

Experimental evidence obtained so far for various elements in a large number of molecules indicates strong correlations between chemical shifts and calculated atomic charges. A typical correlation curve for the 2p level in sulfur obtained from more than 100 compounds containing this

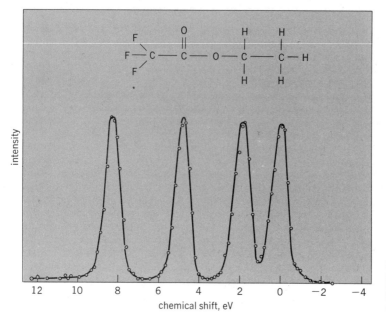

Fig. 7. Electron spectrum from the 1s level of carbon in ethyl trifluoroacetate; binding energy = 291.2 eV.

element is shown in **Fig. 8**. Similar curves are obtained for other elements such as carbon, nitrogen, oxygen, and phosphorus. Chemical shifts are also observed in the electron lines due to the Auger effect. Second-order chemical shifts from groups situated farther away in the molecule (inductive effects) are also observed.

The theoretical importance of the chemical shift effect lies in the study of molecular electronic structure. In other contexts, its usefulness lies precisely in its ability to specify chemical composition, in particular, at surfaces. For example, a metal and its oxide give two distinctly different lines because of the chemical shift effect. It is often easy to follow the rate of oxidation

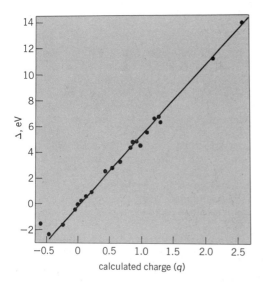

Fig. 8. Binding-energy shifts for the sulfur 2p electrons versus the calculated charge. The points indicate averages from more than 100 compounds.

at a surface, or the adsorption of a gas layer onto it. The chemical composition of a surface or the changes due to various chemical or physical treatments are amenable to detailed examination.
See ATOMIC STRUCTURE AND SPECTRA.

Bibliography. D. Briggs and M. P. Seah, *Practical Surface Analysis by Auger and Photo-Electron Spectroscopy*, 1983; C. R. Brundle and A. D. Baker (eds.), *Electron Spectroscopy, Theory, Techniques and Applications,* vols. 1–5, 1977–1984; A. B. Lever (ed.), *Inorganic Electronic Spectroscopy,* 1984; K. Siegbahn et al., *ESCA Applied to Free Molecules*, 1969; K. Siegbahn et al., *ESCA: Atomic, Molecular and Solid State Structure Studies by Means of Electron Spectroscopy*, 1967; H. Siegbahn and L. Karlsson, Photoelectron spectroscopy: *Handbuch der Physik*, vol. 31, 1982; G. Wendin, *Photoelectron Spectra*, 1981; H. Windawi and F. L. Ho (eds.), *Applied Electron Spectroscopy for Chemical Analysis*, 1982.

NUCLEAR SPECTROSCOPY

Nuclear spectra	**178**
Beta rays	**180**
Gamma rays	**182**
Mössbauer effect	**186**
Neutron spectrometry	**190**
Slow neutron spectroscopy	**196**
Time-of-flight spectrometers	**199**

NUCLEAR SPECTRA
D. J. HOREN

The distribution in energy or momentum of radiations emitted by radioactive nuclei or in a nuclear reaction; also, the graphical display of data from instruments used to measure such radiations (for example, magnetic spectrometers and scintillation detectors).

Radiations can occur when the total energy of a nuclear system (or state) is higher than that of a different configuration (or state) to which a transition can take place. The specific type of radiation which is emitted in such a process is determined by the characteristics of the initial and final nuclear systems as well as the properties of the interaction mechanism. The radiations not only remove energy from the initial system but can also lead to changes in mass, charge, angular momentum (spin), and parity (or symmetry characteristic) of the system.

Nuclear spectra are widely used in both basic and applied research. For the former, it is usually the information which can be inferred from studies of such spectra rather than the radiations themselves which are of primary interest. From experimental determinations of the type of radiation, its energy, and the changes that it causes in angular momentum and party, one is able to deduce information pertaining to the static and dynamic properties of the initial and final nuclear configurations. This, in turn, enables one to gain insight into the structure of nuclei as well as the forces between nucleons. In applied research, it is the radiations themselves or the effects which they produce that are of most importance.

The distribution in energy (momentum) of radiations emitted during transitions between nuclear configurations can be discrete or continuous. Transitions which give rise to discrete spectra are those in which a single type of radiation is emitted. The emitted particle then has a specific energy determined by the two nuclear configurations and type of radiation involved. On the other hand, if two or more radiations are emitted during a transition, the energy will be shared among them. In this case, the energy of each particle emitted can take on a continuum of values between well-defined limits. This then gives rise to a continuous spectrum.

Mathematically a spectrum is described as the number of particles with a given energy (that is, relative intensity) as a function of energy. Graphically, the relative intensity is usually plotted along the ordinate and the energy along the abscissa. A spectrum can be composed of a series of discrete lines (or peaks), a smooth curve, or a combination of these, depending upon the specifics of the transitions involved. In practice, the spectra of particles emitted during nuclear transitions are distorted by the detection instruments used to measure them. For discrete spectra, the instruments are usually calibrated so that the central position of the peak along the abscissa defines the energy of the detected radiation, and the area under the peak is proportional to its relative intensity. The full width at half maximum of the peak is a measure of the effective resolution. A continuous spectrum has an end point corresponding to the energy change of the transition, and it often has a definite shape that is related to the characteristics of the radiation.

In early years, the study of nuclear spectra was mainly confined to naturally occurring radioisotopes (unstable nuclei). The development of nuclear particle accelerators greatly broadened the scope of such research so that nuclear spectra characteristics of over 1600 nuclei have been measured. Some types and characteristics of nuclear spectra are described below, with those associated with the decay of radioisotopes being discussed first.

Beta-ray spectrum. Beta rays are electrons emitted by a nucleus of atomic number Z which, in its ground state, is unstable with respect to one of its neighboring isobars of charge $Z + 1$ or $Z - 1$. Beta decay can also take place from an excited state (isomer) if radiative decay to the ground state is greatly hindered. The charged-particle emission consists of a negative electron to the $Z + 1$ nucleus or a positive electron (positron) to the $Z - 1$ nucleus, and the transition can take place to either the ground or to an excited state of the daughter nucleus.

A transition involving the emission of a beta ray is accompanied by the simultaneous emission of a neutrino, a particle that has neither mass nor charge and is exceedingly difficult to detect. The total energy of such a transition is a fixed quantity, and since this energy is shared between the beta ray and the neutrino, the beta ray can emerge with any energy from zero up to the maximum transition energy. Thus the spectrum of beta rays emitted from an ensemble of nuclei which undero such a transition is continuous. As a result of the statistical manner in which the energy is shared between the beta ray and neutrino, the spectrum of beta-ray intensity plotted versus energy has a bell shape with a broad maximum located at somewhat less than half the maximum energy.

Beta-ray spectra are usually analyzed by a so-called Fermi-Kurie plot. Such a plot converts the experimental intensity-versus-energy spectrum (corrected for source thickness and instrumental effects) into a straight-line distribution which intersects the abscissa at the maximum energy (end-point energy). For negative beta-ray emitters, the end-point energy is equal to the transition energy, except for a nuclear recoil correction. In most instances, this correction is negligible due to the small mass of the beta ray relative to the nucleus. For a positron emitter, the transition energy is equal to the end-point energy plus 1.02 MeV. If the Fermi-Kurie plot of the experimental data, under the assumption of a purely statistical intensity-versus-energy distribution, is not a straight line, the beta-ray transition is said to have a nonstatistical shape and to be of a forbidden type. The degree of forbiddenness is determined by the correction factors needed to straighten out the Fermi-Kurie plot. Whether a beta-ray transition has the allowed or forbidden shape depends upon the characteristics of the initial and final nuclear states that are involved. SEE BETA RAYS.

Alpha-particle spectra. The emission of an alpha particle (equivalent to the nucleus of a mass-4 helium atom) can occur when the state of a nucleus with charge Z, mass A, is unstable with respect to a state of a nucleus with charge $Z - 2$, mass $A - 4$. In order for the alpha particle to emerge, it must overcome the Coulomb-charge-potential barrier of the nucleus and also a centrifugal barrier which depends upon the angular momentum removed. Alpha-particle spectra of radioisotopes are discrete, and the peak energies are less than the corresponding transition energies by an amount equal to the nuclear recoil energy.

Spontaneous fission. Some radioisotopes decay by spontaneous fission; that is, they break up into two fragments (occasionally three). The primary fission fragments and a number of their daughters are themselves radioactive and give rise to nuclear spectra.

Gamma-ray spectra. Gamma rays are emitted when a transition takes place from an excited state to a lower state in the same nucleus. Of the various types of nuclear radiation, gamma rays produce the least amount of nuclear recoil, although in certain kinds of experiments the recoil energy shift is observable. Thus, the gamma-ray energy is almost exactly equal to the energy difference between the states. A process which sometimes competes with gamma-ray emission is internal electron conversion. This process creates holes in the atomic shell structure which, when filled, are accompanied by the emission of x-rays. Hence, it is not unusual to find x-rays in the low-energy regions of gamma-ray spectra.

When gamma rays interact with matter, they produce secondary radiations by means of the photoelectric and Compton effects as well as by pair production (the emission of a positron-electron pair). Instruments (such as ionization chambers, scintillators, and magnetic spectrometers) that measure gamma rays are based upon the detection of these secondary radiations. Of the various types of gamma-ray detectors, it is the scintillator (especially with LiGe crystals) which has been responsible for the gathering of enormous quantities of gamma-ray data. From a determination of gamma-ray energies and intensities, it has been possible to construct nuclear level schemes for many nuclei. Since there is little probability that the level structure of any two nuclei will be identical, precise measurements of gamma-ray transitions can serve as a means to uniquely identify a particular isotope. This feature is used extensively in applied research. SEE GAMMA RAYS.

Nuclear spectra from reactions. There is a myriad of nuclear reactions by which one can produce nuclear spectra of various types (for example, gamma rays, neutrons, protons, and many nuclei). Nuclear reactions can be categorized roughly into three types: (1) elastic and inelastic scattering; (2) transfer reactions in which one or more nucleons (that is, protons or neutrons) are transferred between the projectile and target nucleus; and (3) reactions in which the projectile and target nucleus coalesce to form a compound system which then decays in some manner. In past years, the particles that could be used as projectiles were limited to the lighter elements. However, the development of heavy-ion accelerators has been undertaken to make possible the use of even the heaviest elements as projectiles.

The spectra of particles produced in a nuclear reaction depend upon the target nucleus, the reaction energy release, and the kinematics involved (including geometric and recoil effects).

Usually, more than one type of radiation is emitted, and, in practice, the researcher chooses which particles to measure on the basis of the nuclear structure or reaction properties under study. Whether the spectrum of particles of any given type is discrete or continuous depends upon the specifics of the reaction used.

As an example, consider the case of inelastic scattering. Inelastic scattering takes place when the projectile gives up some of its energy to a target nucleus and is scattered. If the energy transferred to the target is sufficient only to excite discrete levels, the spectrum of the scattered particles will consist of a series of peaks corresponding to each level excited. However, if the energy transferred were sufficient to excite the nucleus up to the point where the density of states is continuous, then the corresponding portion of the spectrum would also be continuous. In the same reaction, nuclear radiations of one type or another (such as gamma rays and neutrons) will also be emitted following the decay of the states which have been excited. In most cases (but not always), the decay of discrete states excited in such reactions takes place by the emission of gamma rays and the spectra involved are also discrete. The decay of excited states in the continuum region usually takes place by neutron or proton emission, and these spectra are usually continuous.

Bibliography. G. F. Bertsch (ed.), *Nuclear Spectroscopy: Proceedings*, 1980; W. E. Burcham, *Elements of Nuclear Physics*, 1986; J. Cerny (ed.), *Nuclear Spectroscopy and Reactions*, pts. A, B, C, and D, 1974; J. H. Hamilton and J. C. Manthuruthil (eds.), *Radioactivity in Nuclear Spectroscopy*, vols. 1 and 2, 1972; K. Siegbahn (ed.), *Alpha-, Beta-, and Gamma-Ray Spectroscopy*, vols. 1 and 2, 1965.

BETA RAYS
Gunnar Backstrom

Fast, charged particles emitted from certain radioactive nuclei. These particles, which are all identical except for the sign of their charge, are classified as positrons (+) and negatrons (−). The latter class is identical with atomic electrons. The kinetic energies range from zero up to 3–5 MeV.

Interaction with matter. Beta rays, often designated β rays or β particles, interact strongly with matter: The particles are generally completely stopped in passing through 0.4 in. (1 cm) of solid material. If various thicknesses of some material are placed across a beam of beta rays, the number of penetrating particles is found to decrease gradually to zero as the thickness increases (**Fig. 1**). There is a definite thickness of absorber which is just sufficient to stop all the beta particles. This quantity, which is called the range of the beta ray, is usually expressed as the product ρx (g/cm^2) of density and thickness, since then this "range" will be approximately independent of the material (**Fig. 2**).

The passage of beta rays through matter is macroscopically observable as heat, due to the kinetic energy dissipated. This radiation also may promote certain chemical reactions and cause structural changes in materials, for instance, the discoloring of glass.

The track of a fast electron, as observed in a photographic emulsion, does not follow a straight line until the particle is stopped, but zigzags in a random way because of collisions with the much heavier nuclei. A beta particle loses its energy by excitation and ionization of atoms and at higher energy also by emitting bremsstrahlung, or brake radiation, when scattered. The first class of energy loss may be pictured as an inelastic collision with a bound electron, which thus reaches a higher energy state. The energy absorbed may or nay not be sufficient to liberate the electron. When a beta particle passes near a nucleus, the electrostatic attraction deflects its path.When so accelerated, the electron emits one or more bremsstrahlung quanta. This type of energy loss increases strongly with the atomic number of the absorber and becomes important in heavy elements at energies of a few MeV.

Apart from the stopping process, which proceeds similarly, negatrons and positrons behave differently in matter. Usually after it has been stopped, the positron combines with an atomic negatron and both vanish (annihilate). The two electron masses appear in the form of the electromagnetic energy of two photons, each of energy 0.511 MeV, which are emitted in opposite direc-

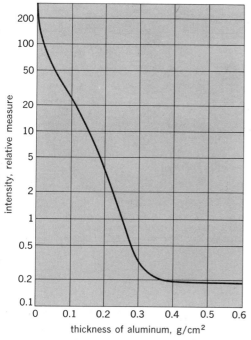

Fig. 1. The intensity of β-rays from indium-116 on passage through aluminum. (*After J. M. Cork, Radioactivity and Nuclear Physics, 3d ed., Van Nostrand, 1957*)

Fig. 2. Dependence of the range ρx of beta rays on their energy. (*After K. Siegbahn, ed., Beta- and Gamma-Ray Spectroscopy, North Holland, 1955*)

tions in order to conserve momentum. Before annihilation the positron-negatron pairs form bound systems, called positronium atoms.

Detection of beta rays. Beta detectors may function so as to produce visible tracks of the particles, as in photographic emulsions and cloud chambers, or may give a signal at the advent of each individual particle. A detector may also indicate the total amount of radiation incident in a definite time. All detectors work on the principle of observing the interaction with matter, in particular ionization and excitation.

Beta-ray spectrometers. These are used for measuring the energy or momentum distribution of beta particles. The energy of an electron is in principle inferred from the way it is influenced by a magnetic field. If a charged particle is moving perpendicularly to the flux of a uniform magnetic field, its orbit is bent into a circle of radius ρ. The momentum is found to be proportional to the product $B\rho$ of magnetic flux density and radius.

A simple and generally used spectrometer is the semicircular type, shown schematically in **Fig. 3**. From the electrons that leave the source in all directions, a thin bundle (represented by an angular spread 2ϕ in Fig. 3) is selected by the entrance slit E to improve resolution. Particles of a definite energy move in congruent circles and are brought to a focus after going through an angle of 180°. Particles of different energies come to a focus at different distances. The particles may be detected by means of a photographic plate along the focal line x, which permits the simultaneous recording of a large energy range. The angular spread 2ϕ allowed by the finite slit width introduces a spread Δx in the position at which the particle strikes the photographic plate. Alternatively a Geiger counter may be located at a fixed distance from the source and the field strength varied to focus one energy after another on a slit in front of the counter. Whereas this instrument can focus a bundle of electrons divergent in the plane of the figure, it does nothing to collect particles which originally make an angle with this plane.

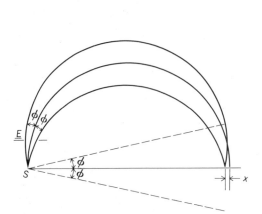

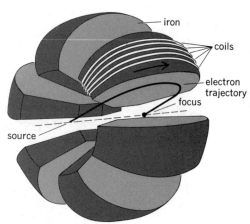

Fig. 3. Principle of the semicircular spectrometer. The magnetic field lines are perpendicular to the paper. The broken lines are diameters of the circular orbits.

Fig. 4. The orange spectrometer, with the very high transmission. (*After K. Siegbahn, ed., Beta- and Gamma-Ray Spectroscopy, North Holland, 1955*)

Focusing in both directions is achieved by using a cylindrically symmetric field with a density decreasing radially as $(\rho)^{-1/2}$ (**Fig. 4**). This spectrometer needs a fixed counter position.

An important figure of merit for a spectrometer is its resolving power, usually given by the width at half-height of a peak due to electrons of a definite energy. Although both instruments described yield relative peak widths down to 0.02%, the latter type realizes the same resolving power with appreciably higher transmission; that is, it accepts a larger percentage of the electrons emitted. Most of the precise work in beta spectroscopy is done with such double-focusing spectrometers.

There exist many types of spectrometers which make use of the focusing properties of magnetic lenses. They have found application especially in cases where high transmission is more important than high resolution.

An instrument with exceptionally high transmission is the so-called orange spectrometer (Fig. 4), which is a number of modified double-focusing spectrometers employing a common source and a common detector.

Bibliography. W. E. Burcham, *Elements of Nuclear Physics*, 1986; M. Morita, *Beta Decay and Muon Capture*, 1973; E. Segre, *Nuclei and Particles*, 2d ed., 1977.

GAMMA RAYS
J. W. OLNESS

Electromagnetic radiation emitted from excited atomic nuclei as an integral part of the process whereby the nucleus rearranges itself into a state of lower excitation (that is, energy content).

Nature of gamma rays. The gamma ray is an electromagnetic radiation pulse—a photon—of very short wavelength. The electric (**E**) and magnetic (**H**) fields associated with the individual radiations oscillate in planes mutually perpendicular to each other and also the direction of propagation with a frequency v which characterizes the energy of the radiation. The **E** and **H** fields exhibit various specified phase-and-amplitude relations, which define the character of the radiation as either electric (EL) or magnetic (ML). The second term in the designation indicates the order of the radiation as 2^L-pole, where the orders are monopole (2^0), dipole (2^1), quadrupole (2^2), and so on. The most common radiations are dipole and quadrupole. Gamma rays range in energy from a few keV to 100 MeV, although most radiations are in the range 50–6000 keV. As such, they lie at the very upper high-frequency end of the family of electromagnetic radiations, which include also light rays and x-rays.

NUCLEAR SPECTROSCOPY

Wave-particle duality. The dual nature of gamma rays is well understood in terms of the wavelike and particlelike behavior of the radiations. For a gamma ray of intrinsic frequency ν, the wavelength is $\gamma = c/\nu$, where c is the velocity of light; energy is $E = h\nu$, where h is Planck's constant. The photon has no rest mass or electric charge but, following the concept of mass-energy equivalence set forth by Einstein, has associated with it a momentum given by $p = h\nu/c = E/c$.

Origin. One of the most frequently utilized sources of nuclear gamma rays is ^{60}Co (that is, the cobalt isotope of $N = 33$ neutrons, $Z = 27$ protons, and thus of atomic mass number $A = N + Z = 60$). As shown in **Fig. 1**, the decay process begins when ^{60}Co (in its ground state, or state of lowest possible excitation) decays to ^{60}Ni ($N = 32$, $Z = 28$) by the emission of a β^- particle. More than 99% of these decays lead to the 2506-keV level of ^{60}Ni; this level subsequently deexcites by an 1173-keV gamma transition to the 1332-keV level, which in turn emits a 1332-keV gamma ray leading to the ^{60}Ni ground state.

The gamma rays from ^{60}Ni carry information not only on the relative excitation of the ^{60}Ni levels, but also on the quantum-mechanical nature of the individual levels involved in the gamma decay. From the standpoint of nuclear physics, the levels of a given nucleus can be described most simply in terms of their excitation energies (E_x) relative to the ground state, and in terms of the total angular momentum (J) and parity (π) quantum numbers given as J^π. For a gamma-ray transition from initial state i to final state f, one obtains $E_x^i - E_x^f = E_\gamma'$, where E_γ' is the measured gamma energy after small (second-order) corrections for nuclear recoil and relativistic effects. Nuclear selection rules restrict the multipole character of the radiation according to the change in the quantum numbers J^π of the initial and final states. In Fig. 1, for example, the transitions must be electric quadrupole (E2), since they connect states of similar parity ($\pi = +$) by radiation of order $L \geq J_i - J_f = 2$. SEE NUCLEAR SPECTRA.

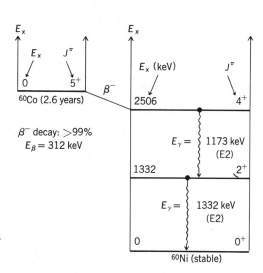

Fig. 1. Energy-level diagram illustrating the gamma decay of levels of ^{60}Ni resulting from beta decay of ^{60}Co.

Use as nuclear labels. Various nuclear species exhibit distinctly different nuclear configurations: the excited states, and thus the gamma rays which they produce, are also different. Precise measurements of the gamme-ray energies resulting from nuclear decays may therefore be used to identify the gamma-emitting nucleus, that is, not only the atomic number Z but also the specific isotope that is designated by A. This has ramifications for nuclear research and also for a wide variety of more practical applications.

The two most widely used detectors for such studies are the NaI(Tl) detector and the Ge(Li) detector. **Figure 2** shows typical gamma spectra measured for sources of ^{60}Co and ^{54}Mn. Full-energy peaks are labeled by the gamma-ray energy, given in keV. The figure of merit for these

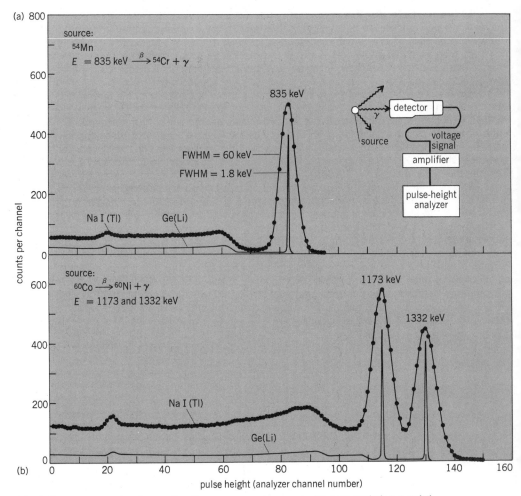

Fig. 2. Gamma-ray spectra from radioactive sources as measured with both NaI(Tl) and Ge(Li) detectors. Inset shows the components of detector apparatus. (a) ^{54}Mn source. (b) ^{60}Co sources.

detectors, defined for a given gamma energy as the full-width-at-half-maximum (FWHM) for the full energy peak, is indicated. Although the more efficient NaI(Tl) detector can clearly distinguish the ^{60}Co and ^{54}Mn gamma rays, it is evident that the Ge(Li) detector, having a line width of only 1.8 keV, is more appropriate for complex nuclei, or for studies involving a greater number of source components. SEE GAMMA-RAY DETECTORS.

Applications to nuclear research. One of the most useful studies of the nucleus involves the bombardment of target nuclei by energetic nuclear projectiles, to form final nuclei in various excited states. For example, ^{48}Ca bombarded by ^{16}O makes ^{60}Ni quite strongly via the ^{48}Ca(^{16}O,4n)^{60}Ni reaction, as well as numerous other final species. Ge(Li) measurements of the decay gamma rays are routinely used to identify the various final nuclei according to their characteristic gamma rays, that is, the 1332- and 1173-keV gamma rays of ^{60}Ni, for example.

Precise measurements of the gamma energies, together with intensity and time-coincidence measurements, are then used to establish the sequence of gamma-ray decay, and thus construct from experimental evidence the nuclear level scheme. Angular correlation and linear polarization measurements determine the radiation character (as M1, E1, E2, or M2 or mixed) and thus the spin-parity of the nuclear levels. These studies provide a very useful tool for investigations of nuclear structure and classification of nuclear level schemes.

Practical applications. In these applications, the presence of gamma rays is used to detect the location or presence of radioactive atoms which have been deliberately introduced into the sample. In irradiation studies, for example, the sample is activated by placing it in the neutron flux from a reactor. The resultant gamma rays are identified according to isotope by Ge(Li) spectroscopy, and thus the composition of the original sample can be inferred. Such studies have been used to identify trace elements found as impurities in industrial production, or in ecological studies of the environment, such as minute quantities of tin or arsenic in plant and animal tissue. *SEE ACTIVATION ANALYSIS.*

In tracer studies, a small quantity of radioactive atoms is introduced into fluid systems (such as the human blood stream), and the flow rate and diffusion can be mapped out by following the radioactivity. Local concentrations, as in tumors, can also be determined.

Doppler shift. If $E_{\gamma 0} = h\nu_0$ is the gamma ray energy emitted by a nucleus at rest, then the energy $E_\gamma = h\nu$ emitted from a nucleus moving with velocity v at angle θ (with respect to the direction of motion) is given by Eq. (1) where c is the velocity of light. In terms of the frequency

$$E_\gamma = E_{\gamma 0}\left(1 + \frac{v}{c}\cos\theta\right) \qquad (1)$$

ν, this expression is entirely analogous to the well-known Doppler shift of sound waves. Experimental measurements of the Doppler shift are used to determine the velocity of the nucleus and, more importantly, to shed light on the lifetime of the nuclear gamma-emitting state. A major advantage of this technique is that the same nuclear reaction which produces the excited nuclear states can also be employed to impart a large velocity to the nucleus.

For example, the velocity of ^{60}Ni nuclei produced via the ^{48}Ca(^{16}O,4n)^{60}Ni reaction at $E(^{16}$O$) = 50$ MeV is $v/c = 0.00204$, and instead of $E_\gamma = 1332$ keV one should observe $E_\gamma = 1359$ keV. The extent of the shift is clearly within the resolving power of the Ge(Li) detector, which may therefore be used to measure E_γ and thus infer v. In most nuclear reactions, v is a known function of time t [that is, $v = v(t)$], and one therefore obtains a distribution of E_γ's whose precise shape may be related to the lifetime of the nuclear state.

Doppler-shift measurements of gamma rays from recoil nuclei produced in nuclear reactions have been routinely used since the mid-1960s to measure nuclear lifetimes of 10^{-9} to 10^{-14} s—a range previously considered inaccessible to study.

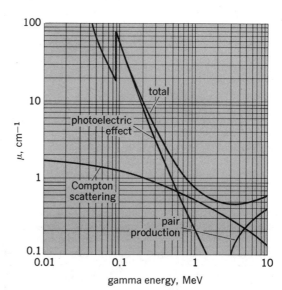

Fig. 3. Graphic representation of partial and total attenuation coefficients for lead as a function of gamma energy. (*National Bureau of Standards*)

Interaction with matter. The energy of a photon may be absorbed totally or partially in interaction with matter; in the latter case the frequency of the photon is reduced and its direction of motion is changed. Photons are thus absorbed not gradually, but in discrete events, and one interaction is sufficient to remove a photon from a collimated beam of gamma rays. The intensity I of a beam decreases exponentially, as in Eq. (2), where x is the path length, I_0 is the

$$I = I_0 e^{-\mu x} \qquad (2)$$

initial intensity, and μ is the linear attenuation coefficient, which is characteristic of the material and the gamma energy.

The dependence of the attenuation coefficient on gamma-ray energy is shown in **Fig. 3** for a lead absorber. For different absorbers, the attenuation is generally greater for the more dense materials. Most attenuation coefficients are tabulated as mass attenuation coefficients μ/ρ where ρ is the material or elemental density.

Bibliography. W. E. Burcham, *Elements of Nuclear Physics*, 1986; J. Cerny (ed.), *Nuclear Spectroscopy and Reactions*, pt. C, 1974; R. D. Evans, *The Atomic Nucleus*, 1955, reprint 1982; E. Segré, *Nuclei and Particles*, 2d ed., 1977; K. Siegbahn (ed.), *Alpha-, Beta-, and Gamma-Ray Spectroscopy*, 1965.

MÖSSBAUER EFFECT
ROLFE H. HERBER

Recoil-free gamma-ray resonance absorption. The Mössbauer effect, also called nuclear gamma resonance fluorescence, has become the basis for a type of spectroscopy which has found wide application in nuclear physics, structural and inorganic chemistry, biological sciences, the study of the solid state, and many related areas of science.

Theory of effect. The fundamental physics of this effect involves the transition (decay) of a nucleus from an excited state of energy E_e to a ground state of energy E_g with the emission of a gamma ray of energy E_γ. If the emitting nucleus is free to recoil, so as to conserve momentum, the emitted gamma ray energy is $E_\gamma = (E_e - E_g) - E_r$, where E_r is the recoil energy of the nucleus. The magnitude of E_r is given classically by the relationship $E_r = E_\gamma^2/2mc^2$, where m is the mass of the recoiling atom. Since E_r is a positive number, the E_γ will always be less than the difference $E_e - E_g$, and if the gamma ray is now absorbed by another nucleus, its energy is insufficient to promote the transition from the nuclear ground state E_g to the excited state E_e.

In 1957 R. L. Mössbauer discovered that if the emitting nucleus is held by strong bonding forces in the lattice of a solid, the whole lattice takes up the recoil energy, and the mass in the recoil energy equation given above becomes the mass of the whole lattice. Since this mass typically corresponds to that of 10^{10} to 10^{20} atoms, the recoil energy is reduced by a factor of 10^{-10} to 10^{-20}, with the important result that $E_r \sim 0$ so that $E_\gamma = E_e - E_g$; that is, the emitted gamma-ray energy is exactly equal to the difference between the nuclear ground-state energy and the excited-state energy. Consequently, absorption of this gamma ray by a nucleus which is also firmly bound to a solid lattice can result in the "pumping" of the absorber nucleus from the ground state to the excited state. The newly excited nucleus remains, on the average, in its upper energy state for a time given by its mean lifetime τ (a quantity dependent on energy, spin, and parity of the nuclear states involved in the deexcitation process) and then falls back to the ground state by reemission of the gamma ray. An important feature of this reemission process is the fact that it is essentially isotropic; that is, it occurs with equal probability in all directions. *See* ENERGY LEVEL; GAMMA RAYS.

Energy modulation. Before this phenomenon of resonance fluorescence can be turned into a spectroscopic technique, it is necessary to provide an appropriate energy modulation of the gamma ray emitted in the initial decay process. An estimate of the energy needed to accomplish this can be calculated from a knowledge of the inherent width or sharpness of the excited-state nuclear level. This is given by the Heisenberg uncertainty principle as $\Gamma = h/2\pi\tau$ (h is Planck's constant and τ is the mean lifetime of the excited state). In the case of ^{57}Fe, a nucleus for which resonance fluorescence is especially easy to observe experimentally, $\Gamma = 4.6 \times 10^{-12}$ keV. In order to modulate the emitted gamma-ray energy, which in this case corresponds to 14.4 keV, one can take advantage of the Doppler phenomenon which states that if a radiation source has a velocity relative to an observer of v, its energy will be shifted by an amount equal to $E = (v/c)E_\gamma$. Setting the required Doppler energy equal to the width of the nuclear level and E_γ equal to the nuclear transition energy leads to the equation below. Relative velocities of this order of magni-

$$v = c\frac{\Gamma}{E_\gamma} = 3 \times 10^{10} \text{ cm/s} \times \frac{4.6 \times 10^{-12}}{14.4} = 0.0096 \text{ cm/s} = 0.0038 \text{ in./s}$$

tude can be used to modulate the gamma ray emitted in a typical Mössbauer transition, that is, to "sweep through" the energy width of the nuclear transition. *See* Doppler effect.

Experimental realization. The experimental realization of gamma-ray resonance fluorescence can be achieved with the arrangement illustrated schematically in **Fig. 1**. In a typical Mössbauer experiment the radioactive source is mounted on a velocity transducer which imparts a smoothly varying motion (relative to the absorber, which is held stationary), up to a maximum of several centimeters per second, to the source of the gamma rays. These gamma rays are incident on the material to be examined (the absorber). Some of the gamma rays (those for which E_γ is exactly equal to $E_e - E_g$) are absorbed and reemitted in all directions, while the remainder of the gamma rays traverse the absorber and are registered in an appropriate detector which causes one or more pulses to be stored in a multichannel analyzer. The electronics are so arranged that the location (address) in the multichannel analyzer, where the transmitted pulses are stored, is synchronized with the magnitude of the relative motion of source and absorber.

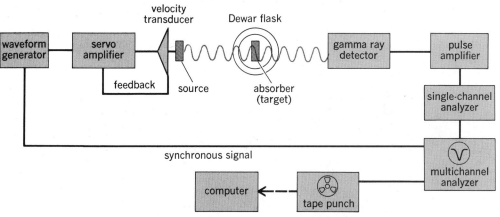

Fig. 1. Experimental arrangement for performing Mössbauer effect spectroscopy. This typical Mössbauer experiment is with ^{57}Fe or ^{119}Sn. (*After R. H. Herber, Mössbauer spectroscopy, Sci. Amer., 255(4):86–95, October 1971*)

A typical display of a Mössbauer spectrum, which is the result of many repetitive scans through the velocity range of the transducer, is shown in **Fig. 2**. Such a Mössbauer spectrum is characterized by a position δ of the resonance maximum (corresponding to a maximum in the isotropic scattering, and thus a minimum in the intensity of the transmitted radiation), a line width Γ, and a resonance effect magnitude ε corresponding to the total area A under the resonance curve.

In the case of the Mössbauer active nuclides ^{57}Fe and ^{119}Sn, among others, two additional features which are of great interest to chemists and physicists may be experimentally elucidated. One of these is the quadrupole coupling which is observed if the Mössbauer nuclide is located in an environment where the electric charge distribution does not have cubic (that is, tetrahedral or octahedral) symmetry. Such a spectrum is shown in **Fig. 3**, in which the magnitude of the quad-

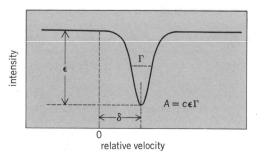

Fig. 2. Mössbauer spectrum of an absorber which gives an upsplit resonance line. The spectrum is characterized by a position δ, a line with Γ, and an area A related to the effect magnitude ε.

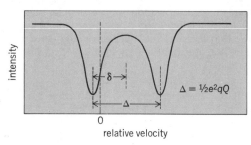

Fig. 3. Mössbauer spectrum of an absorber (containing for example ^{57}Fe or ^{119}Sn) which shows quadrupole splitting Δ.

rupole interaction Δ is equal to $e^2qQ/2$, where e is the electron charge, q is the gradient of the electrostatic field at the nucleus, and Q is the nuclear quadrupole moment. Finally, a Mössbauer spectrum can also give information on the magnitude of the magnetic field H_0 acting on the nucleus through the magnetic hyperfine interaction. This is illustrated in **Fig. 4**, where only a single resonance line would be observed in the absence of a magnetic interaction. SEE HYPERFINE STRUCTURE.

Moreover, all of these parameters—δ, Δ, Γ, A, and H_0—are temperature-dependent quantities, and their study over a range of temperatures and conditions can shed a great deal of light on the nature of the environment in which the Mössbauer nuclide is located in the sample under investigation. More than one hundred Mössbauer transitions, involving 43 different elements, have been experimentally observed and reported.

Application. Mössbauer effect experiments have been used to elucidate problems in a very wide range of scientific disciplines, and only a few examples can be cited as representative of the information extracted from such studies.

Nuclear physics and chemistry. One of the narrowest resonance lines which has been observed is that from the 6.8-microsecond, 6.2-keV gamma transition in ^{181}Ta, and detailed Mössbauer effect measurements using a source of 140-day ^{181}W have shown that the magnetic moment of the spin 9/2 excited state in ^{181}Ta is $+5.35 \pm 0.09$ nanometers and the nuclear quadrupole moment of this state is $+4.4 \pm 0.05) \times 10^{-24}$ cm^{-2}. Such data are of considerable use to nuclear physicists in refining models which describe the fundamental interaction forces in the nucleus. Similarly, the 93.26-keV resonance in ^{67}Zn has been used to determine the magnetic moment of the ½ state (spin = ½, negative parity) in this nuclide and leads to the conclusion that the ½ and 5/2$^-$ states can be considered minus-quasiparticle states. Mössbauer spectroscopy has also been a potent technique for the study of the electromagnetic moments of nuclei, in particular the magnetic dipole moment and the (mean-square) charge radius of nuclear states. The nuclides ^{191}Ir and ^{193}Ir are typical of those which have been used for detailed studies of nuclear parameters. Such nuclear information is frequently difficult to obtain by non-Mössbauer-effect methods.

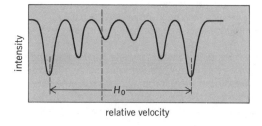

Fig. 4. Mössbauer spectrum of metallic iron showing the splitting of the resonance line by the internal magnetic field (H_0 = 33 T = 330 kG at room temperature).

Recoilless gamma-ray resonance experiments have been able to provide detailed information concerning excited-state lifetimes involved in the nuclear decay process. The lifetime values for the nuclides ^{119}Sn, ^{107}Au, and ^{73}Ge, among others, are largely based on Mössbauer effect measurements.

Recoilless gamma fluorescence spectroscopy has also been used to study the chemical consequences of nuclear decay, and the lifetimes of the Mössbauer transition (typically about 10^{-8} s) provide a convenient time scale to distinguish rapid electronic relaxation processes (typically 10^{-12} to 10^{-14} s) from atomic translation processes (typically slower than 10^{-6} s), and thus study the chemical fate of an atom which results from the decay of a radioactive parent nuclide.

Solid-state physics. Mössbauer effect spectroscopy has made significant contributions to the study of problems in solid-state physics, especially of the nature of the magnetic interactions in iron-containing alloys and the dependence of the magnetic field on composition, temperature, pressure, and other parameters which are of importance in metallurgical processes; solid-state-device fabrication; the structural use of metals and alloys; and numerous related problems of great practical importance.

Combining Mössbauer effect spectroscopy with vibrational spectroscopic studies has led to a clearer understanding of the nature of inter- and intra-molecular forces and the relationship of these forces to the properties of polymeric materials.

It has also been possible, using this technique, to study the effect of high pressure and isotropic compressibility on the chemical properties of materials, especially in the case of experiments with ^{57}Fe, ^{181}Ta, and ^{119}Sn. Such studies have led to the design of high-pressure processes in preparative metallurgy and materials science.

A technique which has also found particular application in metallurgy and the study of catalysts is the use of conversion electron spectroscopy to detect the Mössbauer effect. The advantage of this nondestructive technique is that only the surface layers of the material being investigated are probed by the gamma rays. Thus, in conjunction with the standard transmission type of experimental arrangement shown in Fig. 1, it is possible to differentiate between the atoms on the surface of a material and those in the interior. Such studies frequently can lead to a better understanding of the chemical modification of a surface when solid materials are exposed to a reactive environment as in corrosion and in the poisoning of catalysts. *See* Surface physics.

At the extremely low end of the temperature scale, Mössbauer effect spectroscopy has been useful in examining the nature of those materials which become superconductive at sufficiently low temperatures and the relationship between chemical composition and structure on the one hand and the superconductive transition on the other, as in the dichalcogen layer compounds $TaS_2 \cdot Sn$ and $TaS_2 \cdot Sn_{1/3}$ (both studied using the nuclide ^{119}Sn). Nb_3Sn, a material widely used in the construction of superconducting magnets, has been subjected to detailed Mössbauer effect investigations.

Structural chemistry. The two Mössbauer nuclides most widely exploited by chemists are ^{57}Fe and ^{119}Sn, although a growing body of data resulting from experiments with ^{129}I, ^{99}Ru, ^{121}Sb, and others has been reported. The position of the resonance maximum δ, also called the isomer or chemical shift, can be related to the systematics of the electron configuration of the atom, and extensive isomer shift correlations for iron- and tin-containing compounds have been tabulated. In particular, the isomer shift of tin compounds is readily related to the oxidation state, since it has been observed that all stannous (Sn^{2+}) isomer shifts are larger than that observed for metallic tin (β-Sn), while those for stannic compounds (Sn^{4+}) are smaller than this value. This observation allows an assignment of oxidation state to be made on the basis of the isomer shift parameter, as in the two-dimensional layer compound $SnTa_3S_6$ in which the tin atom is clearly identified as a stannous ion, contrary to expectations based on theory.

The use of Mössbauer spectroscopy to identify the charge state of a given chemical species has also made major contributions to the understanding of the phenomena of valence fluctuations, valence instabilities, and mixed valencies in solids. Such studies are of particular importance in rare-earth chemistry, and the use of the nuclides ^{149}Sm, ^{152}Sm, ^{153}Eu, and ^{151}Eu has been particularly fruitful.

Similarly, the isomer shifts reported for a number of ruthenium compounds, which have been studied using the 89.36-keV resonance in ^{99}Ru, can be correlated systematically with the number of $4d$ electrons involved in the bonding of the metal atom to its nearest-neighbor ligands.

Such Mössbauer effect studies have led to a clearer understanding of the nature of "ruthenium red," a trimeric ammonia ruthenium oxide, and a number of other compounds of this relatively rare transition-metal homolog of iron.

In the field of organometallic chemistry, Mössbauer effect spectroscopy has served to clarify the structure of a number of compounds of iron and tin which are of considerable synthetic and industrial importance, including $Fe_3(CO)_{12}$, $[(\pi C_5H_5)_2Fe(CO)_2]_2$, $[(C_4H_9)_3Sn]_2SO_4$, and the organotin thioglycolates which are used as stabilizers in the plastics industry.

Biological science. Many molecules of biological importance, including hemoproteins, iron-sulfur proteins, and iron storage and transport proteins, offer an ideal system in which Mössbauer effect spectroscopy can be used to elucidate the structure and bonding properties of the metal atom in complex systems. The first measurements on such molecules were reported in 1961, and a very large number of iron-containing systems have been studied since then. Paramagnetic iron compounds can be studied at temperatures below the magnetic ordering point (Néel temperature), and it is thus possible, by means of Mössbauer effect spectroscopy, to determine the sign and magnitude of the magnetic field acting on an iron atom in a complex biological material with molecular weights ranging up to 50,000 or more.

Success has also been achieved in understanding the magnetic basis by which certain microorganisms can orient themselves in the Earth's magnetic field and thus move in the direction of sediments which contain the nutrients essential to their survival. This magnetic orientation, which is of opposite polarity for organisms living in the Northern and the Southern hemispheres of the Earth, is achieved by the incorporation in the microorganism of particles (magnetosomes) which consist of Fe_3O_4 encased in an organic envelope and which function as a compass in the Earth's magnetic flux lines. The new insights in understanding this phenomenon were achieved in part through the application of Mössbauer spectroscopy involving the isotope ^{57}Fe.

It is also possible to study antiferromagnetically coupled iron atoms in biological molecules by carrying out the Mössbauer effect measurements in an external magnetic field over a range of temperatures. Typical of such a study is that of oxidized and reduced putidaredoxin, an iron-sulfur protein (molecular weight = 12,500) which acts as a one-electron transfer enzyme. The Mössbauer experiments on this material clearly showed that in the oxidized material the two iron atoms in the molecule occupy chemically equivalent sites. On one electron reduction, one iron atom remains ferric (Fe^{3+}), while the other becomes ferrous (Fe^{2+}), and the two atoms couple antiferromagnetically to give an electronic ground state of $S = \frac{1}{2}$. Such detailed knowledge of the chemical behavior of the iron atoms in this molecule can elucidate the action of biological catalysts (enzymes) on the molecular level.

Related fields. Mössbauer effect studies have also played a role in studies in many related fields of science, including archeology, geology, engineering studies, theoretical (relativity) physics, chemical kinetics, and biology. The samples of surface material returned from the Moon by the United States Apollo program have been carefully scrutinized by Mössbauer techniques, as have core samples extracted from deep-drilling experiments on the Earth's outer layer. The geographical distribution of ancient Greek pottery has been traced by making use of characteristic Mössbauer effect data, and the pigments used in painting and decorating have been similarly investigated using this technique.

Bibliography. G. M. Bancroft, *Mössbauer Spectroscopy*, 1973; T. E. Cranshaw et al., *Mössbauer Spectroscopy and Its Applications*, 1986; T. C. Gibb, *Principles of Mössbauer Spectroscopy*, 1976; U. Gonser (ed.), *Mössbauer Spectroscopy*, 1975; R. H. Herber (ed.), *Chemical Mössbauer Spectroscopy*, 1984; G. J. Long (ed.), *Mössbauer Spectroscopy Applied to Inorganic Chemistry*, 1984; G. K. Shenoy and F. E. Wagner (eds.), *Mössbauer Isomer Shifts*, 1978.

NEUTRON SPECTROMETRY
JOHN A. HARVEY

A generic term applied to experiments in which neutrons are used as the probe for measuring excited states of nuclides and for determining the properties of these states. The term neutron spectroscopy is also used. The strength of the interaction between a neutron and a target nuclide can vary rapidly as a function of the energy of the incident neutron, and it is different for every

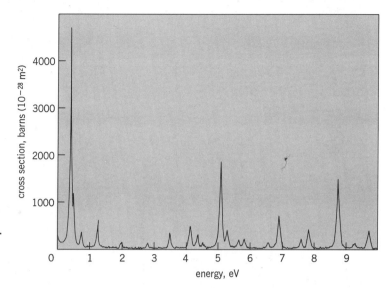

Fig. 1. Neutron total cross section of ^{231}Pa + neutron. The variation in sizes of the resonances and the nonuniform spacing of resonances are apparent.

nuclide. At particular neutron energies the interaction strength for a specific nuclide can be very strong; these narrow energy regions of strong interactions are called resonances. The strength of the interaction, expressing the probability that an interaction of a given kind will take place, can be considered as the effective cross-sectional area presented by a nucleus to an incident neutron. This cross-sectional area is expressed in barns (1 barn = 10^{-28} m^2) and is represented by the symbol σ. The neutron total cross section of the nuclide ^{231}Pa from 0.01 to 10 eV is shown in **Fig. 1**. Even though the neutron has zero charge, neutron energies are measured in electronvolts (1 eV = 1.60×10^{-19} joule). Neutron spectroscopy covers the vast energy range from 10^{-3} eV to 10^3 MeV.

Unbound and bound states of nuclides. Each resonance corresponds to an unbound excited state of the compound nucleus (**Fig. 2**) at an excitation energy that is the sum of the energy of the neutron and the binding energy of the neutron (4–11 MeV) which has been added to the target nuclide. The compound nucleus has a mass number which is one more than that of the target nuclide. Near the ground state of the compound nuclide the spacing of energy levels may be 10^4 to 10^6 eV. However, for a heavy nuclide, such as the compound nuclide ^{232}Pa, the excited states at an excitation energy just above the binding energy of approximately 5.5 MeV are less than 1 eV apart. To observe the individual states, the neutron energy resolution must be smaller than the level spacing. This can be achieved with low-energy neutrons, because they provide the requisite resolution, and there is no Coulomb repulsion to prevent them from entering the target nucleus. Neutron spectroscopy is presently the only technique which can provide this detailed information. For light nuclei (atomic weights ≤ 40) the spacing between the excited states can be many keV, and resonances can be resolved up to neutron energies of several MeV (**Fig. 3**). Lower-energy bound excited states of the compound nucleus below the neutron binding energy can also be studied by gamma-ray spectroscopy, observing the energy of the gamma rays emitted after the capture of neutrons at resonances or at thermal energy (Fig. 2).

Nuclear energy levels of the target nuclide can also be determined by measuring the energy spectrum of neutrons which are inelastically scattered by the target under bombardment from monoenergetic MeV incident neutrons (**Fig. 4**). The energy of an excited state is equal to the difference in energies between the incident and scattered neutrons, $E_n - E_{n'}$. If the incident neutron in Fig. 4 has energy E_{n1}, it has enough energy to excite any of the six lowest levels and emit a neutron of lesser energy than E_{n1}. A neutron of energy E_{n2} could excite only the two lowest levels and emit a neutron of lesser energy than E_{n2}. Information on these same low-energy states can be obtained by measuring the energies and intensities of the gamma rays from the deexcitation of these states excited by inelastically scattered neutrons.

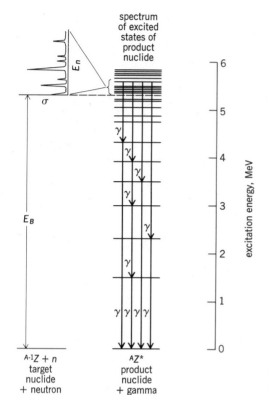

Fig. 2. Energy-level diagram for the product nucleus $^A Z^*$. The asterisk emphasizes that the product nucleus is in an excited state.

Neutron reactions and resonance parameters. The abbreviated notation for neutron reactions, (n,n) and so on, lists the bombarding particle before the comma, and the emitted particle or particles after the comma. The standard symbols are: n (neutron), p (proton), d (deuteron), α (alpha particle), γ (gamma ray), f (fission), and T (total). A more complete description of the reaction lists also the target and product nuclides, for example, $^A Z(n,\gamma)\ ^{A+1}Z$. The reactions most useful for neutron spectroscopy are: the total interaction; elastic scattering (n,n); radiative capture (n,γ); fission (n,f); inelastic scattering (n,n'); charged particle emission (n,p), (n,α), and (n,d); and three-body breakup or sequential decay $(n,2n)$ and (n,np).

The resonances observed in these various reactions can be fitted by a theoretical formula to give parameters of the resonances (E_0, Γ, Γ_f, Γ_γ, Γ_n, and so forth) which correspond to detailed properties of the excited states in the compound nucleus. For example, E_0 is the resonance energy; the fission width, Γ_f, is obtained from the fission cross section; the radiation width, Γ_γ, from the capture cross section, and so forth. The neutron width, Γ_n, can be obtained from the scattering cross section or the total cross section. The total width, Γ, can be obtained if the energy resolution is less than or equal to Γ. In addition, two other properties, the angular momentum of the neutron forming the resonance and the spin, J, of the state can often be determined. For narrow resonances where Γ (in eV) $\leq 0.05 \sqrt{E_0}$ (in eV), it is necessary to consider the Doppler broadening of resonances due to the thermal motion of the target nuclides.

Neutron cross sections. The measurement of a cross section for a particular reaction consists of measuring the number of such reactions produced by a known number of neutrons incident on a known number of target nuclides. When the probability of all neutron interactions with the target nucleus is small, the number of reactions of a particular process i, per unit area and unit time using a beam of neutrons equals $(nv)(Nx)\sigma_i$. The quantity nv is the number of incident neutrons per unit area normal to their direction per unit time, N is the number of target

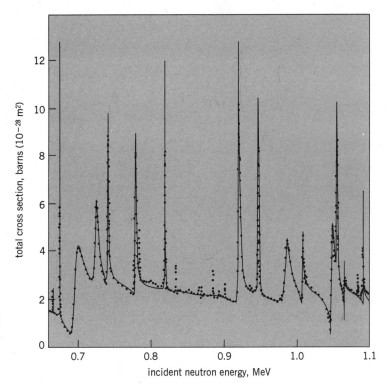

Fig. 3. Experimental neutron total cross section of sulfur compared to a theoretical fit. The fit does not include contributions from small resonances and minor isotopes of sulfur. The asymmetry of some resonances arises from the interference of resonance and potential scattering.

nuclei per unit volume, x is the thickness of the target in the direction of the incident neutrons, and σ_i is the cross section per target nucleus for a particular reaction expressed in units of area. If the probability of all interactions is not small, the incident beam will be attenuated exponentially, as $\exp(-Nx\sigma_T)$, in passing through the sample, where Nx is the number of target nuclei per unit area normal to the beam, and σ_T is the neutron total cross section.

The most common type of neutron cross-section measurement, which can usually be made with the highest neutron energy resolution and usually with the most accuracy, is that of the total cross section. This measurement consists of measuring the transmission of a well-collimated beam of neutrons through a sample of known thickness; the transmission through the sample is simply the ratio of the intensity of the beam passing through the sample to that incident on the sample. The intensity of the incident beam is reduced in passing through the sample because the incident neutrons are absorbed or scattered by the target particles. The total cross section, σ_T, is determined from the equation $\sigma_T = -[\log_e(\text{transmission})]/Nx$.

In order to measure partial cross sections, more elaborate equipment is needed, in general, than for total cross-section measurements, and the measurements are considerably more difficult. For example, to measure the differential elastic scattering cross section, it is necessary to measure the number of elastically scattered neutrons as a function of angle of the scattered neutron relative to that of the incident neutron. MeV-energy neutrons, in addition to being elastically scattered, can also lose energy when scattered from a target nucleus [inelastic scattering, $(n,n'\gamma)$]. Several techniques have been developed for determining inelastic cross sections, both by measuring the energy spectrum of the inelastically scattered neutrons and by measuring the energies and intensities of the gamma rays emitted from the excited nuclei. These cross sections can also be measured as a function of the angle relative to the direction of the incident neutron.

Techniques for neutron spectroscopy. Neutron spectroscopy can be carried out by two different techniques (or a combination): (1) by the use of a time-pulsed neutron source which emits neutrons of many energies simultaneously, combined with the time-of-flight technique to

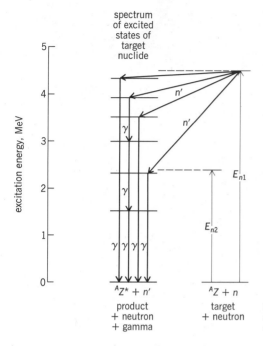

Fig. 4. Energy diagram of the target nuclide showing excitation of levels by the inelastic scattering of MeV neutrons.

measure the velocities of the neutrons; this time-of-flight technique can be used for neutron measurements from 10^{-3} eV to about 200 MeV; (2) by the use of a beam of nearly monoenergetic neutrons whose energy can be varied in small steps approximately equal to the energy spread of the neutron beam; however, useful "monoenergetic" neutron sources are not available from about 10 eV to about 10 keV.

Time-of-flight neutron spectrometers. Time-of-flight neutron spectrometers are the most widely used spectrometers for most neutron cross-section measurements. The time-of-flight technique requires an intense pulse of neutrons which contains neutrons of many energies and a flight path to measure the velocities of the neutrons. Various detectors are placed at the end of the flight path depending on the type of cross-section measurement. Burst widths from 10^{-5} to 10^{-9} s and flight paths from 3 to 3000 ft (1 to 1000 m) have been used. The resolution of a time-of-flight spectrometer is often quoted in microseconds or nanoseconds per meter. The energy resolution ($\Delta E/E$) is equal to $2\Delta t/t$, where Δt is the time width of the neutron burst plus the time spread in the detector, and t is the time of flight of the neutron [t (in microseconds) = 72.3 × path length (in meters)/$\sqrt{E}$ (in eV)]. The time between pulses, the flight path length, and filters in the beam must be selected so that low-energy neutrons from previous pulses do not interfere with the high-energy neutrons from following pulses. By the use of multichannel storage, usually a computer, the complete neutron spectrum (or cross section) can be obtained in one measurement with good energy resolution over a broad energy range.

The most valuable neutron source for neutron time-of-flight spectroscopy is an electron or other charged-particle accelerator capable of producing intense pulses of neutrons of short duration (on the order of 10^{-9} s). Excellent neutron cross-section measurements can be made with these spectrometers from 0.01 eV to 200 MeV. For example, a beam of 140-MeV electrons incident on a tantalum target produces neutrons with an energy distribution that has a peak at about 1 MeV and extends up to about 80 MeV. The peak of the neutron distribution for protons or deuterons incident on a heavy target occurs at a higher neutron energy than for electrons; for deuterons a broad peak occurs at about half the energy of the deuterons. Lower-energy neutrons from these accelerators are obtained by placing a moderator (about 0.8 in. or 2 cm thick) around or near the target. The duration of these moderated neutron pulses in nanoseconds is approximately $2/\sqrt{E}$ (in MeV). The flux distribution of these moderated neutrons approximately follows the relation $E^{-0.8}$ down to thermal neutron energies. With a moderated neutron source the pulse repetition must be sufficiently low, depending on the flight path length, beam filter, and energy range, to prevent overlap of neutrons from successive pulses. Typical high-resolution results obtained using a moderated source are shown in Figs. 1 and 3.

Before the development of short-pulse accelerators, a mechanical chopper rotating at high speed in a well-collimated beam from a moderated fission reactor was used to produce bursts of electronvolt-energy neutrons. The neutron pulses were sufficiently short (about 10^{-6} s) and of sufficient intensity for measurements to be made up to a neutron energy of about 10^4 eV using neutron flight paths up to about 300 ft (100 m) in length. In order to produce pulses of only 10^{-6} s duration, the neutron beam had to be collimated to narrow slits (0.02 in. or 0.05 cm) to match the narrow slits through the chopper. Only when the rotation of the chopper was such that the slits in the rotor lined up with the slits in the collimator was a neutron beam with a broad energy spread passed. A fast-chopper time-of-flight spectrometer is particularly useful for transmission measurements on samples which are available in small amounts, since the sample only needs to be large enough to cover the beams passing through the narrow slits in the collimator.

For time-of-flight measurements in the energy region from about 10 keV to 1 MeV, a pulsed electrostatic accelerator using the ^{7}Li (p,n) reaction is capable of producing neutron pulses of short duration (10^{-9} s) with sufficient intensity for measurements with flight paths of a few meters. By selecting the proton-bombarding energy and a suitable target thickness, neutrons produced in the reaction at a given angle can have well-defined upper and lower energy limits. With no low-energy neutrons and short flight paths, rather high repetition rates of 10^6 Hz can be used and an energy resolution of about 1% can be realized.

The most intense pulsed neutron source used for neutron time-of-flight spectroscopy is that achieved from an underground nuclear explosion. The burst duration is about 80 nanoseconds, and the neutron distribution extends down to about 20 eV. Fission cross-section measurements have been made on very small samples of many radioactive heavy nuclides using such a source and an approximately 980-ft (300-m) flight path length. The availability of this source is obviously rather restricted, but it is unique for measurements on highly radioactive samples.

Neutron time-of-flight measurements have also been made using a pulsed fission reactor where the duration of burst is about 40 μs. Finally, subcritical boosters have been used to multiply the intensity of the neutron pulses from electron accelerators by factors of 10–200, which results in pulse durations of 0.08–4 μs.

Monoenergetic neutron spectrometers. The best technique for obtaining an intense beam of low-energy neutrons ($\leq$10 eV) with an energy spread of only about 1% is to use a crystal monochromator placed in a well-collimated beam of neutrons from a high-flux moderated fission reactor. If a single crystal (such as beryllium, copper, or lead) is properly oriented in a collimated neutron beam, neutrons of a discrete energy, E, will be elastically scattered from a particular set of planes of atoms in the crystal through an angle 2θ given by Bragg's law $n\lambda = 2d \sin \theta$. In this equation, the integer n is the order of the reflection, λ is the neutron wavelength, d is the spacing between the planes of atoms of the particular set in the crystal, and θ is the angle of incidence between the direction of the neutron beam and the set of planes of atoms being considered. The neutron wavelength λ in centimeters equal $0.286 \times 10^{-8}/\sqrt{E}$, where E is in eV. The energy of the diffracted beam can be continuously varied by changing the angle of the crystal. Measurements of many rare-earth nuclides and heavy nuclides have been made with crystal spectrometers up to 10 eV neutron energy. Capture gamma-ray spectra have also been studied as a function of neutron energy, specifically from different neutron resonances.

"Monoenergetic" neutrons in the energy range from a few keV to 20 MeV can be obtained by bombarding various thin targets with protons or deuterons from a variable-energy accelerator such as an electrostatic accelerator. The most useful (p,n) reactions to cover the energy range from a few keV to a few MeV are those on lithium and tritium targets. The (d,n) reaction on deuterium is useful from about 1 to 10 MeV, and the (d,n) reaction on tritium from 10 to 20 MeV. In the energy range up to 1 MeV an energy resolution of about 1 keV is possible, but this resolution is usually not adequate for neutron spectroscopy for neutrons with energies less than 10^4 eV. The measurement of a complete cross-section spectrum up to 1 MeV may require 1000 sequential measurements at slightly different neutron energies. Monoenergetic neutron sources are also useful for measurements such as activation, which cannot be done with the time-of-flight technique.

Applications. Neutron spectroscopy has yielded a mass of valuable information on nuclear systematics for almost all nuclides. The distribution of the spacings between nuclear levels and the average of these spacings have provided valuable tests for various nuclear theories. The properties of these levels, that is, the probabilities that they decay by neutron or gamma-ray emission, or by fission, and the averages and distribution of these probabilities have stimulated much theoretical effort.

In addition, knowledge of neutron cross sections is fundamental for the optimum design of thermal fission power reactors and fast neutron breeder reactors, as well as fusion power reactors now in the conceptual stage. Cross sections are needed for nuclear fuel materials such as ^{235}U or ^{239}Pu, for fertile materials such as ^{238}U, for structural materials such as iron and chromium, for coolants such as sodium, for moderators such as beryllium, for shielding materials such as concrete. The optimum choice of materials for the energy region under consideration is critical to the success of the project and is of great economic significance.

Bibliography. M. Divadeen and N. E. Holden, 4th ed., vol. 1: *Thermal Cross Sections and Resonance Parameters, Part A, Z = 1–60*, 1981; D. I. Garber and R. R. Kinsey, 3d. ed., vol. 2: *Curves*, 1976; J. A. Harvey (ed.), *Experimental Neutron Resonance Spectroscopy*, 1970; S. F. Mughabghab and D. I. Garber (compilers), *Neutron Cross Sections*, 3d ed., vol. 1: *Parameters*, 1981.

SLOW NEUTRON SPECTROSCOPY
BERTRAM N. BROCKHOUSE

The use of beams of slow neutrons, from nuclear reactors or nuclear accelerators, in studies of the structure or structural dynamics of solid, liquid, or gaseous matter. Studies of the chemical or magnetic structure of substances are usually referred to under the term neutron diffraction, while studies of atomic and magnetic dynamics go under the terms slow neutron spectroscopy, inelastic neutron scattering, or simply neutron spectroscopy. The results obtained are to a considerable extent complementary to those obtained by use of optical spectroscopy and x-ray diffraction.

Experiments. In a neutron spectroscopy experiment, a beam of neutrons is scattered by a specimen and the scattered neutrons are detected at various angles to the initial beam. From these measurements, the linear momenta of the incoming and outgoing neutrons (and the vector momentum changes experienced by individual neutrons) can be computed. Experiments are often carried out in a mode in which the vector momentum change Q is held constant and the energy transfer ϵ is employed as the independent variable for the experiment.

In general, just those neutrons which have been scattered once only by the specimen are useful for analysis; the specimen must be "thin" with respect to neutron scattering power as well as to neutron absorption. In practice, the experiments are usually intensity-limited, since even the most powerful reactors or accelerators are sources of weak luminosity when, as here, individual slow neutrons are to be considered as quanta of radiation.

Neutron spectroscopy requires slow neutrons, with energies of the order of neutrons in equilibrium with matter at room temperature, or approximately 0.025 eV. The corresponding de Broglie wavelengths are approximately 0.2 nanometer, of the order of interatomic spacings in solids or liquids. The fast neutrons emitted in nuclear or slow fission reactions can be slowed down to thermal velocities in matter which is transparent to neutrons and which contains light elements, such as hydrogen, carbon, and beryllium, by a process of diffusion and elastic (billiard-ball) scattering known as neutron moderation. By selection of those diffusing neutrons which travel in a certain restricted range of directions (collimation), a beam of thermal and near-thermal neutrons can be obtained. The beam of slow neutrons so produced can be selected as to energy by means of their velocities, by using an electronic or mechanical velocity selector, or by means of their wavelengths, through diffraction from a specially selected and treated single-crystal monochromator. As theoretically expected, the two methods are found to be physically equivalent, but have important differences for the strategy of experiments.

Interpretation. The bulk of the observations can be accounted for in terms of scattering of semiclassical neutron waves by massive, moving-point scatterers in the forms of atomic nuclei and their bound electron clouds. The spatial structure of the scatterers, time-averaged over the duration of the experiment, gives rise to the elastic scattering from the specimen that is studied in neutron diffraction; the spatial motions of the scatterers give rise to the Doppler-shifted inelastic scattering involved in slow neutron spectroscopy.

Quasiparticles. Analysis of the scattering patterns is particularly straightforward when the neutrons can be envisaged as creating or annihilating quantum excitations called quasiparticles in the course of the scattering event. The single-quantum component then includes peaks in the scattering pattern—either sharp peaks, whose energy width reflects the energy resolution of the experiment and instrument, or broadened peaks, whose energy width is interpreted in terms of natural lifetimes for the excitations and of interactions between them. Each peak then yields, through the assumption of conservation of energy and momentum between neutron and quasiparticle, an energy ϵ and a momentum $\mathbf{Q}$ for the (created or annihilated) single quasiparticle involved. Repetition under different conditions permits deduction of the dispersion relation, $\epsilon_q = \epsilon_q(\mathbf{Q})$, of the quasiparticle concerned. Consideration of the intensity of the neutron line allows assignment of the polarization and symmetry type of the excitation concerned—for example, phonon or magnon, longitudinal or transverse, acoustic or optical.

Likewise, through analyses of their widths and shapes, the broadened peaks yield estimates of the lifetimes $\tau_q(\mathbf{Q})$ of the quasiparticle and information on its modes of interaction in the specimen. Inverse lifetimes ($1/\tau$) or frequencies (ϵ/h, where h is Planck's constant) in the range from 10^{10} to 10^{13} Hz are accessible with this method.

Diffusion phenomena. Broadened peaks may occur at zero energy transfer. This quasielastic scattering is usually the result of slow diffusionlike motions of scattering entities through the specimen. What is unusual is that distances involved are of atomic dimensions and times correspondingly short. Thus it is possible to observe self-diffusion in classical liquids on an atomic scale. The phenomena are best considered through the correlation function approach discussed below.

Continuous spectra. For continuous spectra, interpretation is more difficult and indirect. Always evocative in principle (and at times useful in practice) are the Van Hove space-time correlation functions $G(\mathbf{r},t)$. These are natural extensions to space-time of the Patterson pair correlation functions of crystallography and the Zernike-Prins pair distribution functions in liquids. For a monatomic liquid, the self-correlation function $G_s(\mathbf{r},t)$ has a mathematical definition which can be approximately expressed as follows: Given an atom at position $\mathbf{r} = 0$ at time $t = 0$, then $G_s(\mathbf{r},t)$ is the probability that the same atom is at position $\mathbf{r}$ at time t. Similarly, the pair correlation function $G_p(\mathbf{r},t)$ of a monatomic liquid can be approximately defined by the statement: Given an atom at position $\mathbf{r} = 0$ at time $t = 0$, then $G_p(\mathbf{r},t)$ is the number density of other atoms at position $\mathbf{r}$ at time t. In simple cases of continuous spectra, these functions provide a conceptually simple and powerful method of direct analysis, similar to the role played by quasiparticles in the study of line spectra. This is also the case for magnetic inelastic scattering when there is only one species of magnetic atom involved.

When more than one type (j) of atom is involved, the scattering pattern is related to a weighted superposition of correlation functions (G) between pair types (j,j'). In this case, isotopic substitution (for example, deuterium for hydrogen, or ^{60}Ni or ^{62}Ni for natural nickel) can assist in sorting out the contributions. (The weights are proportional to the products of the neutron scattering amplitudes of the pair types.) When the scattering is magnetic, the three spatial components of the vector magnetic moments of the atoms can also be involved through their various self and pairwise correlations.

Applications. Just as (slow) neutron diffraction is the most powerful available scientific tool for study of the magnetic structure of matter on an atomic scale, so slow neutron spectroscopy is the most powerful tool for study of the atomic magnetic and nuclear dynamics of matter in all its phases. The direct nature of the analysis has in some cases added considerable support to the conceptual structure of solid-state and liquid-state physics and thus to the confidence with which the physics is applied. For example, neutron spectroscopy has confirmed the existence of phonons, magnons, and the quasiparticles (rotons) of liquid helium II. Detailed information has been obtained on the lattice vibrations of most of the crystalline elements and numerous simple compounds, on the atomic dynamics of many simple liquids, on the dynamics of liquid helium in different phases, and on the atomic magnetic dynamics of a great variety of ferromagnetic, ferrimagnetic, antiferromagnetic, and modulated magnetic substances.

Bibliography. G. E. Bacon, *Neutron Diffraction*, 3d ed., 1975; A. Larose and J. Vanderwal (eds.), *Scattering of Thermal Neutrons*, 1974; S. W. Lovesy, *Theory of Neutron Scattering From Condensed Matter*, 1984; P. Schofield (ed.), *The Neutron and Its Applications 1982*, 1983; G. L. Squires, *Thermal Neutron Scattering*, 1978.

TIME-OF-FLIGHT SPECTROMETERS
FRANK W. K. FIRK

A general class of instruments in which the speed of a particle is determined directly by measuring the time that it takes to travel a measured distance. By knowing the particle's mass, its energy can be calculated. If the particles are uncharged (for example, neutrons), difficulties arise because standard methods of measurement (such as deflection in electric and magnetic fields) are not possible. The time-of-flight method is a powerful alternative, suitable for both uncharged and charged particles, that involves the measurement of the time t that a particle takes to travel a distance l. If the rest mass of the particle is m_0, its kinetic energy E_T can be calculated from its measured speed, $v = l/t$, using the equation below, where c is the speed of light.

$$E_T = m_0 c^2 \{[1 - (v/c)^2]^{-1/2} - 1\} \approx m_0 v^2/2 \quad \text{if } v \ll c$$

Some idea of the time scales involved in measuring the energies of nuclear particles can be gained by noting that a slow neutron of kinetic energy $E_T = 1$ eV takes 72.3 microseconds to travel 1 m. Its flight time along a 10-m path (typical of those found in practice) is therefore 723 microseconds, whereas a 4-MeV neutron takes only 361.5 nanoseconds.

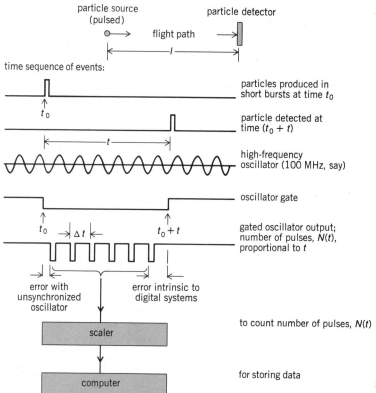

Schematic diagram of a time-of-flight spectrometer.

The time intervals are best measured by counting the number of oscillations of a stable oscillator that occur between the instants that the particle begins and ends its journey (see **illus**.). Oscillators operating at 100 MHz are in common use. If the particles from a pulsed source have different energies, those with the highest energies arrive at the detector first. Digital information from the "gated" oscillator consists of a series of pulses whose number $N(t)$ is proportional to the time-of-flight t. These pulses can be counted and stored in an on/line computer that provides many thousands of sequential "time channels," $t_0, t_0 + \Delta t, t_0 + 2\Delta t, t_0 + 3\Delta t, \ldots$, where t_0 is the time at which the particles are produced and Δt is the period of the oscillator. To store an event in channel $N(t)$, the contents of memory address $N(t)$ are updated by "adding 1."

Time-of-flight spectrometers have been used for energy measurements of uncharged and charged elementary particles, electrons, atoms, and molecules. Their popularity is due to the broad energy range that can be covered, their high resolution ($\Delta E_T/E_T \approx 2\Delta t/t$, where δE_T and Δt are the uncertainties in the energy and time measurements, respectively), their adaptability for studying different kinds of particles, and their relative simplicity. SEE MASS SPECTROSCOPE; NEUTRON SPECTROMETRY.

Bibliography. J. A. Harvey (ed.), *Experimental Neutron Resonance Spectroscopy*, 1970; A. Quayle (ed.), *Advances in Mass Spectrometry*, vol. 8, 1980.

5
MICROWAVE AND RADIO-FREQUENCY SPECTROSCOPY

Microwave spectroscopy	202
Radio-frequency spectroscopy	204
Magnetic resonance	204
Nuclear magnetic resonance (NMR)	209
Nuclear quadrupole resonance	221
Electron paramagnetic resonance (EPR) spectroscopy	222
Cyclotron resonance experiments	226

MICROWAVE SPECTROSCOPY
William Happer

The study of the interaction of matter and electromagnetic radiation in the microwave region of the spectrum. Microwaves are loosely defined as electromagnetic radiation with wavelengths between about 1 mm and 30 cm or frequencies between 1 and 300 GHz. The wavelengths are comparable to the dimensions of experimental apparatus. Experimental techniques make use of ideas from radio-frequency spectroscopy where wavelengths greatly exceed the dimensions of the apparatus, and also techniques from optics where wavelengths are much smaller than the size of the apparatus.

Apparatus. Microwave circuit elements (waveguides, resonant cavities, directional couplers) cannot be characterized by lumped capacitances (electric field regions) or lumped inductances (magnetic field regions), as in the case of radio-frequency circuits. Because of the short wavelengths of microwaves, both electric and magnetic fields are present in most circuit elements. When many wavelengths are present within a microwave circuit element, the momentum of the electromagnetic wave can become well defined, and momentum conservation or phase-matching conditions can have an important bearing on some spectroscopic techniques. An example is the mixing of microwaves and light waves, where the frequency of the mixed wave must equal the sum of the frequencies of the microwave and the light wave, while the momentum of the mixed wave (the inverse wavelength) must equal the vector sum of the momenta of the microwave and the light wave.

The transit times of electrons in ordinary electronic tubes are too long for efficient coupling to the rapidly oscillating fields of the microwave region, and special electronic tubes, klystrons, traveling-wave tubes, magnetrons, and so forth have been designed to overcome transit time limitations. Such tubes are often locked in frequency to a high harmonic of a stable quartz crystal oscillator. Microwave receivers are extraordinarily sensitive, and 10^{-19} W of microwave power can be detected with a good heterodyne system in a 1-Hz bandwidth.

Interaction of microwaves with matter. The interaction of microwaves with matter can be detected by observing the attenuation or phase shift of a microwave field as it passes through matter. These are determined by the imaginary or real parts of the microwave susceptibility (the index of refraction). The absorption of microwaves may also trigger a much more easily observed event like the emission of an optical photon in an optical double-resonance experiment or the deflection of a radioactive atom in an atomic beam. *See* M*olecular beams*.

Microwave energy at a frequency ν is absorbed according to the Bohr frequency condition, Eq. (1), where h is Planck's constant and E_f and E_i are the energies of the final and initial states

$$h\nu = E_f - E_i \qquad (1)$$

of the absorbing system. The initial and final states may be discrete as in the case of rotational states of a molecule, or they may be continuous as in bremsstrahlung in a plasma.

Enhancement of population differences. The characteristic temperature θ of the energy splitting involved in a microwave transition is given by Eq. (2), where k is Boltzmann's

$$\theta = \frac{h\nu}{k} = 0.05 \text{ K} - 15 \text{ K} \qquad (2)$$

constant. At room temperature T, the relative population difference θ/T between the states involved in the transition is a few percent or less. The population difference can be close to 100% at liquid helium temperatures, and microwave spectroscopic experiments are often performed at low temperatures to enhance population differences and to eliminate certain line-broadening mechanisms.

The population differences between the states involved in a microwave transition can be enhanced by artificial means. For example, state selection of atoms or molecules in an atomic beam passing through inhomogeneous magnetic or electric fields can lead to very large population imbalances. Such large population inversions can be prepared so that stimulated emission cross sections for microwaves can exceed the absorption cross section. When the molecules or atoms with inverted populations are placed in an appropriate microwave cavity, the cavity will oscillate spontaneously as a maser (microwave amplification by stimulated emission of radiation). Optical pumping can also lead to large artificial population imbalances between the states involved in a microwave transition.

Applications. The magnetic dipole and electric quadrupole interactions between the nuclei and electrons in atoms and molecules can lead to energy splittings in the microwave region of the spectrum. Thus, microwave spectroscopy has been used extensively for precision determinations of spins and moments of nuclei. The 21-cm hyperfine transition of the hydrogen atom is seen from various parts of the Galaxy, and the field-independent component of the 21-cm line is used in the atomic hydrogen maser as a frequency standard. The field-independent component of the 9.192-GHz hyperfine transition in the cesium atom is used as a time standard in many laboratories. S*EE* H*YPERFINE STRUCTURE.*

Properties of molecules. The rotational frequencies of molecules often fall within the microwave range, and microwave spectroscopy has contributed a great deal of information about the moments of inertia, the spin-rotation coupling mechanisms, and other physical properties of rotating molecules. The rotational frequencies of water and oxygen molecules are responsible for much of the attenuation of microwaves in the atmosphere. Many microwave transitions in molecules are seen from sources in interstellar space, and microwave astronomy has provided much information about the chemistry and molecular composition of various astronomical objects. Inversion frequencies in molecules, for example, the frequency of periodic motion of the nitrogen atom at the apex of the pyramidal NH_3 ammonia molecule through the three hydrogen atoms at the base, often lie in the microwave region. One of the inversion transitions in ammonia was used in the first maser. S*EE* M*OLECULAR STRUCTURE AND SPECTRA.*

Electron-spin resonance. The magnetic resonance frequencies of electrons in fields of a few thousand gauss (a few tenths of a tesla) lie in the microwave region. Thus, microwave spectroscopy is used in the study of electron-spin resonance or paramagnetic resonance. In the simplest cases, the spins may be well isolated in a dilute gas like oxygen (O_2) or cesium vapor. Then the microwave spectrum provides information about the internal spin couplings of a free atom or molecule. For denser gases or condensed phases, the paramagnetic resonance spectrum provides information about interactions between the spins and their environment. Both static interactions, which determine the resonance frequencies of the microwave transitions, and fluctuating random interactions, which determine resonance line widths, are of interest. Typical static interactions are the electrostatic interactions between the spin and the crystal field. Magnetic dipole interactions and spin exchange interactions between spins at neighboring sites can lead to broadening or narrowing of the resonance lines. Particularly strong interactions between neighboring spins occur in ferromagnetic or antiferromagnetic materials, both of which can be investigated by microwave spectroscopy. S*EE* E*LECTRON PARAMAGNETIC RESONANCE (EPR) SPECTROSCOPY;* M*AGNETIC RESONANCE.*

Cyclotron resonance. The cyclotron resonance frequencies of electrons in solids at magnetic fields of a few thousand gauss (a few tenths of a tesla) lie within the microwave region of the spectrum. The effective mass and the cyclotron frequency of an electron in a solid are greatly modified from those of the free electron by the crystal interactions. Microwave spectroscopy has been used to map out the dependence of the effective mass on the electron momentum. S*EE* C*YCLOTRON RESONANCE EXPERIMENTS.*

Plasmas. Gaseous plasmas are strong emitters and absorbers of microwaves. The main coupling mechanisms of charged-particle motion to microwave radiation are bremsstrahlung and cyclotron radiation. Thomson scattering is also important in dense plasmas. Microwaves are completely reflected from plasmas where the electron density n is so high that the plasma resonance frequency f given by Eq. (3) exceeds the microwave frequency. Electron densities and tempera-

$$f = 8980^{1/2} \; Hz \cdot cm^{-3/2} \tag{3}$$

tures in plasmas can be determined by microwave spectroscopy.

Cosmic microwave radiation. Thermal microwaves at a temperature of about 3 K permeate all space and can be detected by sensitive microwave receivers. These microwaves are believed to be the cooled radiation left over from the big bang at the beginning of the universe.

Bibliography. G. W. Chantry, *Modern Aspects of Microwave Spectroscopy*, 1980; M. A. Heald and C. B. Wharton, *Plasma Diagnostics with Microwaves*, 1965, reprint 1978; C. Kittel, *Introduction to Solid State Physics*, 5th ed., 1976; D. G. Lominadze, *Cyclotron Waves in Plasma*, 1981; C. P. Poole, Jr., *Electron Spin Resonance*, 2d ed., 1982; T. L. Squires, *Introduction to Microwave Spectroscopy*, 1963; T. M. Sugden and C. N. Kenney, *Microwave Spectroscopy of Gases*, 1965; C. H. Townes and A. L. Schawlow, *Microwave Spectroscopy*, 1955, reprint 1975.

RADIO-FREQUENCY SPECTROSCOPY
POLYKARP KUSCH

The branch of spectroscopy concerned with the measurement of the intervals between atomic or molecular energy levels that are separated by frequencies from about 100 kHz to 1000 MHz (10^5–10^9 s^{-1}), as compared to the frequencies that separate optical energy levels of about 6×10^{14} s^{-1}. The importance of radio-frequency spectroscopy lies in the fact that certain specific properties of the nucleus, such as spin, magnetic dipole moment, and electric quadrupole moment, play a relatively major role in determining the intervals between closely lying energy levels; the results of this branch of spectroscopy have been of great importance in determining nuclear properties. For a discussion of radio-frequency spectroscopy by the molecular-beam magnetic-resonance method SEE MOLECULAR BEAMS. SEE ALSO MAGNETIC RESONANCE; MICROWAVE SPECTROSCOPY.

MAGNETIC RESONANCE
CHARLES P. SLICHTER

A phenomenon exhibited by the magnetic spin systems of certain atoms whereby the spin systems absorb energy at specific (resonant) frequencies when subjected to alternating magnetic fields. The magnetic fields must alternate in synchronism with natural frequencies of the magnetic system. In most cases the natural frequency is that of precession of the bulk magnetic moment **M** of constituent atoms or nuclei about some magnetic field **H**. Because the natural frequencies are highly specific as to their origin (nuclear magnetism, electron spin magnetism, and so on), the resonant method makes possible the selective study of particular features of interest. For example, it is possible to study weak nuclear magnetism unmasked by the much larger electronic paramagnetism or diamagnetism which usually accompanies it.

Nuclear magnetic resonance (that is, resonance exhibited by nuclei) reveals not only the presence of a nucleus such as hydrogen, which possesses a magnetic moment, but also its interaction with nearby nuclei. It has therefore become a most powerful method of determining molecular structure. The detection of resonance displayed by unpaired electrons, called electron paramagnetic resonance, is also an important application. These two phenomena, as well as other related resonance phenomena, are discussed in this article.

Origin. Because **M** has its origin in circulating currents or intrinsic spins, there is always an angular momentum **J** associated with it. The vector quantities **M** and **J** are related by Eq. (1), where γ is called the gyromagnetic ratio.

$$\mathbf{M} = \gamma \mathbf{J} \tag{1}$$

For a system to exhibit a magnetic resonance it must possess a magnetic moment, possess angular momentum, and experience torques. The magnetic moment may arise from nuclei of atoms, from orbital electronic motion, from electronic spins, or from moving nuclear charges during molecular rotation. The angular momentum arises from the same sources. The torques may arise from externally applied magnetic fields; from magnetic dipole fields exerted by neighboring nuclei, atoms, or molecules; from electric fields acting on, for example, a nuclear electric quadrupole moment or the nonspherical electron cloud of an atom; or from electron exchange coupling. In any given case it is necessary to decide which interactions are large and which are small. Thus for a paramagnetic atom possessing a nuclear magnetic moment a distinction might be made between small applied static fields, in which the coupled nucleus and electron angular momenta act as a unit, and large magnetic fields, which decouple them so that they act independently. An effective angular momentum and magnetic moment can often be defined, giving an effective γ (in analogy to the Landé g factor of optical spectroscopy).

Several types of resonances have been observed; these differ in one or more of the three basic requirements. However, the principal features can be understood by assuming that the torques arise from an effective static applied magnetic field **H**. The torque $\mathbf{M} \times \mathbf{H}$ causes the angular momentum to change with time according to Eq. (2). The resultant motion of **M** is a

$$\frac{d\mathbf{J}}{dt} = \mathbf{M} \times \mathbf{H} \text{ or } \frac{d\mathbf{M}}{dt} = \gamma \mathbf{M} \times \mathbf{H} \qquad (2)$$

precession at angular frequency γH about the direction of **H**. At thermal equilibrium **M** is parallel to **H**, and no precession occurs. Application of an alternating magnetic field $H_x \cos \omega t$ perpendicular to **H** causes **M** to tilt away from **H** with a consequent absorption of energy, provided the resonant condition $\omega = \gamma H$ is satisfied. In practice the absorption takes place over a narrow range of frequency on both sides of γH. The magnetization M_x parallel to H_x obeys Eq. (3), where $\chi'(\omega)$

$$M_x = H_x \chi'(\omega) \cos \omega t + H_x \chi''(\omega) \sin \omega t \qquad (3)$$

and $\chi''(\omega)$ are the real and imaginary parts of the complex magnetic susceptibility $\chi = \chi'(\omega) - j\chi''(\omega)$ $(j = \sqrt{-1})$, and characterize dispersion and absorption respectively. For a typical resonance, $\chi''(\omega)$ attains maximum value for a region of frequencies near $\omega = \gamma H$.

For many cases it is necessary to analyze magnetic resonance by quantum theory. Consider a system composed of many (weakly interacting) identical parts (atoms or nuclei), each with angular momentum quantum number F (total angular momentum $= \sqrt{F(F + 1)}\hbar$, where $\hbar$ is Planck's constant h divided by 2π). The spatial quantization in the magnetic field H gives $2F + 1$ equally spaced energy levels labeled by the quantum number $M_F = (F, F - 1, \ldots, -F)$ and energy spacing between adjacent levels of $\gamma \hbar H$. The field $H_x \cos \omega t$ produces transitions with the selection rule $\Delta M_F = \pm 1$. To satisfy the Bohr frequency condition, $\hbar \omega = \gamma \hbar H$, in agreement with the classical result, $\omega = \gamma H$.

According to quantum theory, the probability of transition from any energy level A to any other B is the same as that from B to A; thus a net absorption of energy requires that the population of the lower energy states be greater than that of the upper. The reverse situation in which the upper states are more populated leads to an induced emission and is the basis for the operation of the solid-state maser. At thermal equilibrium, as a result of spin-lattice relaxation, the lower energy states are more heavily populated in accordance with the classical Maxwell-Boltzmann statistics (ordinarily it is unnecessary to use either Fermi-Dirac or Bose-Einstein statistics). When H_x becomes sufficiently large, the level populations become disturbed from thermal equilibrium. The population difference between states joined by H_x decreases, a phenomenon known as saturation, because it causes χ' and χ'' to diminish with increasing H_x. Population differences between pairs of states other than A and B may simultaneously be increased, or even inverted, as in the three-level maser. The intensity of the alternating field necessary to produce saturation depends on the width of the absorption line and on the spin-lattice relaxation time (wider lines or shorter times require larger H_x).

Observation. Experimentally it is possible to detect magnetic resonance by measuring the absorption of magnetic energy of a circuit containing the magnetic material or by measuring the change in inductance or resonant frequency of the circuit. The two methods measure $\chi''(\omega)$ or $\chi'(\omega)$ respectively. The resonant condition $\omega = \gamma H$ may be produced by varying ω, or more customarily, by changing H. In some experiments one tilts **M** away from the direction of **H** by alternating fields of short duration and then observes voltages induced by the subsequent free precession of **M**. This method is particularly useful for studying relaxation times.

Nuclear magnetic resonance (NMR). The nuclei of many atoms possess angular momentum (spin) and nonvanishing magnetic moments. The former may be characterized by an angular momentum quantum number I (integer or half integer) of the nuclear particles. As far as is known, stable nuclei with an even number of neutrons and even number of protons have zero spin and magnetic moment, hence are incapable of exhibiting magnetic resonance. *SEE NUCLEAR MOMENTS.*

Nuclear resonances have been observed in insulators, metals, paramagnetic salts, antiferromagnetic substances, and other solids, and in gases and liquids. Often, to observe NMR a sample is placed between the poles of an electromagnet (**Fig. 1**) which in addition to the main winding carries a small auxiliary winding or sweep. A coil connected to an oscillator surrounds the sample, as does a second coil at right angles to both the oscillator coil and the sweep winding, to avoid direct coupling. The oscillator frequency is fixed and the sweep circuit is used to vary the magnetic field strength continuously. When a resonance frequency of the sample is reached, a signal induced in the second coil is detected and amplified. Typical resonance frequencies in a field of 10,000 gauss lie in the radio-frequency region (1–45 MHz). For example, the ^{1}H nucleus shows a resonance frequency of 42.6 MHz at this field strength. ^{13}C nuclei give a much weaker signal, further decreased in a sample containing this isotope in its natural abundance of 1.1%.

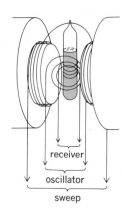

Fig. 1. Arrangement of a sample and coils in a nuclear induction apparatus. (*After J. D. Roberts, Nuclear Magnetic Resonance, McGraw-Hill, 1959*)

Nuclei with quadrupole moments. If a nucleus has a spin ≥ 1, it generally has a nonvanishing electric quadrupole moment (there are good grounds for believing that all nuclei have zero electric dipole moments). The electrical interaction between the nucleus and electric potentials $V(x,y,z)$ from other charges depends on the nuclear orientation (specified by the direction of nuclear spin) and on the spatial second derivatives of the potential $\partial^2 V/\partial x^2$, $\partial^2 V/\partial x\, \partial y$, and so on, at the position of the nucleus. For potentials of spherical, tetrahedral, or cubic symmetry, the interaction energy is independent of orientation and may be disregarded. When the electric quadrupole interaction is nonzero but nevertheless much weaker than the static magnetic interaction, the unique resonance condition $\omega = \gamma H$ is changed, and the resonance line splits into $2I$ components centered around γH. A convenient method of determining the nuclear spin is thereby provided.

The name pure quadrupole resonance is used when one dispenses with the static field H and observes the reorientation of the nucleus among its various quantized orientations with respect to the electric potential alone. Classically, the nonspherical nuclear charge experiences a torque which causes precession.

Applications. Nuclear magnetic resonance has been used widely to measure nuclear magnetic moments, electric quadrupole moments, and spins. Because the resonance lines may be on occasion very sharp (1 cycle wide at 40 MHz), nuclear resonance is frequently employed to measure magnetic fields with great precision.

The extensive use of NMR in molecular structure determinations arises from the slight shift in the resonance frequency of an atom—commonly that of a proton—due to the environment of neighboring atoms. Because the magnitude of this shift depends on the type of environment, it is called the chemical shift. **Figure 2** shows the NMR spectrum of ethyl alcohol, CH_3CH_2OH. The three main resonance frequencies are due to protons in the OH, CH_2, and CH_3 groups, respectively, and the spacing between them (which varies with the field strength) shows the chemical shift characteristic of the protons in these three typical structural groups.

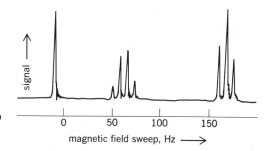

Fig. 2. Proton resonance spectra of ethyl alcohol at 40 MHz. (*After J. D. Roberts, Nuclear Magnetic Resonance, McGraw-Hill, 1959*)

The separate peaks at each frequency are due to spin-spin splitting. Often, the presence of n protons will split the frequency of a given, structurally different proton into $n + 1$ peaks, in direct analogy to ordinary spectral lines. The triplet in the NMR spectrum of ethyl alcohol results from the splitting of the frequency of the CH_3 protons by the two adjacent CH_2 protons; the quadruplet is at the typical CH_2 frequency and is split into four peaks by the three protons of the CH_3 group. SEE MOLECULAR STRUCTURE AND SPECTRA.

Because the time required for nuclear transitions is relatively long, substances undergoing fast reactions show altered NMR spectra, and rates of such rapid processes as ionization or intramolecular rotation can be measured. SEE NUCLEAR MAGNETIC RESONANCE (NMR).

Electron-nuclear double resonance (ENDOR). In this technique the magnetic resonance of a nucleus is detected by observing that of a nearby electron. The magnetic coupling between the nuclear and electronic magnetic moments gives rise to a back reaction on the electron resonance when the nuclei are brought into resonance. The sample under study is placed in a conventional electron resonance apparatus. In addition, an oscillator drives a coil to produce alternating magnetic fields at the sample under study, the frequency of alternation being in an appropriate range to produce nuclear transitions. Typically, the static magnetic field is adjusted to produce electron resonance. With the static magnetic field and the frequency of the electron resonance apparatus held fixed, the frequency of the nuclear resonance oscillator is then swept. As it passes through the resonant frequency of any nucleus that is coupled to the electron, a change in the electron absorption occurs.

The electron resonance and the nuclear resonance both represent transitions between energy levels of the combined system of electron and nucleus, each resonance being between a pair of levels. The strength of the electron resonance depends on the difference in population between its two levels. Thus, if one of these levels takes part in both resonances, the nuclear resonance may influence the electron resonance by changing the population of the common energy level.

Since the individual quanta absorbed in an electron resonance are much larger in energy than those absorbed in a nuclear resonance, the double resonance often permits the detection of the resonant absorption of a smaller number of nuclei than it would be possible to detect by a direct observation of the nuclear resonance. Since ENDOR requires both electron and nuclear resonances, it is applicable only to systems possessing both. It has been used, for example, to study the nuclear resonance associated with impurity atoms in semiconductors and that associated with F-centers in alkali halides, as well as to measure nuclear moments of rare elements. Important results have included the determination of the electronic structure of point imperfections and the measurement of nuclear magnetic moments and hyperfine anomalies.

Paramagnetic resonance. Magnetic resonance arising from electrons in paramagnetic substances or from electrons in paramagnetic centers in diamagnetic substances is called paramagnetic resonance. For applied fields of several thousand gauss the electron paramagnetic resonance (EPR) experiments are done at microwave frequencies, commonly at 3-cm (1.2-in.) or at 1-cm (0.4 in.) wavelengths. In some instances, nuclear resonance apparatus has been used with correspondingly lower applied fields. The most sensitive apparatus detects approximately 10^{12} electron spins for a line 1 gauss broad, a sensitivity far greater than that obtained by nonresonant methods, such as those utilizing paramagnetic susceptibility.

Resonances have been observed in atoms of the iron group, rare earths, and other transition elements; in paramagnetic gases; in organic free radicals; in color centers in crystals (such as F- and V-centers); in metals (conduction electron spin); and in semiconductors (both conduction electron and impurity center spins).

When two paramagnetic ions or molecules approach one another, the spins become coupled via the exchange interaction. Much weaker couplings than the usual exchange interactions within atoms, in chemical bonds, or in ferromagnets produce pronounced effects. For additional information on this phenomenon and its applications SEE ELECTRON PARAMAGNETIC RESONANCE (EPR) SPECTROSCOPY.

Ferromagnetic resonance. In the case of both nuclear and paramagnetic resonance, the spins of neighboring atoms are nearly randomly oriented with respect to one another. In contrast, the electron spins in one domain of a ferromagnet are nearly all parallel for temperatures sufficiently below the Curie point. The alignment may be described in terms of the exchange coupling between neighboring spins, or equivalently in terms of the Weiss molecular magnetic field $\mathbf{H}_w$.

It is simplest to consider the case where magnetization is uniform throughout the sample. Neglecting relaxation effects, the equation of motion is still Eq. (2), but an effective field is substituted for $\mathbf{H}$, consisting of the applied static and alternating fields, the demagnetizing corrections (from the electron magnetic dipolar fields), and the effects of crystalline anisotropy. Because the Weiss molecular field is always parallel to $\mathbf{M}$, $\mathbf{H}_w$ exerts no torque and plays no role as long as the magnetization is uniform throughout the sample. (The exchange energy between spins does not change as long as their relative orientation does not change.)

The crystalline anisotropy can be shown to be equivalent to a magnetic field $\mathbf{H}_A$ along the direction of easy magnetization as long as $\mathbf{M}$ points nearly in that direction.

Because of the demagnetizing effects, the resonant frequency depends on sample geometry. For an infinite plane perpendicular to the applied field, the resonant angular frequency ω is given by $\omega = \gamma\sqrt{BH}$, but for a sphere, $\omega = \gamma H$.

The large demagnetizing and exchange fields are the principal difference between ferromagnetic and paramagnetic resonance. The demagnetizing field has components $-N_x M_x$, $-N_y M_y$, and $-N_z M_z$, where N_x, N_y, and N_z are the demagnetizing coefficients. Thus, suppose the static field and $\mathbf{M}$ lie along the z direction. Application of the alternating field tilts $\mathbf{M}$ away from z, changing the effective z field. If the change brings the spins closer to resonance, $\mathbf{M}$ may tilt out more. It is possible for such a nonlinear effect to be unstable for sufficiently large alternating fields. This instability is utilized in the Suhl ferromagnetic amplifier.

Antiferromagnetic resonance. The two sublattices of spins in an antiferromagnet are strongly coupled together by exchange forces. If both magnetizations ($\mathbf{M}_1$ and $\mathbf{M}_2$) are tilted together, away from the normal direction of magnetization in the crystal (call this the z direction), the only change in energy results from the anisotropy fields. However, because the anisotropy fields are reversed in direction for the two lattices, the magnetizations $\mathbf{M}_1$ and $\mathbf{M}_2$ tend to precess in opposite directions, bringing about a change in the exchange energy. An external field along the z direction aids one anisotropy field but opposes the other. The resonant angular frequency ω for a sphere is given by Eq. (4), where H is the applied field, H_A the equivalent anisotropy field,

$$\omega = \gamma(H \pm \sqrt{H_A(H_A + 2H_E)}) \quad (4)$$

and H_E the equivalent exchange field. The plus and minus signs refer to two opposite directions of rotating magnetic fields which may be used to observe the resonance. If H_E is 10^6 oersted, and H_A is 10^4 oersted, the corresponding frequency is 3×10^{11} Hz.

Ferrimagnetic resonance. Magnetic resonance in ferrites is called ferrimagnetic resonance. Ferrites are the natural generalization of antiferromagnets, containing two or more sublattices which may differ in magnetization. The basic coupling terms are still anisotropy fields, exchange fields, and the applied fields. The resonant angular frequency ω for the case of two sublattices is given by Eq. (5), where η is a parameter measuring the relative sizes of the two

$$\omega = \gamma\left[H - \frac{\eta H_E}{2} \pm \sqrt{\left(\frac{\eta H_E}{2}\right)^2 + H_E H_A(2 - \eta) + H_A^2}\right] \quad (5)$$

magnetization vectors. Taking $\mathbf{M}_1$ to be the smaller magnetization, η is defined by Eq. (6). This

$$\mathbf{M}_1 = (1 - \eta) \mathbf{M}_2 \qquad (6)$$

equation assumes the magnetizations to be at saturation and the two sublattices to have the same γ (deviations might differ from spin-orbit coupling). SEE MOLECULAR BEAMS.

Bibliography. Atta-ur-Rahman, *Nuclear Magnetic Resonance: Basic Principles*, 1986; A. Carington and A. D. McLachlan, *Introduction to Magnetic Resonance with Applications to Chemistry and Chemical Physics*, 1979; C. Dybowski and R. Litcher, *Nuclear Magnetic Resonance*, 1987; E. Kundla et al., *Magnetic Resonance and Related Phenomena*, 1980; C. P. Poole and H. A. Farach, *Theory of Magnetic Resonance*, 2d ed., 1986; R. T. Schumacher, *Introduction to Magnetic Resonance*, 1970; F. Seitz and D. Turnbull (eds.), *Solid State Physics*, vol. 2, 1956; C. P. Slichter, *Principles of Magnetic Resonance*, 2d ed., 1980.

NUCLEAR MAGNETIC RESONANCE (NMR)
GEORGE SLOMP

A phenomenon exhibited by a wide variety of atoms which is based upon the existence of magnetic forces (moments) attributed to quantized nuclear spins. These nuclear moments, when placed in a magnetic field, give rise to distinct nuclear Zeeman energy levels between which spectroscopic transitions can be induced by radio-frequency radiation. When the applied radio-frequency matches the required transition frequency for these nuclei, the interaction which is detected as resonance results. Plots of these transition frequencies, called spectra, furnish important information about molecular structure and sample composition. The method is nondestructive and can be applied to chemical substances of living systems. When tuned to the protons of water, images of the human anatomy, much like x-ray images, can be made without any radiation hazard.

Spectroscopy. Spinning nuclei with nonzero angular momentum behave like tiny magnets. When substances are placed in the magnetic field of a large magnet, these tiny nuclear magnets present in the sample assume certain allowed orientations (Zeeman energy levels) with respect to that external magnetic field. The nuclei can be reoriented only by adding energy of the exact transition frequency. This transition frequency is unique and varies widely for different kinds of atoms. A plot of all the frequencies absorbed by a sample is called a spectrum. SEE MAGNETIC RESONANCE; MOLECULAR STRUCTURE AND SPECTRA.

Nuclei excluded from consideration are those with zero angular momentum or spin ($I = 0$) and therefore zero magnetic moment (for example, the important ^{12}C and ^{16}O isotopes). High-resolution methods generally exclude nuclei with I greater than $\frac{1}{2}$, as they possess electrical quadrupole moments which interact with electric field gradients so as to broaden the magnetic resonance signals and prevent resolution of closely spaced resonance lines. High-resolution techniques, therefore, have been limited primarily to the nuclear species of spin $\frac{1}{2}$ (for example, ^1H, ^{13}C, ^{19}F, and ^{31}P). SEE NUCLEAR MOMENTS.

As the separation between the nuclear Zeeman levels is directly proportional to the strength of the perturbing magnetic field, the transition frequency can be varied for a given nucleus by merely changing the applied magnetic field. In this regard, nuclear magnetic resonance spectroscopy is unlike other spectroscopic methods, in which the investigator is unable to control the frequency of the spectral transition. Thus an NMR spectrum may be secured by varying the magnetic field to bring the separation of the Zeeman levels into correspondence with a constant irradiating frequency; or the alternative experimental method may be used, in which a constant magnetic field is employed and the irradiating frequency is varied over the range of spectroscopic frequencies. SEE ZEEMAN EFFECT.

A spectrum is interpreted by assigning all the resonance signals to the different nuclei in the molecular structure. The assignment is considered reasonable when all the lines have the appropriate multiplicity and the chemical shifts are verified by related examples.

Chemical shift. It is this unique field-frequency relationship that makes possible the application of NMR spectroscopy to molecular structure studies. Organic molecules usually contain many of the same kinds of atoms. Although identical nuclei have the same frequency dependence upon the magnetic field, a difference in the chemical environment can modify the applied magnetic field, so that all such nuclei in the same sample do not experience the same local magnetic field, and thus they do not all resonate at exactly the same frequency. Each nucleus in the sample is, in effect, telegraphing out information about its environment from which molecular structure information can be deduced. The small spectral shift in the transition frequencies between two such chemically nonequivalent nuclei is referred to as the chemical shift. Being small and directly proportional to the total applied field, this parameter is recorded in the relative units of parts per million (ppm) relative to the resonance frequency of a standard substance (**Fig. 1**).

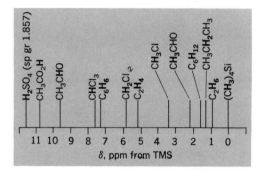

Fig. 1. Chemical shifts portrayed schematically for several representative compounds. Decreasing values of δ correspond to increasing magnetic field in a constant-frequency spectrometer. The scale calibration is obtained from the resonance signal of a small amount of tetramethylsilane (TMS) placed in the sample tube to provide a zero reference point.

It is convenient to subdivide the chemical shift parameter into a diamagnetic term and a paramagnetic term. Diamagnetism induced by an applied magnetic field is a well-known phenomenon and is attributed to currents in the molecular electrons that were first described by W. E. Lamb, Jr. Diamagnetic shielding decreases the field intensity at the nucleus and thereby decreases the separation between the nuclear Zeeman levels. A higher applied field or a lower frequency is now required to attain resonance, and the signal moves to the right (upfield) in the spectrum. Considered simply, this part of the chemical shift is proportional to the electron density in that segment of a molecule in which the magnetic nucleus is found, and therefore reflects in an approximate manner the charge polarization of the molecular electrons. A paramagnetic shift (downfield), attributed to increased magnetic field intensity at the nucleus, is observed insome cases as a result of diamagnetic currents existing in remote anisotropic groups of a molecule. Aromatic systems with their associated ring currents constitute typical examples of such anisotropic groups which enhance the magnetic field in certain regions of space external to the aromatic ring. Finally, in molecules of certain symmetries the magnetic field can remove the quenching of orbital angular momentum associated with electrons involving p orbitals in completed subshells. There is evidence that this paramagnetic interaction may be significant in ^{13}C and ^{19}F magnetic resonance studies. However, theoretical estimates of the magnitude of the several terms in the chemical shift parameter involve considerable difficulty, and the relative importance of the various shielding mechanisms is not completely resolved.

Because nuclei contained in various functional groups have their own characteristic resonance frequencies. NMR has become an important spectroscopic tool for molecular structure determination. Many tables of chemical shifts are available for identification purposes. Figure 1 schematically portrays the distribution of proton chemical shift values for a few selected compounds. Low diamagnetic shielding is observed for the electropositive protons in the two acid compounds. The chemical shifts of less acidic methyl groups are found at high fields. The shift to lower fields with the addition of an electronegative group is exhibited by the series CH_3Cl, CH_2Cl_2, and $CHCl_3$. Finally, the relatively low field position of the benzene resonance is explained as noted before by a paramagnetic shift resulting from π-electron ring currents. Thus the chemical shifts tell what functional groups may be present in the sample.

Spin coupling. The NMR spectrum of ethyl bromide portrayed in **Fig. 2**a is presented as an example of a moderately high-resolution spectrum in which the resonance peaks of the chemically nonequivalent methylene (CH$_2$) and methyl (CH$_3$) protons are separated by a chemical shift of 1.77 ppm. The relative intensities in these two peaks of 2 and 3 reflect the number of hydrogens in the methylene and methyl groups, respectively.

With additional improvement in resolution, each of the ethyl bromide peaks subdivides into the multiplet structure shown in Fig. 2b. The methylene resonance is observed to split into a quartet of lines, whereas the methyl peak is replaced by a triplet of lines. Resulting from a nuclear spin-spin interaction between the two sets of protons, the multiplet pattern can be rationalized on the basis of the allowed orientation of the methylene and methyl protons as shown schematically in the figure. Thus, the magnetic field experienced by the methylene protons is perturbed by the four different distinguishable spin orientations exhibited by methyl protons.

Furthermore, the 1:3:3:1 statistical weights for these orientations are reflected in the intensity pattern of the methylene multiplet. In a like manner the two protons in the methylene group induce a 1:2:1 triplet in the methyl peak. Normally the coupling constant, which measures the multiplet splittings due to spin-spin interactions, attenuates rapidly for protons separated by more than two or three chemical bonds, and only neighboring protons interact significantly. Thus, in addition to the identification of functional groups from the chemical shifts, the multiplicity in the splitting patterns and the magnitude of the coupling constant tells how the groups are arranged in the molecule.

It is not always possible to interpret spectra in the manner indicated by Fig. 2, where the splitting patterns can be explained on the basis of a first-order perturbation of one spin system by a second, neighboring group. Specifically, whenever the spin-spin coupling constant becomes comparable to or larger than the chemical shift parameter, higher-order mixing of the spin states occurs to give spectra of considerably greater complexity. As an example, the spectrum of 1-bromo-2-chloroethane is given in **Fig. 3**. This spectrum is derived from a molecule differing only

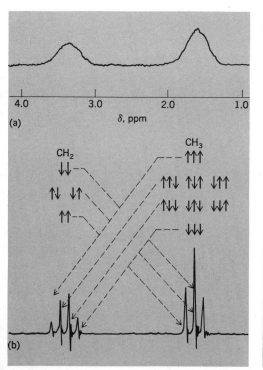

Fig. 2. Nuclear magnetic spectra of ethyl bromide (CH$_3$CH$_2$Br), with schematic representation of nuclear spin orientations. (a) At a moderate resolution. (b) At a high resolution.

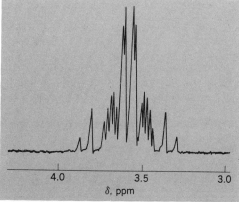

Fig. 3. High-resolution spectrum of 1,2-bromochloroethane (ClCH$_2$CH$_2$Br), exhibiting higher-order split.ings and complexities that are due to magnetic nonequivalence.

slightly from that considered in Fig. 2: yet the spectral features do not resemble the simple pattern shown in Fig. 2b. The similarity between the bromine and chlorine atoms results in chemical similarity between the two methylene groups, and the chemical shift between these two sets of protons is reduced to a value comparable with the intramolecular spin-spin coupling constants. Higher-order splitting features are often observed in proton NMR spectra, and exact interpretation usually requires detailed numerical analysis with a computer. Further complexity is introduced into spectral features whenever the spin-spin coupling values between the two sets of chemically equivalent nuclei are unequal. This element of complexity, which is referred to as magnetic nonequivalence, is found in Fig. 3, where the inequalities in the coupling constants between protons in the two methylene groups are not eliminated by averaging over the several rotameric conformations existing for this molecule. Were the two methylene groups in a 1,2-disubstituted ethane to have the same chemical shift (either by coincidence or from molecular symmetry in the event that both substituents are identical), then all splittings would vanish and a single resonance line would be observed. Spin-spin interactions between nuclei which are both chemically and magnetically equivalent do not affect the spectral features, and coupling constants for such interactions therefore become unobtainable.

Theoretical interpretation of coupling constants indicates that nuclear spin-spin interactions are transmitted through the molecular electrons. Direct magnetic interactions between nuclei through space are observed in solids to be relatively large, but these coupling terms average to zero in the liquid state under the influence of rapid molecular tumbling. As a result of the quantized orientation of magnetic moments associated with the spin and the orbital angular momentum of electrons, magnetic coupling mechanisms involving the molecular electrons do not average to zero with rapid molecular reorientation. Thus, spin-spin coupling values contribute to a better understanding of the electronic structure of molecules, especially in the areas of electron spin correlation and valence theory.

Spin decoupling. Spectra with higher-order splittings and overlapping multiplets are often simplified by a technique known as spin decoupling. In this procedure the coupling is removed by irradiating the coupled partner with a high-intensity radio-frequency field adjusted to its exact resonance frequency. This changes the polarization of the perturbing nuclear spin system and wipes out the coupling.

The resulting simplifications provide chemical shift data from spectra which are more easily interpreted. Furthermore, information derived with this technique can also be used in obtaining the relative signs of spin-spin coupling constants, which can assume either positive or negative values.

The resonant frequency of the perturbing nucleus can be found by the reverse process, in which the irradiating frequency is swept while the perturbed nucleus is under continuous observation.

Nuclear Overhauser effect. The intensities of some resonance signals may be changed by the nuclear Overhauser effect (NOE). When one group of nuclei is irradiated, not only are the signals from the coupled nuclei simplified (decoupled) but other nuclei, nearby in space, are also perturbed, causing their signals to change in intensity (Overhauser enhancement). Since the magnitude of this enhancement decreases with the internuclear distance ($1/d^6$), this technique is a powerful test for proposed molecular structures.

Lanthanide shift reagents. Another development aids in spectrum interpretation. Certain organolanthanide reagents can be added to the sample to expand and untangle otherwise overlapped NMR spectra. The phenomenon is understood as a pseudocontact shift, which means that the lanthanide reagent coordinates with an electron-rich site of the sample molecule, for example, an oxygen atom, and the deshielding effect of the lanthanide ion magnetic dipole, like a beacon, is aimed at this atom. Other nuclei in the sample molecule, for example, protons or carbons, which are observable by the NMR method will therefore be deshielded (shifted downfield in the spectrum) by an amount depending on their position (distance and angle) relative to that beacon. It is possible to quantitate these shifts, and computer programs have been written to fit molecular structures to these measured shifts. N. S. Angerman and coworkers used this method to determine the three-dimensional structure of the important antimalarial chloroquine in acetone solution, to be compared with the molecular structure in the crystalline state as determined by x-ray crystallography.

Quantitative analysis. The technique of quantitative analysis by NMR is based on the fact that the area under each peak in an NMR spectrum is directly proportional to the number of atomic nuclei causing the absorption peak. The method is applicable to many kinds of atoms. Once the peaks in the spectrum have been assigned to the atoms in the sample, various types of quantitative analyses are possible, including analysis of numbers of substituent groups in the molecule, analysis of the molecular composition of mixtures, elemental analyses, and molecular weight determinations. When the proper instrument settings have been determined, analyses can be made rapidly and conveniently. The method is nondestructive and is especially useful on dynamic systems. The results have good accuracy and precision.

NMR analysis, based on the intensity of magnetic resonance signals, is applicable to more than 100 different nuclei. The determination of hydrogen (^{1}H) is most used because its easy to do, is the most sensitive, and the applications encompass the whole field of organic chemistry. Analyses for ^{13}C, ^{11}B, ^{31}P, ^{2}H, ^{15}15N, ^{29}Si, and ^{17}O, where the inherent sensitivity is less, are also possible. Often, nuclei for which no instruments are available can be determined indirectly if they replace or shift the position of a hydrogen signal in the same molecule.

For brevity, the discussion emphasizes application to hydrogen, leaving the reader to visualize specialized applications involving other nuclei.

The variables determining intensity of an NMR signal have thoroughly studied. They can be grouped to include (1) the number of resonating nuclei in the sample, (2) certain design characteristics of the instrument, and (3) appropriate quantum constants for the nuclei. The first variable can be singled out and the others eliminated by performing comparative measurements under carefully selected experimental conditions. The extent of absorption, revealing the number of nuclei present, is then expressed in arbitrary units rather than as the familiar percentage absorption or transmission units used in other forms of spectroscopy. Intensities are measured with respect to one another in the same sample or are referred to those of a reference sample.

Unlike other types of electromagnetic radiation, radio-frequency radiation employed in magnetic resonance experiments can be regarded as constant, for less than about 1% of it is absorbed. Thus the Beer-Lambert law can be disregarded.

To cancel instrumental effects, the spectrometer must have high stability and must be operated in such a way as to avoid errors from saturation. Stability is achieved in internal-locked spectrometers by audiomodulation circuits. Saturation occurs when the radiation power is too high and too many nuclei are excited, significantly decreasing the population of the ground state. The problem arises because nuclei in different molecular environments saturate at different rates. Saturation is therefore minimized by a choice of low radiation power in exchange for some loss in sensitivity. Part of the difference is made up by using more concentrated samples and employing repetitive scans.

Relative intensities are obtained from the areas under the NMR signals in the spectrum. Areas can be measured with a planimeter, or the peaks can be cut out and weighed. However, most modern instruments have automated integrating capability. If the peaks are very sharp, as they frequently are in NMR, an electronic or digital integrator must be used to avoid errors arising from exceeding the response time of the graphic recorder. With integrators, the areas may be recorded as a step integral presentation directly above the spectrum, as seen in **Fig. 4**. Since peaks are not all the same width, peak heights can be used only if variation in widths is taken into account.

The areas of strong signals can be measured more accurately than weak ones because there is less relative contribution from background noise. Thus, the reliability of the data depends on the performance of the spectrometer, on concentration of the sample, and on the shape of the absorption (how the area is distributed).

Proton counting. The method has most often been applied to proton counting, the determination of the relative numbers of hydrogen atoms of each structural type present in the sample. The data are invariably used in molecular structure determination. In a typical case the overall precision of the count (based on areas measured from the step integral) expressed as standard deviation was 0.6% for concentrated solutions, 2.8% for 0.4 M solution, where the absorption occurred as a single peak, and about 10% at the 0.1 M level. For example, for toluene, the aryl/methyl-hydrogen ratio was 1.666 ± 0.013 (calculated 1.667); for ethanol the methyl/methylene-hydrogen ratio was 1.498 ± 0.010 (calc 1.500) and the hydroxyl/methylene-

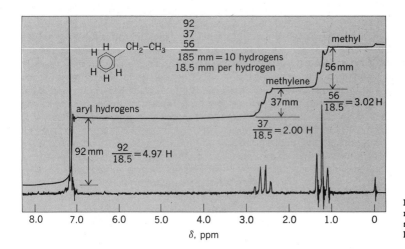

Fig. 4. Step integral representation and NMR spectrum for ethyl benzene.

hydrogen ratio was 0.504 ± 0.006 (calc 0.500). In another study of 26 samples, the standard error was 0.3% (relative standard deviation from theory).

The method is sometimes used to discover signals which are very broad and consequently too weak to be detected in the absorption spectrum.

A reverse application has been the determination of the location and extent of substitution of other isotopes for hydrogen in isotope-labeled compounds by measuring the decrease in proton signal intensities. The method has frequently been applied to ^{2}H- and ^{13}C-enriched molecules, and the precision and accuracy of the results compare favorably with those from mass spectrometry. Active hydrogen is determined similarly, allowing the sample to exchange with deuterium oxide. In a study of seven compounds, the average error in this determination was only 1.6%.

Mixtures. A second type of application is the analysis of the molecular composition of mixtures. Examples include mixtures from chemical reactions, pharmaceutical and natural product mixtures, mixtures of structural isomers, and equilibrium mixtures (where NMR is the method of choice). For the analysis of an n-component mixture, resolved peaks are needed for $n - 1$ of the components, and the number of contributing nuclei per molecule must be known. Spectra of the pure components are not usually required. If the analysis does not involve closely spaced signals, high resolution will not be required. Better precision is in fact obtained if the resolution is deliberately lowered by not spinning the sample or not degassing the sample to remove the paramagnetic broadening effects of oxygen.

The precision of the results varies with the application. In an analysis of a small amount of by-product in chlorphenesin carbamate, the precision was ±0.5% (standard deviation) and the accuracy was between −0.26 and +0.35% mean error.

Each component need not be identified. In the analysis of petroleum fractions, aromatic, paraffinic, and naphthenic content are determined without identifying the individual components of the mixture. The relative tacticity (the state of being stereochemically tactic) of polymers can be determined without knowing the molecular structure. Moisture content of starches, ground meats, soaps, breakfast cereals, and paper products is regularly determined by low-resolution NMR for quality control of production. Seeds with high oil content have been selected by NMR analysis and later used for propagation—a tribute to the nondestructive quality of the method. One laboratory performs more than 700 moisture analyses during a single day.

The classical example of an equilibrium system is acetylacetone. The enol content, 79.1%, by NMR analysis is accepted as more accurate than the 76% value obtained earlier by the bromine titration method. The method has frequently been applied to inorganic systems. For example, for sodium tripolyphosphate, the pyrophosphate composition is 21 mole % by NMR.

If the intensities are compared with those of an absolute reference standard, absolute assays are possible. In this way, the method has been used to determine total percent of hydrogen or fluorine in a sample. The latter is sometimes difficult to determine by combustion procedures. The standard can be either admixed or observed separately, but must be selected to have a similar response.

The results are as good as most routine combustion analyses on samples of 10% hydrogen or less. In a study of 57 samples, the average error was ±2% relative and the precision was ±0.5%. On eight samples, the average error in fluorine determination was ±1.3% relative.

A related application is the measurement of certain physical constants. Molecular weights and average molecular weights, and the degree of unsaturation expressed as iodine number of unsaturated fats and oils, have been determined by NMR methods using a reference standard for calibration.

Instrument development. The first commercial NMR spectrometer was sold in 1953. Second-generation instruments were transistorized and had internal-lock stabilization. The introduction of a totally new third generation of instruments with integrated circuits and minicomputers in the early 1970s caused a quantum jump in NMR spectroscopy. In the old continuous-wave (CW) method the narrow radiation frequency was varied slowly to scan the frequency-calibrated spectrum. The new idea in pulsed NMR is to excite the whole spectrum at once and let the computer untangle the time-based interference patterns of signals coming from the detector by using a mathematical Fourier transformation (FT). The savings in time and increase in sensitivity make for about two orders of magnitude of improvement in sensitivity, permitting the analysis of much smaller samples or nuclei of much less sensitivity such as carbon.

Once instruments with computer-based data systems became common, it was only logical that the on-board computer be improved and given many more tasks. Instead of using one pulse of radiofrequency to orient (tip) the nuclei prior to collection of the interference patterns, all sorts of complex pulse trains were applied around any axis, and this allowed scientists to steer the nuclei in many interesting ways. This capability led to two-dimensional NMR and to imaging.

The most recent trend has been to higher magnetic fields attainable only with superconducting magnets employing special wire for windings and cooled with liquid helium to remove the electrical resistance. At higher magnetic fields the differences among the unique frequencies are increased and the lines of the spectra are better dispersed. Increasing the strength of the magnetic field also changes the populations of the energy levels, making the method more sensitive.

Carbon-13 NMR. Chemists have long been interested in the magnetic resonance of the mass-13 isotope of carbon (^{13}C) because carbon is the most fundamental atom in nature and life. Compared to protons, the ^{13}C signals are narrower and the range of chemical shifts is about 20 times wider, allowing more detailed analysis of the structural features of fairly large molecules. But this attractive applicability is accompanied by certain experimental difficulties that have slowed the development of carbon-13 nuclear magnetic resonance (CMR).

Sensitivity. Carbon-13 behaves much like ^{1}H in NMR, but two important differences make its signals much more difficult to detect. Compared to ^{1}H, the ^{13}C nuclei are much weaker magnets. The magnet strength (magnetogyric ratio γ) is only about ¼ that of protons, and the NMR experiment depends on the cube of this term; hence the net sensitivity is actually $\frac{1}{63}$.

The low natural abundance (1.1%) of the ^{13}C isotope in a carbon-containing sample is a mixed blessing. Because the rest of the carbon (^{12}C), having no magnetic moment, is not observable by NMR, the sensitivity is further decreased by a factor of about 100, making the total loss in sensitivity, compared to ^{1}H, about $\frac{1}{6700}$. Thus the signals from ^{13}C are very weak. However, because of the low natural abundance, the chances of finding two adjacent ^{13}C atoms in a molecule are so small that ^{13}C-^{13}C spin coupling is negligible, and only the coupling with protons need be considered.

Several techniques have been combined to overcome this great sensitivity loss and make CMR practical. The most obvious development was to make bigger magnets so that larger samples could be studied. This increase in sample volume gave an 11-fold increase in sensitivity. A. Allerhand described experiments with even larger (20-mm or 0.8-in. outside diameter) sample cells.

Another development, suggested by D. Grant, is the decoupling of all the protons in the sample. For example, the ^{13}C resonance of the six identical carbons of benzene is actually split up into 15 lines because of the coupling of each carbon to one attached hydrogen, two α-hydrogens, two β-hydrogens, and one γ-hydrogen. Thus there is a 15-fold distribution of the net signal intensity from the six ^{13}C nuclei, and all the observed lines are that much weaker in intensity. By broad-band irradiation of the whole region where hydrogens normally resonate, all these hydrogens can be decoupled at once. The ^{13}C multiplet is collapsed, and all the intensity is concentrated into the single sharp line. This gives a "win back" of 2× for doublets, 3× for triplets, and so on, up to 15× for multiplet signals like those from benzene.

To get complete decoupling of a multiplet to a singlet, the decoupling radiation must be applied at exactly the resonance frequency of the coupled proton. Since there are usually many kinds of coupled protons in the sample having many different proton resonance frequencies to be irradiated at the same time, J. Roberts conceived of the idea of using a broad band of frequencies that would cover the whole proton resonance range. For his first experiment he successfully used a phonograph recording to audiomodulate the decoupler frequency. Modern instruments use noise modulation of a square wave for this purpose.

Another development was the observation that the nuclear Overhauser effect factor for ^{13}C is 2.98, considerably more than for ^{1}H. This enhancement comes from interaction of the ^{13}C nuclei with nearby hydrogens and varies somewhat with structure. Blanket irradiation of the hydrogens (as described above) gains back as much as a threefold improvement in the intensity of the carbon signals through the nuclear Overhauser effect enhancement.

There is still a lack of sensitivity by about a factor of 10–100. Another way to improve the sensitivity is to add repetitive spectra and store the result in a small digital computer. In this way the signals increase linearly while the noise, being random, increases only as the square root of the number of spectra added. Thus addition of four spectra doubles the sensitivity, nine sums gives a threefold improvement, and so on. With only a little additional gadgetry the system can be made to run unattended. To get adequate sensitivity on small amounts of material, several thousand spectra have to be added, requiring several days' time.

Pulsing FT spectrometers. In 1965, L. Johnson suggested that pulsing was the answer to the elapsed time problem. In this way the data could be collected rapidly, and a lot of time would not be wasted sweeping through blank regions of the spectra and collecting nothing but noise. This method requires a digital computer for experimental control, data collection and Fourier transformation of the results. The advent of small dedicated computers made this procedure very attractive for CMR.

In 1969 available results mistakenly indicated that ^{13}C relaxation times would be very long compared to those of protons. This would require special procedures to avoid the necessity of a long delay between pulses, while waiting for the nuclei to relax so they could be excited (tipped) again. In 1970 Allerhand showed that only a few special types of carbons have such long spin-lattice relaxation times, and investigations employing rapid summing of ^{13}C spectra by pulsing techniques were launched in earnest.

Applications of CMR. Carbon atoms are significantly different from protons, and it might be expected that this would show up in the NMR. Unlike hydrogen, carbon is rarely found on the surface of a molecule and is rarely bothered by inter-molecular interactions. Hence, large molecules and viscous or plastic samples do not present a problem in CMR. Therefore, one of the most important uses of CMR is in the determination of molecular structure. The method has been useful in the study of large molecules such as polymers of synthetic or biological origin.

Since the low natural abundance inhibits chances of two such nuclei being adjacent in the same molecule, coupling is not observed in the spectrum, and so the utility of CMR in structure determination comes mostly from the chemical shift information in the NMR spectrum. Furthermore, since carbon has a full set of valence electrons the paramagnetic type of shielding effects become dominant. These shifts are a result of the electronic structure of the carbon atom. Thus doubly bonded carbon atoms resonate in the low-field region of the spectrum, triply bonded atoms in the middle, and singly bonded atoms at high field, with a total spread for most molecules of about 200 parts per million. The shifts also depend to a lesser extent on the electronic contributions from other atoms in the molecule as was observed for ^{1}H NMR.

Some theoretical treatments have been advanced in which the ^{13}C chemical shift in question can be calculated from the additive contributions of substituents in the α, β, γ, and δ positions. One effect which is particulary useful is the steric compression shift. Thus Grant and his students found that a γ-substituent, when oriented so that it is nearby in space, causes the carbon

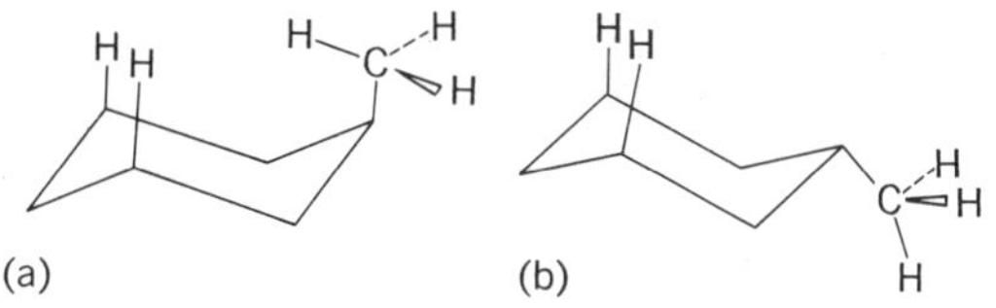

Fig. 5. Two possible configurations of substituents on a cyclohexane molecule, showing proximity of 3- and 5-hydrogens to (a) an axial and (b) an equatorial methyl group.

resonance to shift upfield because of compression effect on its electrons. An axial methyl carbon on a cyclohexane ring (**Fig. 5**a) is shielded by the two axial hydrogens at C-3 and C-5 and resonates at a higher field compared to an equatorial methyl carbon (Fig. 5b) which does not have this interaction. Numerous examples were found among ortho-substituted aromatics, steroids, and sugars. The effect also appears in alkenes where cis carbons (**Fig. 6**a) are more shielded than trans (Fig. 6b). In general an observed upfield shift detects any carbon which can exist, even partly on a time-shared basis, in a gauche or eclipsed orientation with respect to another carbon or heteroatom, and the effect is a powerful tool for making stereochemical assignments by ^{13}C NMR.

Considerable progress has been made on the semiempirical correlation of ^{13}C shifts with structural features. Since carbon is quite free of intermolecular interactions, substituent effects are reproducible from one kind of molecule to another. and model compounds can be used in the analysis of complex spectra. A structure is considered reasonable when all the chemical shifts are verified by closely related examples.

Very large molecules and solids usually give broad proton resonance lines because the neighboring magnets interact. But with natural-abundance CMR, nearest-neighbor ^{13}C atoms are at least 1 nm apart, so broadening is only about 10–15 Hz, an amount which is insignificant compared to the width of CMR spectra. Hence the method has shown much success for the determination of molecular structure of synthetic polymers and large biomolecules. With certain restrictions the method can even be applied to solids.

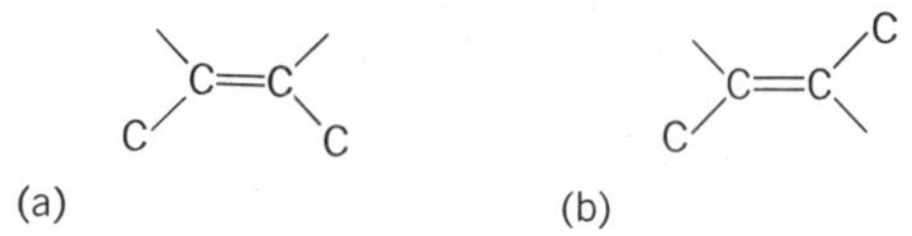

Fig. 6. Alkene configurations showing proximity of (a) cis substituents versus (b) trans substituents.

Carbon counting. Since pulsed ^{13}C spectra are usually well-resolved singlet signals with the couplings intentionally removed, the technique is often applied to count carbon atoms. Further quantitation, employing integrals of the carbon signals in a spectrum, is usually inaccurate because of saturation problems which are more severe than with protons, and because of variations of the amount of Overhauser enhancement for variously protonated carbons. Where quantitative data are necessary, the Overhauser effect can be removed by gating the decoupler or by use of certain paramagnetic additives, and saturation can be minimized by careful selection of pulse widths and repetition rate so that in favorable cases an accuracy up to ±5% can be obtained.

Proton counting. After the broad-band proton-decoupled carbon spectrum consisting of single lines has been obtained, as described above, it is often profitable to repeat the experiment, allowing some proton coupling to return. If the decoupler is turned off, all the multiple splittings return and they are so wide that components from adjacent signals (that were singlets before) overlap badly, making them difficult to identify. The width of the splittings depends on how accurately the decoupler frequency matches that of the proton. Hence decoupling the protons with a single frequency placed just outside of the proton resonance range (single frequency off-resonance decoupling) allows a small amount of coupling to return which is narrow enough to minimize overlaps and yet wide enough to count the lines and deduce the number of hydrogens on that carbon atom. The accurate count of primary, secondary and tertiary hydrogens attached to carbons (methyl, methylene, and methine groups) obtained in this way is very useful in molecular structure determination.

Tracer studies. Another practical application of CMR is the use of artificially enriched samples for tracer studies of reaction mechanisms and biosynthesis pathways. Since ^{13}C is not radioactive, it is easy to handle, and the high resolving power of the ^{13}C NMR spectrometer as a detection system is another advantage. SEE TRACE ANALYSIS.

Carbon-13 labeling has also been used for studies of mechanisms of rearrangement reactions. A variation of the tracer technique depends on the isotope shift of ^{13}C resonance frequency when attached hydrogens are replaced with deuterium. Thus without elaborate synthesis deuterium can be exchanged at certain points to identify the attached carbon atoms.

The fast CMR techniques are being applied to studies of the mechanism of rapid free-radical reactions where the appearance of emission- (upside down) or enhanced-absorption lines in the NMR spectrum yields evidence on mechanisms of thermal decomposition of peroxides and azo compounds.

Dynamic effects. Methods have been developed to measure the relaxation times of the different carbon nuclei in the sample by special pulsing sequences. It was shown that the time required for an excited nucleus to dispose of its excess energy to the surroundings (spin-lattice relaxation time T_1) is another very useful parameter for identifying the lines in a CMR spectrum. The T_1 value has also been used to characterize detailed motions of molecules such as ring conversion in substituted cyclohexanes, rotational barriers in substituted benzenes, and segmental motions in large molecules.

The biologists are anxious to characterize the segmental motions in enzymes and have already shown, for example, that the 17 different lysine residues in the protein cytochrome c have eight different relaxation times. Thus some lysines are much freer to move around than others, and this information can be correlated with the overall shape of the molecule.

Two-dimension NMR. In traditional one-dimensional experiments the nuclear spins are tipped by a radio-frequency pulse, causing them to precess around the z axis of the magnet. The precision frequency of the various nuclear magnets, detected by a radio receiver, is recorded as a digitized voltage with respect to time (**Fig. 7**). The Fourier transform of this free induction decay gives a one-dimensional NMR spectrum which is typically plotted as intensity versus frequency.

As computers became more sophisticated, scientists conceived of more elaborate tasks for them to perform. Single pulses have been replaced with elaborate pulse trains in which multiple pulses, sometimes at both the proton and carbon resonance frequencies, separated by very carefully measured delay times are directed at the sample. In this way the nuclear moments can be steered until they are exactly in the xy plane, crosswise to the field of the magnet, and there a second interaction is allowed to take place for a certain period of time. The system is said to evolve during this time. Next the receiver is turned on and the resonance frequencies are recorded as usual as a free induction decay (**Fig. 8**). If the pulse sequences were designed so that the

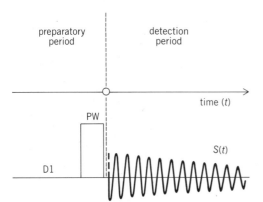

Fig. 7. Schematic for a basic pulse sequence. After a delay time, $D1$ for the nuclei to forget whatever may have been done to them previously (relaxation delay), a carefully timed radio-frequency pulse of width PW is applied to tip these nuclear magnets away from their equilibrium alignment along the z axis. The receiver is turned on, and the precession of the nuclei is detected in the transverse plane as an interferogram decaying with time, $S(t)$.

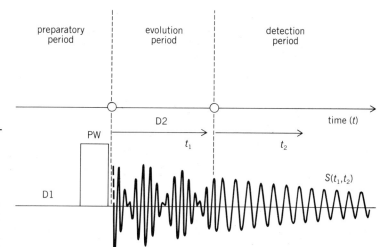

Fig. 8. Schematic for a two-dimensional experiment. An evolution period D2 is inserted in the pulse sequence, during which time the system is perturbed in some additional time-dependent way. The detected signal differs accordingly, carrying information from both time domains, $S(t_1, t_2)$.

resulting free induction decay has been influenced by a second nuclear phenomenon during the evolution time, two experiments are combined into one. For example if the hydrogen nuclei so prepared have interacted with the nuclear moments of carbon which were also tipped into the same plane (**Fig. 9**), the free induction decay has information about both proton and carbon nuclei, and it is a simple matter to analyze what went on during this evolution time and add a second dimension to the NMR experiment. Measurement of two time-dependent phenomena in the same experiment (in this example both the proton and the carbon resonance frequencies) requires repeating the experiment with carefully incremented evolution times. A total of 512 increments would give 512 free induction delays. If each free induction delay were 1024 points long, the result would be a 512 × 1024 matrix of points, showing one time result horizontally and the second time result vertically. Now the matrix need only be Fourier-transformed in both these directions to present frequency spectra in both directions and to provide a two-dimensional NMR experiment.

A two-dimensional plot would have the proton spectra plotted in one direction and the carbon spectrum in the other. The plot would then be a surface with intensity peaks rising up toward the observer and would most conveniently be recorded as a contour plot, familiar to geographers (**Fig. 10**). Each peak in the matrix has a proton- and a carbon-frequency.

The power of the method is in this orthogonal correlation. In this proton-carbon correlation each peak in the matrix shows that the protons which gave rise to a signal at that frequency in the proton spectrum must be attached to a carbon in the molecule which gave rise to a signal at that frequency in the carbon spectrum. This greatly facilitates the interpretation of NMR spectra, since if one assignment is known the other follows.

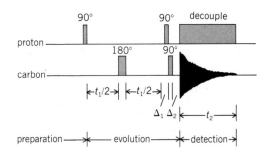

Fig. 9. Schematic pulse sequence for a proton-carbon chemical shift correlation. The protons are pulsed 90° to prepare the system. During the evolution time the carbons are also pulsed into the *xy* plane, where the two nuclei interact. The resulting free induction delay contains information about which protons interact with which carbons.

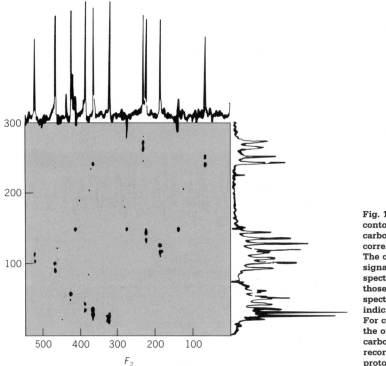

Fig. 10. Two-dimensional contour plot from a proton-carbon chemical shift correlation experiment. The contours correlate the signals in the carbon spectrum (along F_2) with those in the proton spectrum (along F_1) indicating the attachments. For convenience the one-dimensional carbon spectrum is recorded along F_2 and the proton spectrum along F.

Many other types of two-dimensional experiments can be performed to make interpretation of the spectra and the molecular structure of the sample possible.

Magnetic resonance imaging. A modification of the two-dimensional NMR experiment can provide plots in which the intensity at each spot in the matrix is color-coded (or gray-shaded) instead of being presented as a contour plot. Magnets are available which are large enough to accommodate a person serving as the sample. By the use of gradients in the magnetic field, the sensitivitiy can be limited to a cross-sectional slice anywhere through the person, and the proton intensity data presented as a two-dimensional color intensity plot is a magnetic resonance image of the anatomy at this cross section. These images have advantages over x-ray–computed tomograms. They show much better contrast on soft tissue, especially on tumors, edema, and infarcts, and involve no harmful radiation.

Bibliography. J. W. Akitt, *NMR and Chemistry: An Introduction to the Fourier Transform-Multinuclear Era*, 1983; Atta-ur-Rahman, *Nuclear Magnetic Resonance: Basic Principles*, 1986; T. Axenrod and G. Ceccarelli (eds.), *NMR in Living Systems*, 1986; A. Bax, *Two-Dimensional Nuclear Magnetic Resonance in Liquids*, 1982; E. D. Becker, *High Resolution NMR: Theory and Chemical Applications*, 2d ed., 1980; C. Dybowski and R. Litchter, *NMR Spectroscopy*, 1987; F. Kasler, *Quantitative Analysis by NMR Spectroscopy*, 1973; G. C. Levy (ed.), *NMR Spectroscopy: New Methods and Applications*, 1981; G. C. Levy, R. L. Lichter, and G. L. Nelson, *Carbon-13 Nuclear Magnetic Resonance Spectroscopy*, 2d ed., 1980; A. P. Marchand, *Stereochemical Applications of NMR*, 1982; M. L. Martin, J. J. Delpuech, and G. J. Martin, *Practical NMR Spectroscopy*, 1980; J. D. Memory and N. K. Wilson, *NMR of Aromatic Compounds*, 1982; D. Shaw, *Fourier Transform N.M.R. Spectroscopy*, 2d ed., 1984; C. P. Slichter, *Principles of Magnetic Resonance*, 2d ed., 1980; P. Sohar, *Nuclear Magnetic Resonance Spectroscopy*, vols. 1–3, 1983–1984; F. W. Wehrli and T. Wirthlin, *Interpretation of Carbon-13 NMR Spectra*, 1983.

NUCLEAR QUADRUPOLE RESONANCE
Hans Dehmelt

A selective absorption phenomenon observable in a wide variety of polycrystalline compounds containing nonspherical atomic nuclei when placed in a magnetic radio-frequency field. Nuclear quadrupole resonance (NQR) is very similar to nuclear magnetic resonance (NMR), and was originated in the late 1940s by H. G. Dehmelt and H. Krüger as an inexpensive (no stable homogeneous large magnetic field is required) alternative way to study nuclear moments. SEE MAGNETIC RESONANCE; NUCLEAR MAGNETIC RESONANCE (NMR).

Principles. In the simplest case, for example, ^{35}Cl in solid Cl_2, NQR is associated with the precession of the angular momentum of the nucleus, depicted in the **illustration** as a flat ellipsoid of rotation, around the symmetry axis (taken as the z axis) of the Cl_2 molecule fixed in the crystalline solid. (The direction of the nuclear angular momentum coincides with those of the sym-

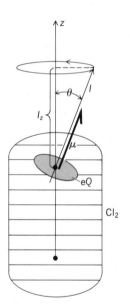

Interaction of ^{35}Cl nucleus with the electric field of a Cl_2 molecule.

metry axis of the ellipsoid and of the nuclear magnetic dipole moment μ.) The precession, with constant angle θ between the nuclear axis and symmetry axis of the molecule, is due to the torque which the inhomogeneous molecular electric field exerts on the nucleus of electric quadrupole moment eQ. This torque corresponds to the fact that the electrostatic interaction energy of the nucleus with the molecular electric field depends on the angle θ. The interaction energy is given by Eq. (1), where ρ is the nuclear charge density distribution and ϕ is the potential of the molecular electric field. Its dependence on θ is given by Eq. (2), where ϕ_{zz} is the axial gradient of the

$$E = \int \phi \rho \, dV \qquad (1) \qquad E = eQ\phi_{zz}(3\cos^2\theta - 1)/8 \qquad (2)$$

(approximately) axially symmetric molecular electric field. The quantum-mechanical analog of this expression is Eq. (3), where I and m denote the quantum numbers of the nuclear angular momen-

$$E_m = eQ\phi_{zz}[3m^2 - I(I+1)]/4I(2I-1) \qquad (3)$$

tum and its z component I_z. The absorption occurs classically when the frequency of the rf field ν

and that of the processing motion of the angular momentum coincide, or quantum-mechanically when Eq. (4) is satisfied, where m and m' are given by Eqs. (5) and $m' - m = \pm 1$, corresponding to the magnetic dipole transitions.

$$h\nu = |E_{m'} - E_m| \quad (4)$$

$$\begin{aligned} m,m' &= 0 \pm 1, \pm 2 \ldots \pm I & \text{for integer } I \geq 1 \\ m,m' &= \pm \tfrac{1}{2}, \pm \tfrac{3}{2} \ldots \pm I & \text{for half-integer } I \geq \tfrac{3}{2} \end{aligned} \quad (5)$$

It is not necessary that the rf field direction is perpendicular to z; a nonvanishing perpendicular component suffices. This eliminates the necessity of using single crystals and makes it practical, unlike in the NMR of solids, to use polycrystalline samples of unlimited mass and volume. In fact, a polycrystalline natural sulfur sample of 2.7 quarts (3 liters) volume was used in work on the rare (0.74% abundance) ^{33}S isotope.

Techniques. The original NQR work was done on approximately 3 in.^{3} (50 cm^3) of frozen *trans*-dichloroethylene submerged in liquid air using a superregenerative detector. The oscillator incorporated a vibrating capacitor driven by the power line to sweep a frequency band about 50 kHz wide over the approximately 10-kHz-wide ^{35}Cl and ^{37}Cl resonances near 30 MHz, and an oscilloscopic signal display was used. This work demonstrated the good sensitivity and rugged character of these simple, easily tunable circuits which are capable of combining high rf power levels with low noise. Their chief disadvantage is the occurrence of side bands spaced by the quench frequency which may confuse the line shape. For nuclear species of low abundance it becomes important to use nuclear modulation. In the ^{33}S work zero-based magnetic-field pulses periodically smearing out the absorption line proved satisfactory.

Application. NQR spectra have been observed in the approximate range 1–1000 MHz. Such a range clearly requires more than one spectrometer. Most of the NQR work has been on molecular crystals. While halogen-containing (Cl, Br, I) organic compounds have been in the forefront since the inception of the field, NQR spectra have also been observed for K, Rb, Cs, Cu, Au, Ba, Hg, B, Al, Ga, In, La, N, As, Sb, Bi, S, Mn, Re, and Co isotopes. For molecular crystals the coupling constants $eQ\phi_{zz}$ found do not differ very much from those measured for the isolated molecules in microwave spectroscopy. The most precise nuclear information which may be extracted from NQR $eQ\phi_{zz}$ data are quadrupole moment ratios of isotopes of the same element, since one may assume that ϕ_{zz} is practically independent of the nuclear mass. As far as ϕ_{zz} values may be estimated from atomic fine structure data, for example, for Cl$_2$ where a pure p-bond is expected and the molecular nature of the solid is suggested by a low boiling point and so forth, fair Q values may be obtained. However, it has also proved very productive to use the quadrupole nucleus as a probe of bond character and orientation and crystalline electric fields and lattice sites, and a large body of data has been accumulated in this area. SEE MICROWAVE SPECTROSCOPY.

Bibliography. I. P. Biryukov, M. G. Voronkov, and I. A. Safin, *Tables of Nuclear Quadrupole Resonance Frequencies*, 1969; T. P. Das and E. L. Hahn, *Nuclear Quadrupole Resonance Spectroscopy*, 1958; H. G. Dehmelt, Nuclear quadrupole resonance (in solids), *Amer. J. Phys.*, 22:110–120, 1954, and *Faraday Soc. Discuss.*, 19:263–274, 1955; G. K. Semin, T. A. Babushkina, and G. G. Yakobson, *Nuclear Quadrupole Resonance in Chemistry*, 1975; J. A. S. Smith (ed.), *Advances in Nuclear Quadrupole Resonance*, vols. 1–5, 1974–1983.

ELECTRON PARAMAGNETIC RESONANCE (EPR) SPECTROSCOPY
S. I. WEISSMAN

The study of magnetic resonance spectra of materials which show paramagnetism because of the magnetic moment of unpaired electrons. EPR spectra are usually presented as plots of the absorption or dispersion of the energy of an oscillating magnetic field of fixed radio frequency versus the intensity of an applied static magnetic field. SEE MAGNETIC RESONANCE.

EPR spectroscopy has been used for detection and identification of paramagnetic materials, for determinations of electronic structure, for studies of interactions between molecules, and for measurements of nuclear spins and moments. Among the wide variety of paramagnetic substances to which EPR spectroscopy has been applied are free radicals (including free atoms), impurity centers, and compounds of the transition elements, rare earths, and actinides. EPR spectra have been obtained from gases, liquids, and solids. Much of the work has been done with the oscillating magnetic field either in the vicinity of 9×10^9 Hz (X band) or 24×10^9 Hz (K band). Measurements at other frequencies lying between 10^6 and 10^{11} Hz have been performed.

Spectra characteristic of individual paramagnetic molecules, uncomplicated by magnetic interactions with neighboring paramagnetic molecules, may be obtained only from dilute solutions in diamagnetic solvents. The required degree of dilution depends on the nature of the magnetic molecules. Concentrations lower than 10^{16} molecules per cubic centimeter frequently must be used for solutions of organic free radicals, whereas concentrations as high as 10^{20} molecules per cubic centimeter may sometimes be tolerated for solutions of inorganic ions.

In some cases spin-lattice relaxation (exchange of magnetic energy with thermal motions of the environment) obscures the spectra. The effects are especially pronounced in inorganic magnetic ions and frequently require the use of low temperatures. The spectra of organic free radicals, on the other hand, are not severely affected and may usually be obtained at ordinary temperatures.

Solids. Maximum information is yielded by EPR spectra of solid solutions in single crystals. The spectrum of a single species may contain scores of lines, their positions and intensities varying with orientation of the specimen relative to the static magnetic field. The many-lined structure results in part from interactions of the orbital motion of the electrons with the electric fields of the environment and from interactions of the magnetic moments of the electrons with nuclear magnetic moments. The latter effect (hyperfine interaction) is the sole cause of the splittings in the EPR spectra of most organic free radicals. Most magnetic ions exhibit pronounced anisotropy in their EPR spectra (**Fig. 1**) because of the anisotropic nature of the orbitals (wave functions) which give rise to their magnetism. Most organic free radicals and a few ions with highly symmetrical charge distributions or with highly quenched orbital magnetism exhibit little anisotropy in their EPR spectra.

From analysis of hyperfine interactions with nuclei whose spins and magnetic moments are known, details of the distribution of electrons about the nuclei may be determined. The average value of the cube of the reciprocal of the distances between electrons and nuclei, the orientation of the orbits relative to the crystal axes, and the density of unpaired electrons about the various nuclei in a free radical may be evaluated from the hyperfine structure.

Relative values of nuclear moments of isotopes may be found from the relative splittings produced by them in the same chemical environment. If only one isotope is available, its nuclear magnetic moment may be obtained from EPR spectra only if suitable properties of the electronic orbits are known. Nuclear spins, on the other hand, may be found, simply by counting hyperfine components.

Liquids. Much work has been done, particularly with organic free radicals, in liquid solutions, where less information is obtainable by EPR spectroscopy than in crystals. Only averages of orientation-dependent properties can be observed in motions in liquids. The nature of the average is dependent on the rapidity of the motions. Well-resolved EPR spectra may be obtained in liquids only if the variations of positions of lines with orientation of molecular axes are not great. Many organic free radicals fulfill this requirement and yield lines only a few tenths of an oersted (1 Oe = 79.6 A/m) broad in liquid solutions. Highly characteristic spectra ranging from those containing only one line (such as the semiquinone of chloranil) to others containing more than 100 lines (triphenylmethyl) have been recorded. The spectra of organic free radicals are symmetrical about a center (**Fig. 2**). At fields of 3200 oersteds (25 kA/m), the centers usually lie within a few oersteds of the position of the resonance of the spin of a free electron, that is, one which has no spin-orbit interaction.

Hyperfine interactions are responsible for the complexity of the EPR spectra of most free radicals. Most of the splittings observed thus far have been produced by ^{1}H, which has a nuclear spin quantum number of ½ ($I = ½$). Splittings by ^{2}H, ^{11}B, ^{13}C, ^{14}N, and ^{15}N have also been studied. The contribution to the splitting by each kind of proton (in a free radical the protons differ in chemical environment) is determined by observation of the EPR spectra of radicals with appropriate substitutions of ^{1}H by ^{2}H ($I = 1$).

Rates of electron transfer. Migration of electrons among different molecules may produce measurable effects on the EPR spectra. In favorable cases, electron spin rates and mecha-

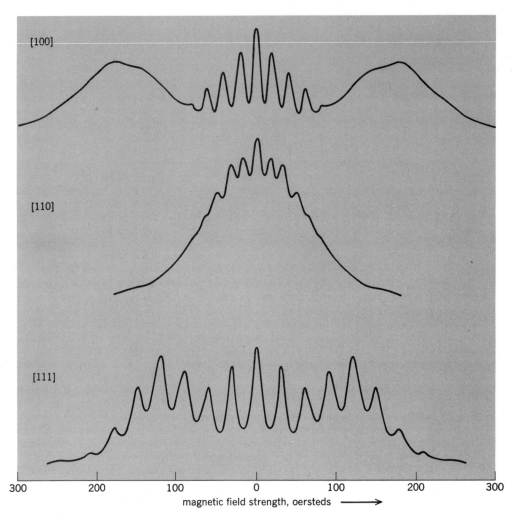

Fig. 1. Electron paramagnetic resonance spectra showing energy absorption versus magnetic field strength for a dilute solution of FeF_6^{3-} in a single crystal of Na_2KGaF_6. The indices [111], [110], and [100] give the direction of the magnetic field relative to the crystal axes. (*Courtesy of L. Helmholz*)

nisms may be detected. In a stable free radical with resolved hyperfine structure, each line is associated with the frequency of precession of the electron spin in the presence of a particular arrangement of nuclear spins. When the electron jumps to a molecule with a different arrangement of nuclear spins, its spin precesses at a different frequency. Jumps occurring with mean time $1/\bar{\nu}$ between them add breadth $\bar{\nu}$ to the spectral lines as long as $\bar{\nu}$ is small compared with the separation of the various frequencies of precession. The method has been applied to measurements of electron-transfer reactions with second-order rate constants in the range 10^6–10^9 liters/(mole)(s).

When $\bar{\nu}$ becomes large compared with separation of lines, a new spectrum appears. The hyperfine structure of the new spectrum reveals the nature of the groups of atoms which accompany the electron in its migrations.

Motions effects. The effects described above arise from modulation of the hyperfine interaction accompanying electron transfer from one molecule to another. Reorientation of a single paramagnetic molecule produced similar effects through modulation of the anisotropic part of the hyperfine interaction. The effect has been exploited for probing motional processes in large molec-

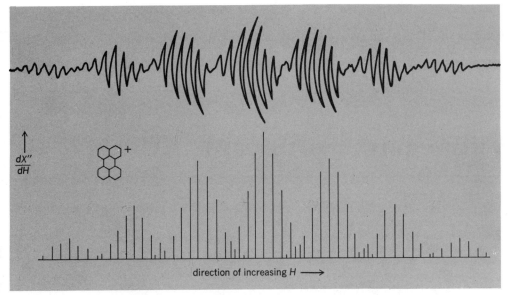

Fig. 2. Spectrum of dX''/dH versus H (X'' is the energy absorption, H is the magnetic field strength) for a liquid solution of the free radical perylene positive ion. Vertical lines give the positions and intensities of the spectral lines as calculated from molecular orbital theory. (*Courtesy of E. de Boer and S. I. Weissman*)

ular structures, particularly ones of biological importance. Radicals containing the nitroxide group have been inserted into hemoglobins, enzymes, membranes, and other complex systems with little impairment of biological function. Due to large anisotropy in the nitrogen hyperfine interaction, the EPR spectra are sensitive to reorientation of the radical. Characteristic times for reorientation in the range 10^{-9} to 10^{-5} s may be measured.

Multiple resonance methods. Greatly enhanced resolution may be obtained in many cases through simultaneous irradiation with several frequencies. The most useful of these methods is electron nuclear double resonance (ENDOR). The material is simultaneously irradiated at one of its EPR resonant frequencies and by a second oscillatory field whose frequency is swept over the range of nuclear frequencies. The intensity of the EPR response is measured as a function of the second frequency. Changes in EPR intensity accompany passage through nuclear resonance. Resonances as narrow as 10 kHz are observed. The spectra are far more easily interpreted than are conventional EPR ones, owing to the fact that each variety of nucleus yields only a doublet in ENDOR, independent of the number of such nuclei. The triphenylmethyl radical which contains 15 protons (3 para, 6 ortho, 6 meta) has 196 lines in its EPR spectrum, and only 3 doublets in its ENDOR spectrum. In crystalline materials, proton ENDOR has permitted precise location of hydrogen atoms.

Transient methods. The experimental methods described above rely on steady-state responses; that is, absorption or dispersion at each part of the spectrum is recorded only after all transients have subsided. Just as in nuclear magnetic resonance, the transients often contain more information than the steady-state responses. Echoes associated with electronic paramagnetism have now been observed in a variety of systems. They are particularly useful in materials with inhomogeneously broadened lines. A succession of two microwave pulses evokes a conventional echo; and a succession of three pulses, stimulated echoes. From the way in which the echo intensities vary with time intervals between pulses, relaxation parameters and hyperfine splittings may be obtained. Developments in solid-state technology, such as wide-banded microwave amplifiers and very rapid digitizers, have provided enhanced sensitivity of detection, as well as richer information. *See* николаа magnetic resonance *(NMR).*

Bibliography. A. Abragam and B. Bleaney, *Electron Paramagnetic Resonance of Transition Ions*, 1986; L. R. Dalton et al. (eds.), *EPR and Advanced EPR Studies of Biological Systems*, 1985; J. E. Harriman (ed.), *Theoretical Foundations of Electron Spin Resonance*, 1978; L. Kevan, *Time Domain Electron Spin Resonance*, 1979; L. Kevan and L. D. Kispert, *Electron Spin Double Resonance Spectroscopy*, 1976; P. F. Knowles et al., *Magnetic Resonance of Biomolecules: An Introduction to the Theory and Practice of NMR and ESR*, 1976; C. P. Poole, *Electron Spin Resonance: A Comprehensive Treatise on Experimental Techniques*, 2d ed., 1983; M. C. Symons, *Chemical and Biochemical Aspects of Electron Spin Resonance Spectroscopy*, 1978.

CYCLOTRON RESONANCE EXPERIMENTS
WALTER M. WALSH

The measurement of charge-to-mass ratios of electrically charged particles from the frequency of their helical motion in a magnetic field. Such experiments are particularly useful in the case of conducting crystals, such as semiconductors and metals, in which the motions of electrons and holes are strongly influenced by the periodic potential of the lattice through which they move. Under such circumstances the electrical carriers often have "effective masses" which differ greatly from the mass in free space; the effective mass is often different for motion in different directions in the crystal. Cyclotron resonance is also observed in gaseous plasma discharges and is the basis for a class of particle accelerators.

The experiment is typically performed by placing the conducting medium in a uniform magnetic field H and irradiating it with electromagnetic waves of frequency ν. Selective absorption of the radiation is observed when the resonance condition $\nu = qH/2\pi m^*c$ is fulfilled, that is, when the radiation frequency equals the orbital frequency of motion of the particles of charge q and effective mass m^* (c is the velocity of light). The absorption results from the acceleration of the orbital motion by means of the electric field of the radiation. If circularly polarized radiation is used, the sign of the electric charge may be determined, a point of interest in crystals in which conduction may occur by means of either negatively charged electrons or positively charged holes.

For the resonance to be well defined, it is necessary that the mobile carriers complete at least $½\pi$ cycle of their cyclotron motion before being scattered from impurities or thermal vibrations of the crystal lattice. In practice, the experiment is usually performed in magnetic fields of 1000 to 100,000 oersteds (1 Oe = 79.6 A/m) in order to make the cyclotron motion quite rapid ($\nu \sim 10 - 100$ gigahertz, that is, microwave and millimeter-wave ranges). Nevertheless, crystals with impurity concentrations of a few parts per million or less are required and must be observed at temperatures as low as 1 K in order to detect sharp and intense cyclotron resonances.

The resonance process manifests itself rather differently in semiconductors than in metals. Pure, very cold semiconductors have very few charge carriers; thus the microwave radiation penetrates the sample uniformly. The mobile charges are thus exposed to radiation over their entire orbits, and the resonance is a simple symmetrical absorption peak.

In metals, however, the very high density of conduction electrons present at all temperatures prevents penetration of the electromagnetic energy except for a thin surface region, the skin depth, where intense sheilding currents flow. Cyclotron resonance is then observed most readily when the magnetic field is accurately parallel to the flat surface of the metal. Those conduction electrons (or holes) whose orbits pass through the skin depth without colliding with the surface receive a succession of pulsed excitations, like those produced in a particle accelerator. Under these curcumstances cyclotron resonance consists of a series of resonances $n\nu = qH/2\pi m^*c$ ($n = 1, 2, 3, \ldots$) whose actual shapes may be quite complicated. The resonance can, however, also be observed with the magnetic field normal to the metal surface; it is in this geometry that circularly polarized exciting radiation can be applied to charge carriers even in a metal.

Cyclotron resonance is most easily understood as the response of an individual charged particle; but, in practice, the phenomenon involves excitation of large numbers of such particles. Their net response to the electromagnetic radiation may significantly affect the overall dielectric behavior of the material in which they move. Thus, a variety of new wave propagation mechanisms may be observed which are associated with the cyclotron motion, in which electromagnetic energy is carried through the solid by the spiraling carriers. These collective excitations are generally referred to as plasma waves. In general, for a fixed input frequency, the plasma waves are observed to travel through the conducting solid at magnetic fields higher than those required for cyclotron resonance. The most easily observed of these excitations is a circularly polarized wave, known as a helicon, which travels along the magnetic field lines. It has an analog in the ionospheric plasma, known as the whistler mode and frequently detected as radio interference. There is, in fact, a fairly complete correspondence between the resonances and waves observed in conducting solids and in gaseous plasmas. Cyclotron resonance is more easily observed in such low-density systems since collisions are much less frequent there than in solids. In such systems the resonance process offers a means of transferring large amounts of energy to the mobile ions, which is a necessary condition if nuclear fusion reactions are to occur.

Bibliography. C. Kittel, *Introduction to Solid State Physics*, 5th ed., 1976; P. M. Platzman and P. A. Wolff, *Waves and Interactions in Solid State Plasmas*, 1973.

MASS SPECTROSCOPY

Mass spectroscope	**230**
Mass spectrometry	**235**

MASS SPECTROSCOPE
Alfred O. Nier and Harry E. Gove
H. E. Gove is author of the section Tandem Accelerator Mass Spectrometers.

An instrument used for determining the masses of atoms or molecules found in a sample of gas, liquid, or solid. It is analogous to the optical spectroscope, in which a beam of light containing various colors (white light) is sent through a prism to separate it into the spectrum of colors present. In a mass spectroscope, a beam of ions (electrically charged atoms or molecules) is sent through a combination of electric and magnetic fields so arranged that a mass spectrum is produced. If the ions fall on a photographic plate which after development shows the mass spectrum, the instrument is called a mass spectrograph; if the spectrum is allowed to sweep across a slit in front of an electrical detector which records the current, it is called a mass spectrometer.

Operation. A typical mass spectroscope has a continuously pumped vacuum chamber, commonly called the spectrometer tube, into which the gas or vapor to be investigated flows at such a rate that the equilibrium pressure in the chamber is of the order of 10^{-6} mm of mercury (10^{-8} lbf/in.^{2} or 10^{-4} pascal). **Figure 1** is a schematic drawing of a type of mass spectrometer tube widely used for making gas and isotope analyses. The pumping system consists of a mechanical vacuum forepump followed by either an oil or mercury diffusion pump and a cold trap maintained at dry ice or liquid air temperature. A sufficient quantity of the gas to be analyzed is placed in a vacuum chamber, whose pressure is approximately 0.05 mmHg (10^{-3} lbf/in.^{2} or 7 Pa). A pinhole having a diameter of approximately 0.001 in. (0.025 mm) permits this gas to leak continuously into the mass spectrometer tube.

A heated tungsten or rhenium filament, not shown in the diagram, produces an electron beam normal to the plane of the diagram, as shown. The electrons in the beam collide with the molecules of gas present and knock off one or more electrons, thus creating positive ions.

In a monatomic gas such as argon, multiply charged positive ions, designated as Ar^+ (singly charged), Ar^{2+} (doubly charged), and so on, are formed. In the case of polyatomic molecules, ionized fragments may also be formed. For example, for methane, CH_4, the ions CH_4^+, CH_3^+, CH_2^+, CH^+, C^+, and H^+ are found. In some cases negative ions are also created as a result of an electron capture process. Thus for CO, in addition to the ions CO^+, C^+, O^+, the ions C^- and O^- are observed.

An electric field resulting from the application of a potential difference of several volts between A and B draws the ions through the slit in plate B. Further energy is given the ions by allowing them to fall through an electric potential of several hundreds or thousands of volts applied between plates B and G. In such an instrument setup, plate G is grounded.

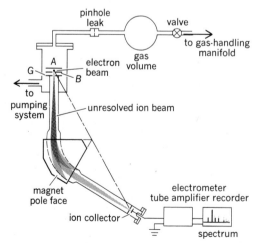

Fig. 1. Schematic diagram of mass spectrometer tube. Ion currents are in the range 10^{-10} to 10^{-15} A and require special electrometer tube amplifiers for their detection. In actual instruments the radius of curvature of ions in a magnetic field is 4–6 in. (10–15 cm).

The beam of ions travels downward and because of the finite width of the slits in plates B and G diverges slightly, as shown. The beam passes between the poles of a magnet as indicated. The magnetic field is perpendicular to the plane of the diagram. In the magnetic field the ions experience a force at right angles to their direction of travel given by Eq. (1), where f is the force,

$$f = Bev \qquad (1)$$

B the magnetic field intensity, e the charge on the ion, and v its speed, all in SI units. This force results in a circular trajectory, whose radius r is found by equating the force to the product of mass and acceleration according to Newton's laws of motion, as in Eq. (2).

$$Bev = mv^2/r \qquad (2)$$

Equation (2) may be put in the more convenient form of Eq. (3) by equating the kinetic

$$\tfrac{1}{2}mv^2 = eV \qquad (3)$$

energy gained to the potential energy lost and expressing v in terms of potential difference V through which the ion fell. By combining Eqs. (2) and (3), solving for r, and substituting units which are more convenient for actual calculation, Eqs. (4) are obtained, where r' is now in inches,

$$r' = 57(MV/e)^{1/2}B \qquad (4a) \qquad\qquad r = 144\,(mV/e)^{1/2}/B \qquad (4b)$$

r is in cm, B in gauss (1 gauss = 10^{-4} tesla), V in volts, e is the charge of the ion measured in terms of the number of electrons removed (or added) during ionization, and m is the mass measured in atomic mass units; that is, for hydrogen $m = 1$, for the most abundant isotope of oxygen, ^{16}O, $m = 16$, and so on.

Figure 1 shows the paths of ions having three different masses. Only the intermediate group has the proper mass to reach and pass through the collector slit and to be measured.

If the source of ions, the apex of the wedge-shaped field, and the collector slit all lie on a straight line, the diverging beam of ions is focused as shown. This property of a wedge-shaped magnetic field is analogous to the focusing property of a convex lens in optics.

From Eqs. (4) it is clear that if either B or V is varied continuously, a mass spectrum may be swept. **Figure 2** is a recording of such a spectrum for krypton. The abundances of the several isotopes are proportional to the recording pen deflection.

Routine gas or isotope analyses may be made on 10^{-3} in.^{3} (10^{-2} cm^3) of gas at normal pressure and temperature (NTP), and in special cases as little as 10^{-9} in.^{3} (10^{-8} cm^3) of gas at NTP suffices. Some solids, such as alkali metals or alkali earths, give off ions directly when heated; therefore a filament coated with a salt containing these elements may serve as a source of ions. With this method, analyses have been made on samples as small as 3×10^{-14} oz (10^{-12} g). Electric sparks may also serve as sources of ions in solid analyses. Because of the unsteady nature of a spark, photographic recording of the spectra is generally employed.

In order to obtain higher resolution and a more definite relation between mass and position

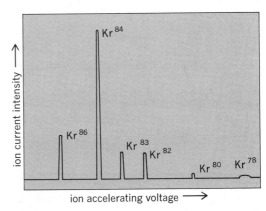

Fig. 2. Mass spectrum of krypton isotopes obtained with an instrument such as that shown in Fig. 1.

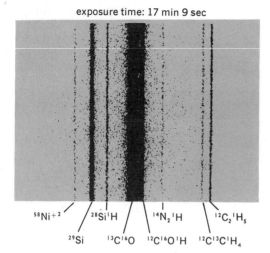

Fig. 3. Mass spectrum observed at mass number 29 when several substances are present simultaneously in a mass spectrograph. Resolution is sufficient to completely separate two close lines such as $^{12}C^{13}C^1H_4$ and $^{12}C_2{}^1H_5$, which differ in mass by only 1 part in 75,000.

in a spectrum, an electric field region is employed in series with the magnetic field region. With this arrangement, called double focusing, resolutions of the order of 1 part in 50,000 are obtained, and masses can be determined to 1 part in 10^7, or better. **Figure 3** is a mass spectrum obtained with a double-focusing mass spectrograph.

Applications. Mass spectroscopes are used in both pure and applied science. Atomic masses can be measured very precisely. Because of the equivalence of mass and energy, knowledge of nuclear structure and binding energy of nuclei is thus gained. The relative abundances of the isotopes in naturally occurring or artificially produced elements can be determined. Thus nuclear processes occurring either in nature or in the laboratory may be investigated. Isotopic analyses of elements such as lead, argon, or strontium, which may result from the radioactive decay of other elements, are of particular interest because they make possible the determination of the geological age of the minerals from which the elements are extracted. Other geological effects which cause variations in relative abundances of isotopes may also be investigated.

Empirical and theoretical studies have led to an understanding of the relation between molecular structure and the relative abundances of the fragments observed when a complex molecule, such as a heavy organic compound, is ionized. When a high-resolution instrument is employed, the masses of the molecular or fragment ions can be determined so accurately that identification of the ion can frequently be made from the mass alone. Thus, although $C_2H_4N_4$ and $C_2H_4N_2O$ have molecular weights of approximately 84 when made up of the most abundant isotope in each of the elements concerned, the deviation from integral numbers of the constituents is sufficiently different to cause the exact masses to be 84.04359 and 84.03236 atomic mass units (amu), respectively (^{12}C = 12.00000 amu). If the isotopic nature of the elements is also taken into account, additional ions will be observed whose relative abundances depend upon the relative abundances of the isotopes. This fact provides another powerful tool in determining the identification of the ion in question. These methods, supplemented by others, frequently make the determination of the molecular formula of a compound relatively easy. Once this has been accomplished, the actual structure can usually be found from a consideration of the relative abundances of the fragment ions.

Mass spectrometers also make possible isotopic analyses of compounds which have reacted chemically with other compounds containing elements with artificially altered isotopic abundance ratios. Thus the instruments make possible tracer studies of chemical or biochemical reactions.

Because chemical compounds may have mass spectra as unique as fingerprints, mass spectroscopes are widely used in industries such as oil refineries, where analyses of complex hydrocarbon mixtures are required. For further information on applications SEE BETA RAYS.

Miscellaneous types. In the instruments thus far discussed, a mass spectrum is obtained by making use of the fact that ions of different mass are deflected through different angles in passing through a magnetic field. In one important modification a superimposed electric and magnetic field is used, resulting in a trochoidal ion path. This has certain advantages, especially in that high resolution is obtained in a very compact apparatus. Various time-of-flight mass spectrometers are also employed. In some of these, ions are accelerated in pulses or by potentials varying sinusoidally with time and then sent through a system of grids where potentials also vary with time. A separation according to mass is effected because ions of different mass require different times to traverse the arrangement. In still others, use is made of the cyclotron principle; that is, the time for an ion to traverse a complete circle in a magnetic field is independent of the energy of the ion and depends only on its mass. In the quadrupole spectrometer, or *Massenfilter*, ions pass along a line of symmetry between four parallel cylindrical rods. An alternating potential superimposed on a steady potential between pairs of rods filters out all ions except those of a predetermined mass.

Tandem accelerator mass spectrometers. Mass spectrometers of the type described above have limitations in certain circumstances. For example, they may not be able to resolve molecules from equivalent mass number atoms, although the double-focusing mass spectrometers described above generally have sufficient mass resolution to avoid this problem. Such high resolution (1 part in 50,000 or better) is achieved, however, with a sacrifice in efficiency. A more serious problem is the separation of the wanted atom from its isobars. For example, ^{14}C and ^{14}N differ in mass by 1 part in 100,000.

Both these problems are overcome by the use of a tandem electrostatic accelerator as a mass spectrometer (**Fig. 4**). The system starts with negative ions (neutral atoms to which an extra electron is attached) produced in a sputter ion source. These are mass-analyzed to about 1 part in 200 and injected into the low-energy end of a tandem electrostatic accelerator. They are accelerated to the central positive terminal whose voltage can be adjusted up to 12 MV. In the terminal they pass through a differentially pumped argon gas stripper and emerge with several electrons removed. These positive ions are then accelerated through the second half of the machine and exit

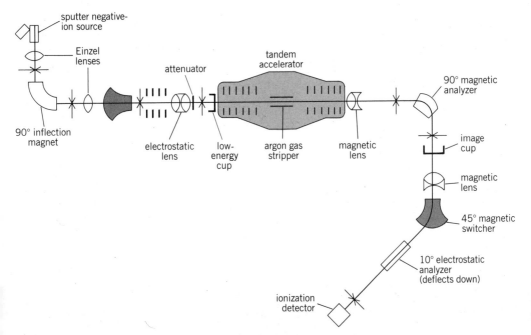

Fig. 4. Typical arrangement of a tandem accelerator mass spectrometry system.

Radionuclides studied in tandem accelerator mass spectrometer

Radio-nuclide	Half-life, years	Stable isotopes	Stable isobars	Chemical form	Charge state	Energy, MeV	Limit of detection
^{10}Be	1.6×10^6	^{9}Be	^{10}B	BeO	3+	33†	7×10^{-5}
^{14}C	5730	^{12,13}C	^{14}N*	C	4+	40	0.3×10^{-15}
^{26}Al	7.2×10^5	^{27}Al	^{26}Mg*	Al	5+	48	10×10^{-15}
^{32}Si	108	^{28,29,30}Si	^{32}S	SiO$_2$	5+	55	7×10^{-12}
^{36}Cl	3.0×10^5	^{35,37}Cl	^{36}Ar*, ^{36}S	AgCl	7+	80	0.2×10^{-15}
^{129}I	15.9×10^6	^{127}I	^{129}Xe*	AgI	5+	30	0.3×10^{-12}

*Does not form negative ions.
†BeO from source.

at the high-energy end. Here they are subjected to a series of magnetic and electrostatic deflections and finally pass into an ionization detector. Molecules are dissociated in the terminal stripper, and isobars are separated by their different rates of energy loss in the ionization detector due to their different atomic number.

Such systems are used to measure cosmogenic radioisotopes in a wide variety of naturally occurring material as well as stable isotopes in matrices of much more abundant elements. The **table** lists some of the radionuclides that have been measured. The limit of detection is listed in the last column. This limit is the ratio of the number of atoms of the radionuclide to the number of atoms of the stable isotopes. For ^{14}C and ^{36}Cl, the detection limit approaches 1 part in 10^{16}. Such detection limits never have been, and probably never will be, achieved by conventional mass spectrometers.

Radioactive isotopes are normally measured by detecting their decay products. For example, the amount of ^{14}C in a carbonaceous artifact can be used to determine the age of that artifact. To make such a measurement by radioactive decay counting, because of the relatively long half-life of 5730 years, requires carbon samples in the range of 1 to 10 grams. The tandem accelerator technique can measure the ^{14}C contents of a carbon sample by using 10 micrograms to a few milligrams of carbon. There are innumerable examples of carbonaceous artifacts which it would be of great interest to date but whose carbon content or the sample size is much too small to permit radioactive decay counting. The accelerator technique makes their age measurement possible.

Many of the other radioisotopes listed in the table have much larger half-lives than ^{14}C and thus are even more difficult to measure by decay counting. The accelerator mass spectrometry technique, which does not require that the isotope be measured by radioactive decay, can readily be used.

^{14}C has now been measured by tandem accelerators in a wide variety of important archeological, anthropological and geological samples. ^{36}Cl has been measured in meteorites, lunar rocks, ice cores, groundwater, and many other naturally occurring materials. ^{10}Be has been measured in lake and ocean cores, a Moon core, rainwater, ice, and manganese modules. ^{129}I, which is the longest-lived cosmogenic radioisotope, has been measured in meteorites and in lunar rocks and groundwater.

The accelerator mass spectrometry technique has also been applied to the measurement of stable isotopes. In particular, platinum and iridium have been measured in samples of geological and anthropological interest.

Accelerator-based ultrasensitive mass spectrometry finds applications in quite different areas of research than conventional mass spectrometry. The two techniques complement one another. SEE MASS SPECTROMETRY.

Bibliography. J. H. Benyon and A. G. Brenton, *An Introduction to Mass Spectrometry*, 1982; H. E. Duckworth et al., *Mass Spectroscopy*, 2d ed., 1986; A. E. Litherland, Ultrasensitive mass spectrometry with accelerators, *Annu. Rev. Nucl. Part. Sci.*, 30:437–473, 1980; F. W. McLafferty, *Interpretation of Mass Spectra*, 3d ed., 1980; F. W. McLafferty, *Tandem Mass Spectrometry*, 1983; M. E. Rose and R. A. Johnstone, *Mass Spectrometry for Chemists and Biochemists*, 1982.

MASS SPECTROMETRY
Maurice M. Bursey

An analytical technique for identification of chemical structures, determination of mixtures, and quantitative elemental analysis, based on application of the mass spectrometer. Organic and inorganic molecular structure determination is based on the fragmentation pattern of the ion formed when the molecule is ionized; further, because such patterns are distinctive, reproducible, and additive, mixtures of known compounds may be quantitatively analyzed. Quantitative elemental analysis of organic compounds requires exact mass values from a high-resolution mass spectrometer; trace analysis of inorganic solids requires a measure of ion intensity as well. See Mass Spectroscope.

Methods of ion production. For analysis of organic compounds the principal methods are electron impact, chemical ionization, field ionization, field desorption, and particle bombardment.

Electron impact. When a gaseous sample of a molecular compound is ionized with a beam of energetic (commonly 70-V) electrons, part of the energy is transferred to the ion formed by the collision, as shown in reaction (1). For most molecules the production of cations is favored

$$A\text{—}B\text{—}C + e \rightarrow A\text{—}B\text{—}C^+ + e + e \tag{1}$$

over the production of anions by a factor of about 10^4, and the following discussion pertains to cations. The ion corresponding to the simple removal of the electron is commonly called the molecular ion and normally will be the ion of greatest m/e ratio in the spectrum. In the ratio m is the mass of the ion in atomic mass units and e is the charge of the ion measured in terms of the number of electrons removed (or added) during ionization. Occasionally the ion is of vanishing intensity, and sometimes it collides with another molecule to abstract a hydrogen or another group. In these cases an incorrect assignment of the molecular ion may be made unless further tests are applied. Proper identification gives the molecular weight of the sample.

The remaining techniques were devised generally to circumvent the problem of the weak or vanishingly small intensity of a molecular ion.

Chemical ionization. Here the ions to be analyzed are produced by transfer of a heavy particle (H^+, H^-, or heavier) to the sample from ions produced from a reactant gas. Frequently the reactant gas is methane at pressures of 0.2–2.0 torr (27–270 pascals). As above, the initial process in methane upon electron impact is ionization to yield CH_4^+ ions; some of these have enough energy to fragment to CH_3^+ + H. These ions in turn react with neutral methane as in reactions (2) and (3). The resulting ions are strong Brönsted acids and react by proton transfer as

$$CH_4^+ + CH_4 \rightarrow CH_5^+ + CH_3 \tag{2} \qquad CH_3^+ + CH_4 \rightarrow C_2H_5^+ + H_2 \tag{3}$$

in reactions (4) and (5) to ionize the molecule of interest. The $C_2H_5^+$ ion also reacts by hydride

$$CH_5^+ + A\text{—}B\text{—}C \rightarrow H\text{—}A\text{—}B\text{—}C^+ + CH_4 \tag{4} \qquad C_2H_5^+ + A\text{—}B\text{—}C \rightarrow H\text{—}A\text{—}B\text{—}C^+ + C_2H_4 \tag{5}$$

abstraction, as in reaction (6). Other gases may also be used for ionization; H_2, H_2O, NH_3, and

$$C_2H_5^+ + A\text{—}B\text{—}C\text{—}H \rightarrow A\text{—}B\text{—}C^+ + C_2H_6 \tag{6}$$

isobutane are common. In these the reactive ions are H_3^+, H_3O^+, NH_4^+, and $C_4H_9^+$ [the conjugate acid of $(CH_3)_2C\text{=}CH_2$], respectively. The thermochemistry of these proton transfer reactions and many others is summarized by reference to **Fig. 1**. The proton affinity is the negative of the enthalpy change for the solvation of a proton by the compound, as in reaction (7). It follows that

$$A\text{—}B\text{—}C + H^+ \rightarrow A\text{—}B\text{—}C\text{—}H^+ \tag{7}$$

A—B—C—H$^+$ protonates molecules with greater proton affinities in the absence of kinetic complications; such reactions are exothermic. Recently, proton affinities have been determined both by high-pressure techniques akin to chemical ionization and by the ion cyclotron resonance technique. Thus H_3^+ transfers the most energy when it protonates a molecule; NH_4^+ transfers the least of the four examples. If the energy transferred (typically 10–50 kcal/mole or 40–200 kilo-

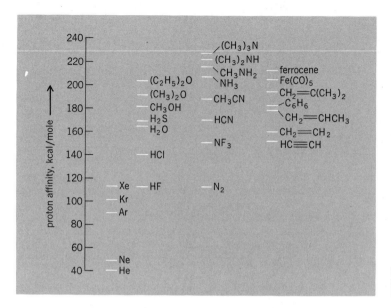

Fig. 1. Representative gas-phase proton affinities of molecules.

joules/mole) is great enough, fragmentation can occur, but there is much less in chemical ionization than in electron impact mass spectra.

In addition to heavy-particle transfer of other types (transfer of CH_3^+, $C_2H_5^+$, Cl^-), another important method is charge exchange. For a gas with an ionization potential greater than that of the molecule of interest, reaction (8) is exothermic, and the excess energy is transferred as internal

$$G^+ + A\text{—}B\text{—}C \rightarrow A\text{—}B\text{—}C^+ + G \qquad (8)$$

energy. He^+, Ne^+, and Ar^+ transfer large amounts of internal energy and molecular ions are weak; CO and NO are useful gases for this process. Occasionally a combination of chemical ionization and charge exchange is used, with mixtures of Ar and H_2O yielding information about both the molecular weight and important fragments, for example.

The term negative chemical ionization is usually intended to include all ionization processes yielding negative ions under source pressure conditions characteristic of chemical ionization. In some cases ions may be formed by a true chemical ionization process like proton transfer, as in reaction (9), in which OH^- is the reagent ion, produced from H_2O or $N_2O + CH_4$. In others

$$RCOOH + OH^- \rightarrow RCOO^- + H_2O \qquad (9)$$

the reagent gas only mediates the energy of free electrons to low values compatible with capture by neutral molecules to yield negative molecular ions. Molecules with several electronegative atoms have large cross sections for this latter process. Thus the direct analysis of polychlorinated aromatics, and the analysis of small peptides after derivatization by pentafluorobenzoylation, are possible in the femtogram region.

New applications of ions formed by chemical ionization (atmospheric pressure ionization; plasma chromatography, in which the drift time of the ion through a flowing gas is measured) also have femtogram sensitivities.

Field ionization and field desorption. For less volatile material, the sample is ionized when it is in a very high field gradient (several volts per angstrom) near an electrode surface. **Figure 2** illustrates the distortion of the molecular potential well so that an electron tunnels from the molecule to the anode. The ion thus formed is repelled by the anode. Typically, the lifetime of the ion in the mass spectrometer source is much less (10^{-12} to 10^{-9} s) than in electron impact. Because little energy is transferred as internal energy and the ion is removed rapidly, little fragmentation occurs, and the molecular weight is more easily determined. Field ionization is also

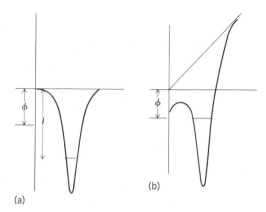

Fig. 2. Potential well of a molecule: (a) undistorted, with ionization potential I near a metal with work function ϕ in the absence of a field gradient; and (b) in a strong field gradient. Near the metal surface the external field raises the most weakly bonded electron to the Fermi level so that tunneling through the small barrier may occur.

used in the time-resolved study of ion fragmentation and rearrangement called field ionization kinetics. This technique permits determination of the fragmentation products at specific times from about 10^{-11} to 10^{-8} s after ionization by energy-focusing of ions at different points in the field ionization source. In this way simple fragmentations uncomplicated by rearrangements of hydrogen or other atoms in the molecular ion may be observed at the shortest times yet used for sampling ions, and complex rearrangements of ions may be defined from studies over a range of times.

Cations, most commonly Li^+ but often other alkali metals, may be field-desorbed from a salt coating on a wire and pass through a gaseous or adsorbed sample M. MLi^+ ions are produced in some cases where simple field desorption fails.

Electrohydrodynamic ionization. A high electric-field gradient induces ion emission from a droplet of a liquid solution, that is, the sample and a salt dissolved in a solvent of low volatility. An example would be the sample plus NaI dissolved in glycerol. Spectra include peaks due to cationized molecules of MNa^+, $MNa(C_3H_8O_3)_n^+$, and $Na(C_3H_8O_3)_n^+$.

Rapid heating methods. A sample heated very rapidly may vaporize before it pyrolyzes. Techniques for heating by raising the temperature of a source probe on which the sample is coated by 200 K/s have been developed. Irradiation of an organic sample with laser radiation can move ions of mass up to 1500 daltons into the gas phase for analysis. This technique is the most compatible with analyzers which require particularly low pressures such as ion cyclotron resonance. Time-resolved spectra of surface ejecta are proving to be the most useful kinds of laser desorption spectra available.

Particle bombardment techniques. A solid sample or a sample in a viscous solvent such as glycerol may have ions sputtered from its surface by bombardment with accelerated electrons, ions, or neutrals. Bombardment by electrons is achieved simply by inserting a probe with the sample directly into the electron beam of an electron impact source (in-beam electron ionization). The energetic ions may likewise be the plasma in a chemical ionization source (in-beam chemical ionization), or alternatively, ions in a beam of 2.5 kV Ar^+ directed to a surface coated with the sample (without the coating, this is the inorganic surface analysis technique of secondary ion mass spectrometry). With charge exchange between accelerated Ar^+ (most commonly 6–8 keV) and thermal Ar, a beam of accelerated Ar atoms is produced; desorption of surface layers of glycerol solutions of biological molecules up to 15,000 molecular weight or inorganic salts $(Cs_x I_{(x-1)})^+$ up to 20,000) produces useful spectra by fast atom bombardment. The simplicity of application of this broadly useful technique is associated with the long duration of spectra because of the constantly regenerated surface of the solution after bombardment, and not so much (as originally thought) with the avoidance of surface charging by using atoms instead of molecules; beams consisting partially or wholly of accelerated Ar^+ ions produce spectra from viscous solutions in techniques more precisely called liquid secondary ion mass spectrometry (SIMS), or liquid-target SIMS. The most powerful method of this variety is the californium-252 plasma desorption source. It must be noted that nomenclature is not unified in this area; some authors also refer to ion-beam chemical ionization as plasma desorption.

The ^{252}Cf source produces fission fragments which penetrate a thin foil on which a film of sample has been coated. The fission fragment deposits energy by electronic excitation in a cylindrical track with a 20-nanometer diameter. Molecular, $(M + 1)^+$, and $(M - 1)^-$ ion ejection from the surface then occurs as the energy is dissipated through the remaining lattice in a shock wave. Molecular ions of mass up to 4000 daltons have been observed with this technique.

Multiphoton ionization of gaseous molecules. This is a subject of much academic interest, particularly on the range of mechanisms important in the process. If absorption of several photons from a laser produces an excited state of the molecule, molecules accumulate in this state so that absorption of further photons produces especially abundant ions (resonant multiphoton ionization) in contrast to the case where no excited state is reached as an intermediate (nonresonant multiphoton ionization). This tunability for a certain energy to achieve selective ionization of a particular molecule suggests that analysis of mixtures for especially targeted components would be possible. The difficulty is that for even modestly complex molecules the number of states in rotationally and vibrationally broadened bands is large, and the search for selectivity for target molecules is leading to use of supersonic nozzle introduction of the sample, where extreme cooling will yield states with insignificant population of higher rotational and vibrational levels, and ideally, electronic levels of nearly line width; thus interference due to excitation of bands of contaminants of the target will be greatly reduced.

Assignment of empirical formula. If the spectrum is a high-resolution spectrum, the deviation of the molecular weight from an integral value is used to determine the elemental composition (for example, $^{12}C_{12}{}^1H_{10}$ has molecular weight 154.0782; $^{12}C_9{}^1H_{14}{}^{16}O_2$, 154.0994; $^{12}C_{10}{}^1H_{16}{}^{16}O$, 154.1358; and $^{12}C_{11}{}^1H_{22}$, 154.1721). If it is low-resolution, then the worker takes advantage of the natural abundances of isotopes (for example, ^{12}C, 98.9%; ^{13}C, 1.1%; ^{35}Cl, 75.4%; ^{37}Cl, 24.6%; ^{79}Br, 50.6%; and ^{81}Br, 49.4%) to calculate at least part of the empirical formula by application of the binomial coefficients. In **Fig. 3** the intensity of the $(M + 1)$ peak is 11% that of the M peak. Now only ions containing one ^{13}C ion contribute to the $(M + 1)$ peak, and the probability that any C chosen randomly is ^{13}C is 1.1%. If 10 C atoms are examined as a group, the probability that one is ^{13}C is 10 times greater, or 11%; hence the molecule contained 10 carbons.

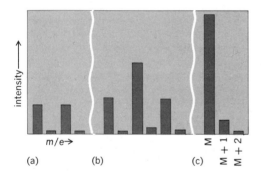

Fig. 3. Use of isotope distribution to determine elemental composition. (a) Molecular ion for a compound containing one Br atom. (b) Molecular ion for a compound containing two Br atoms. (c) Molecular ion for a compound containing neither Br nor Cl.

Fragmentation patterns. Since the amount of energy transferred to the molecule is much more than that required simply to ionize it or to break some of the bonds in the remaining ion, some of the molecules fragment after ionization, by competing consecutive decompositions, as indicated in reactions (10)–(15). These ions are separated by mass in the spectrometer and

$$A—B—C^+ \rightarrow A—B^+ + C\cdot \quad (10) \qquad A—B—C^+ \rightarrow A—B\cdot + C^+ \quad (11)$$

$$A—B—C^+ \rightarrow A\cdot + B—C^+ \quad (12) \qquad A—B^+ \rightarrow A + B^+ \quad (13)$$

$$B—C^+ \rightarrow B^+ + C \quad (14) \qquad A—B—C^+ \rightarrow A—C^+ + B \quad (15)$$

produce the mass spectrum. Although fragments at almost every mass are usually produced, the most favorable routes for decomposition, which give the most intense peaks in the spectrum, are characteristic of the functional groups in the molecule. Therefore, the intense ions, particularly those at the high-mass end of the spectrum, are used in the assignment of structure to compounds. The common mechanistic rationalizations used for interpretation assume that the charge is localized at the functional group, because the electron lost is most likely to be a π-electron or a nonbonding electron. Other less empirical explanations, notably the statistical approach of the quasi-equilibrium theory of mass spectra, have been outlined in the literature.

The reactions of functional groups bearing electronegative atoms are typically loss of the electronegative atom or the group containing it [X = Cl, Br, I, OR, SR, acyl, in reaction (16)] or

$$R\text{—}X \rightarrow R^+ + X\cdot \quad (16)$$

loss of a substituent group on the α-carbon [R=H or alkyl, X = OR, NR_2, SR, Cl, Br, in reaction (17); R= H, alkyl, or aryl, X = H, OH, OR, NR_2, Cl, Br, I, alkyl, or aryl, in reaction (18)]. These are

$$R\text{—}CH_2\text{—}X \rightarrow R\cdot + CH_2\text{=}X^+ \quad (17) \qquad R\text{—}\overset{\overset{O^{+\cdot}}{\|}}{C}\text{—}X \rightarrow R\text{—}C\text{=}O^+ + X\cdot \quad (18)$$

simple fragmentations [reactions (10)–(12)] corresponding to losses of a radical; the remaining ion is an even-electron ion. Except for the molecular ion, which is an odd-electron ion, the principal ions in a spectrum are even-electron ions unless special structural requirements are met. In these less common cases, prominent odd-electron ions, formed by the loss of a molecule, are found.

The most general case of odd-electron ion formation is illustrated in reaction (19), where in

$$\quad (19)$$

general a γ-hydrogen is abstracted by a multiply bonded electronegative atom with rupture of the

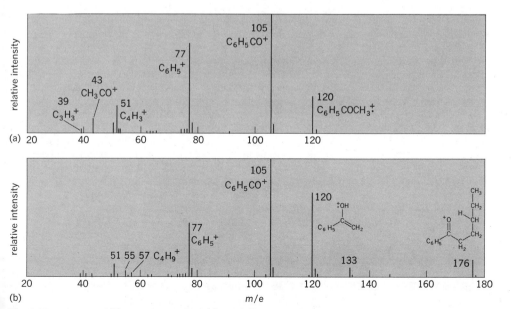

Fig. 4. Mass spectra of (a) acetophenone and (b) valerophenone, using electron impact.

β bond; this reaction is called the McLafferty rearrangement. Other cases of odd-electron ion formation involve interactions of ortho substituents on aromatic systems and certain decompositions of cyclic molecules, which may or may not be rearrangements [reaction (15)] Typical electron-impact mass spectra of volatile compounds are shown in **Fig. 4**. The spectrum of acetophenone is characterized only by simple bond-cleavage reactions, but there is an intense peak in the spectrum of valerophenone corresponding to a McLafferty rearrangement in which C_4H_8 is lost from the molecule.

Even-electron ions, such as many fragment ions found in electron-impact spectra and also (M + 1) ions from chemical ionization, decompose most often to smaller even-electron ions by loss of a small molecule. Illustrations are given by reactions (20) and (21), from the chemical

$$R_1C(\overset{\overset{\bar{O'}}{\|}}{\underset{+}{-}}\overset{H}{\underset{}{N}}H)-R_2 \longrightarrow R_1C\equiv O^+ + H_2N-R_2 \tag{20}$$

$$R_a-C(\!=\!O)-N(CH(R_b))-C(=O)\cdots H-\overset{+}{N}H_2-R_c \longrightarrow R_aC(=O)-N(CH(R_b))-C=O^+ + \overset{+}{N}H_3R_c \tag{21}$$

ionization of small peptides. Complex rearrangements are more competitive with simple cleavages in ions of low internal energy than in ions of higher internal energy; so in chemical ionization spectra, though only a few pathways for loss of small molecules may exist, some of the pathways may be found to be quite remarkable. Occasionally the fragmentations involve rearrangements still so poorly understood that the fragment peaks are not yet of much use in establishing the structure of the molecule.

Typical spectra for the various techniques discussed are shown in **Fig. 5** for the involatile compound creatine; both the molecular ion in Fig. 5a and the (M + 1) ions in Fig. 5b – d lose water by a rearrangement process. It is not universally true that the chance of finding the molecular ion increases as one progresses from electron impact to chemical ionization, field ionization, and field desorption, but these results are typical.

Further useful information about fragments is obtained, first, from the potentials required to ionize the molecule and to form fragments and, second, from observations of shifts of peaks with isotopic labeling techniques. For example, in reaction (19), substitution of D for H at the β position does not increase the mass of the product ion, but when H at the γ position is replaced by D, the mass of the product is shifted to higher mass by one unit. Finally, correlations with trends observed in ordinary chemical reactions yield information. For example, in both aromatic and aliphatic systems, orders of reactivities depending on substitution patterns show correspondence; in some aromatic systems the correlation is quantitative. If an unknown compound is suspected of belonging to a certain group of compounds, then its structure may often be established by comparison of its spectrum with spectra of compounds of known structure. This technique is particularly valuable in the study of natural products. The shift technique, the interpretation of substitution patterns when the molecular ion and some fragments are shifted by the same amount, for example, 30 units by a methoxy group, is helpful in studying alkaloid spectra. In addition to correlation with typical solution phenomena, relation of fragmentations to pyrolytic reactions and to photochemical processes is studied. The McLafferty reaction is analogous to a well-known photochemical reaction of ketones, the Norrish type II rearrangement, in which an olefin molecule is lost from the alkyl ketone while the enol form of a smaller ketone is produced by the influence of light. Other ionization techniques (by large electric-field gradients, by photons, or by other ions) produce different types of spectra useful in structural analysis.

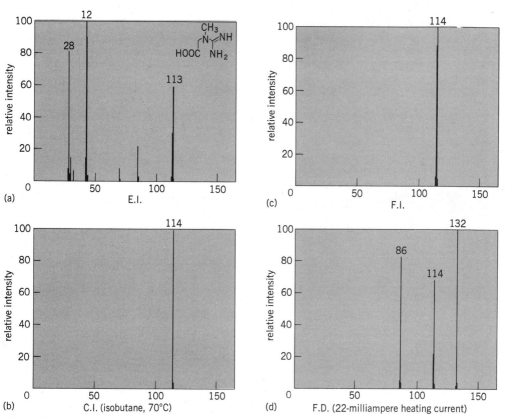

Fig. 5. Mass spectra of creatine (molecular weight 131) obtained by (a) electron impact, (b) chemical ionization, (c) field ionization, and (d) field desorption with moderate heating. Both the molecular ion (not observed in a) and the $M + 1$ ion in b, c, and d undergo a loss of water. Field desorption does not always produce the largest ion related to the molecular weight on comparing the four techniques, but this result is typical. (*After H. M. Fales et al., Anal. Chem., 47:207–219, 1975*)

Metastable ions. Correlation of spectra with structure requires the additional use of further decomposition products beyond the first fragmentation steps, but it is difficult to define the origin of products possibly arising by more than one path, for example, B^+ in reactions (13) and (14). A clue to origin of fragments is given by metastable ions: If in reaction (13) some AB^+ ions decompose after they are accelerated but before they are magnetically deflected, a broad peak (**Fig. 6**) appears at an m/e value numerically equal to m_B^2/m_{AB}. In Fig. 6 this broad peak is seen at $m/e = 56.5$. The value 56.5 equals $77^2/105$; therefore, the metastable peak corresponds to the reaction $105 \rightarrow 77$ and indicates that at least part of the $C_6H_5^+$ ion is formed by the loss of CO from $C_6H_5CO^+$. The order of the decomposition steps is suggested by the collection of metastable ions and bears on the organization of the parts of the molecule into the whole.

Study of metastable ions and other ions which can be made to decompose between acceleration and mass analysis has been carried out through ion kinetic energy spectroscopy (IKES) and mass-analyzed ion kinetic energy spectroscopy (MIKES), also called direct analysis of daughter ions (DADI). In the IKES technique, which may be studied with a conventional double-focusing mass spectrometer, those ions which decompose between the accelerating region and the electric sector will form ionic fragments with only a fraction of the kinetic energy imparted to the original ions. The fraction is the ratio of the mass of the fragment to that of the original ion. Scanning the

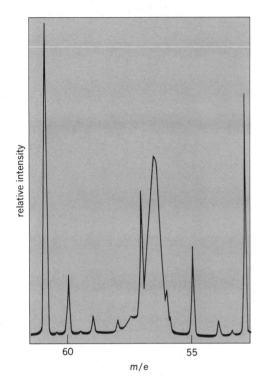

Fig. 6. Scale-expanded portion of acteophenone spectrum showing broad metastable ion at *m/e* 56.5.

voltage of the electric sector and detecting the ions which leave it provides a scan of ions according to their kinetic energy. Frequently intensities of ions differ substantially for even closely related isomers; thus this technique is useful for fingerprinting difficult compounds.

In MIKES, the electric sector and the magnetic sector in the double-focusing mass spectrometer are reversed, so that the ions, after acceleration, enter the magnetic sector and are analyzed according to mass before they enter the electric sector. By setting the magnetic field, ions of a certain mass are chosen for study; those which decompose before entering the electric sector have only the fraction of the kinetic energy according to the ratio of fragment and original mass noted above, and by sweeping the electric sector, the amounts of daughter ions from this one particular precursor may be observed. The intensities of ions are useful for structure determination, although they are somewhat affected by the distribution of energies within the fragmenting ion. Much information about energetics of decomposing ions has been gained from studies of the width of peaks due to their products in MIKES, for these are always broadened because kinetic energy T is released in the fragmentation. It has been shown that T is a characteristic of molecular structure; for example, the McLafferty product ion formed by the loss of C_2H_4 from the molecular ion of butyrophenone, as in reaction (19), fragments by loss of CH_3, releasing 48 meV (4.6 kJ/mole) of T. Since the molecular ion of acetophenone releases 7 meV (0.7 kJ/mole) energy, the McLafferty product does not have the acetophenone structure. Further studies of rearrangements have led to analyses of the partitioning of energy between internal and kinetic energy in fragmentation, as a function of geometry of the activated complex and other parameters. The study of metastable peak shape as a function of ion lifetime, or of ion internal energy, provides information on ion structure.

Unimolecular MIKE spectra, that is, MIKE spectra of metastable ions, are somewhat dependent on ion internal energy. A less energy-dependent method, and therefore a surer guide to structure, is that of collisionally activated decomposition (CAD), occasionally also known as collisionally induced dissociation (CID). In this, a collision gas is admitted at a low pressure to the region between the magnetic and electric sector of the reversed instrument. As mass-selected ions travel from the magnetic sector, they collide with the gas molecules, and a fraction of their kinetic energy is transformed into internal energy sufficient to cause rapid decomposition of the ions. The spectra are recorded as in mass-analyzed ion kinetic energy. Charge-exchange processes may also occur on collision, as in reactions (22)–(24). If the electric sector is set at twice,

$$A\text{—}B\text{—}C^{2+} + N \rightarrow A\text{—}B\text{—}C^{+} + N^{+} \qquad (22)$$

$$A\text{—}B\text{—}C^{+} + N \rightarrow A\text{—}B\text{—}C^{2+} + N + e \qquad (23)$$

$$A\text{—}B\text{—}C^{+} + N \rightarrow A\text{—}B\text{—}C^{-} + N^{2+} \qquad (24)$$

one-half, or minus the voltage which transmits normal ions, the products of reactions (22)–(24) can be detected. Thus mass spectra of ions produced by special routes are generated, and the utility of these spectra for interpretive purposes is being explored.

The spectra have been found to depend on the angle by which the products are deflected from their path in the absence of collision, deflects of only a half degree to one degree producing major changes in the spectra in the direction of more energy content in the ions (angle-resolved mass spectrometry).

Collision-induced dissociations in MIKE spectrometers involve ion accelerations of 4–8 kV. A similar experiment can be carried out in a triple quadrupole instrument, in which the acceleration is 2–100 eV. In the former, energy transfer is through electronic modes; in the latter, through vibrational modes. Because the kinetic energy of the ion strongly influences the collisionally activated decomposition product distribution in the latter experiment, studies of energy-resolved mass spectra of ions by this technique are of much interest, especially in comparison to the angle-resolved mass spectra.

These techniques for producing mass spectra from mass-resolved ions have become of paramount importance to establishing the structures of small ions, particularly in combination with more advanced methods of measuring heats of formation of ions and onset energies for ion formation.

Applications of computers. The principal applications of computers to analysis have been to data acquisition and structure interpretation. High-resolution spectra contain so much data that the rapid acquisition and presentation of data in a form easily assimilated by the operator has been adapted to the computer. Similarly, in cases where a gas chromatograph effluent passes through the source of the mass spectrometer, the generation of even low-resolution data is so rapid that a dedicated computer is appropriate. A variety of data displays are useful for interpretation. The plot of total ion current versus time produces a reconstructed chromatogram; the plot of ions of a single mass versus time, called a mass fragmentogram, is useful in identifying compound classes among the gas chromatogram peaks if the appropriate mass is chosen, or in identifying compounds directly if some other mass like the molecular weight of a desired component is chosen.

For interpretation two approaches have been used: library searching and training. Library searches of large collections (over 25,000 spectra) by comparison of the spectrum with known spectra yield degrees of closeness of agreement of the unknown and known spectra. Various algorithms for spectral comparison, using the 10 most intense peaks in the spectrum or the two most intense peaks in each 14-mass-unit segment, for example, have been devised, and the minimum amount of information to be supplied for a good chance of identification has been studied. The other approach involves several pattern recognition approaches in which various features of the spectra are correlated with structural characteristics by methods independent of formal theories of mass spectral interpretation; these include learning-machine and factor-analysis approaches. Hybrid techniques, in which the self-trained computer approach is augmented by selected tests derived from the fragmentation theory noted previously, have also been devised.

Analysis of mixtures. For a given instrument, the mass spectrum of a compound serves as a fingerprint of a compound under standard conditions; it varies with temperature, ionizing voltage, and parameters of the construction of the instrument. A mixture of gases may be analyzed because spectra are additive; hydrocarbon mixtures have been so analyzed quantitatively since the 1940s.

A further powerful technique for analysis of mixtures containing unknowns is mass spectral analysis of gas chromatographic effluents either by direct hookup of the gases exiting from the chromatograph, including carrier gas, to the introduction system of the spectrometer or by analysis of trapped effluent fractions.

For electron-impact or charge-exchange ionization of the sample, it is necessary to remove the carrier gas preferentially from the sample; several devices based upon preferential diffusion or effusion have been developed to achieve this. Gas chromatography using methane as carrier can be used directly as a method of introducing reagent gas and sample simultaneously for chemical ionization. Increased sensitivity can be achieved by monitoring the intensities of only a few peaks in the mass spectrum as the chromatograph effluents pass through the source, using rapid voltage switching between these peaks, instead of scanning the whole spectrum repeatedly; this technique is called multiple ion detection.

Similarly, mass spectrometers may be coupled to liquid chromatographs. There are three important methods for interfacing the instruments: moving belt, direct liquid introduction, and thermospray. In direct liquid introduction the effluent from the liquid chromatograph is split, and a small fraction is introduced directly into the source of the mass spectrometer, where it is vaporized; occasionally auxiliary ionization by electron ionization or chemical ionization is used. In thermospray the sample enters the source through a nozzle heated to 250–350°C (482–662°F), and rapid desolvation of solutes produces ions for analysis. Less common methods for interfacing include electrospray and monodisperse aerosol generation interface for liquid chromatography (MAGIC). Because of the new power of chromatography using supercritical fluids, studies on interfacing supercritical fluid chromatographs with mass spectrometers have been undertaken.

Ion cyclotron resonance spectroscopy, previously used to study reactions between ions and molecules (and thus the thermodynamics and kinetics of reactions in the absence of solvent), has become of special importance here because of the development of Fourier-transform ion cyclotron resonance (FT/ICR) or Fourier transform mass spectrometry (FTMS) techniques; extremely high resolution (>500,000) and very rapid acquisition of spectra (millisecond range) suggest FT/ICR applications to wall-coated open tubular column capillary gas chromatography, in which 1-s-wide effluent peaks present problems for other scanning mass spectrometric detection methods. The technique also has no upper mass limit inherent in instrument geometry.

Collision-induced dissociation forms part of a new technique in which an unseparated mixture, for example, a biological sample, is admitted to the source, perhaps by direct probe. Under chemical ionization conditions, (M + 1) ions of each component are produced. Each of these ions is transmitted in its turn through the magnetic sector, and a collision-induced spectrum of the component is obtained. Mass spectra are thus obtained without prior separation.

Analysis of inorganic solids. The analysis of solid inorganic samples can be made either by vaporization in a Knudsen cell arrangement at very high temperatures or by volatilization of the sample surface so that particles are atomized and ionized with a high-energy spark (for example, 20,000 eV). The wide range of energies given to the particles requires a double-focusing mass spectrometer for analysis. Detection in such instruments is by photographic plate; exposures for different lengths of time are recorded sequentially, and the darkening of the lines on the plate is related empirically to quantitative composition by calibration charts. The method is useful for trace analysis (parts per billion) with accuracy ranging from 10% at higher concentration levels to 50% at trace levels. Methods for improving accuracy, including interruption and sampling of the ion beam, are being studied.

Secondary ion mass spectrometry is most commonly used for surface analysis. A primary beam of ions accelerated through a few kilovolts is focused on a surface; ions are among the products sputtered from this surface, and they may be directly analyzed in a quadrupole filter. SEE LASER SPECTROSCOPY. SECONDARY ION-MASS SPECTROMETRY (SIMS); TRACE ANALYSIS.

Bibliography. F. Adams, R. Gijbels and R. VanGrieken, *Inorganic Mass Spectrometry*, 1987; R. G. Cooks, K. L. Busch, and G. L. Glish, Mass spectrometry: Analytical capabilities and potential, *Science*, 222:273–281, 289–291, October 21, 1983; S. Facchetti, *Mass Spectrometry of Large Molecules*, 1985; J. H. Futrell, *Gaseous Ion Chemistry and Mass Spectrometry*, 1986; A. G. Harrison, *Chemical Ionization Mass Spectrometry*, 1983; K. Levsen, *Fundamental Aspects of Organic Mass Spectrometry*, 1978; F. W. McLafferty, *Interpretation of Mass Spectra*, 3d ed., 1980; G. R. Waller, *Biochemical Applications of Mass Spectrometry*, 1972, suppl., 1980; F. A. White and G. M. Wood, *Mass Spectrometry: Applications in Science and Engineering*, 1986.

7

ANALYTIC TECHNIQUES

Trace analysis	246
Spectrophotometric analysis	247
Spectrochemical analysis	253
Emission spectrochemical analysis	256
Flame photometry	260
Combustion	260
Fluorometric analysis	262
Activation analysis	265
Proton-induced x-ray emission (PIXE)	267
Secondary ion mass spectrometry (SIMS)	268

TRACE ANALYSIS
ANDREW T. ZANDER

The determination of the elemental constituents of a sample in which these constituents make up approximately 0.01% of the sample or less. There is no sharp boundary between nontrace and trace constituents. The lower limit is set by the sensitivity of the available analytical methods and, in general, is pushed downward with progress in analytical techniques. A large number of different physical and chemical techniques have been developed for the measurement of the elemental composition at the microgram-per-gram and nanogram-per-gram level, thereby constituting the field of trace analysis.

Goals. Included in the goals of trace analysis are the determination of the bulk or total concentrations of the trace elements; the preconcentration of microconstituents into a small, essentially matrix-free sample to improve detectability or remove matrix interferences; the determination of local concentrations using probe techniques in order to establish the topographical distribution of trace elements in solid samples; and the determination of major and minor species in a minute initial sample. In the past much emphasis has been placed on analyses for bulk concentration, but there has been increased interest in the determination of local concentrations.

Methods. All trace analytical methods can be divided into three component steps: sampling, chemical or physical pretreatment, and instrumental measurement. Depending upon the type of information desired in an analysis and the requirements of sensitivity, precision, and other performance figures of merit, an appropriate instrumental technique is selected. Therefore, it is essential to know the capabilities and limitations of the various methods of instrumental measurement. Once they are known, appropriate steps can be taken in the sampling and pretreatment steps to provide sufficient microconstituent free of interferences and in the appropriate form for the final measurement. In a number of methods the pretreatment step may be omitted, and in others the sampling and measurement occur simultaneously. In spite of possible deviations, these steps are interrelated and require different degrees of emphasis, depending upon the individual analytical situation. In many cases the analyst uses a particular physical or physicochemical method in which manifestations of energy provide the basis of measurement. These methods are indirect in the sense that the emission or absorption of radiation or transformation of energy must be related in some way to the mass or concentration of the species that are being determined. The establishment of these relations almost invariably requires calibration, with the use of standards of known content of the constituent in question.

Techniques. Methods having applicability to trace analysis range from the more classical chemical methods of colorimetric and absorption spectrophotometric analysis to modern instrumental approaches. The wide diversity of methods is apparent from the number of different approaches and the attendant classes and subclasses in the accompanying **table**. Within any one of these categories there are still more specialized techniques.

Of the various criteria used in the selection of an appropriate trace analytical method, sensitivity, accuracy and precision, and selectivity are of prime importance. Other important considerations, such as scope, sampling and standards requirements, cost of equipment, and time of analyses, are of great practical significance.

Sensitivity and detection limits. The application of any analysis technique to trace problems is governed to a large extent by the analytical sensitivity that can be achieved for the species of interest in a given material. Sensitivity reflects the ability to discern a small change in concentration or amount of species of interest. Detection limit is a closely related, but distinct, term used to indicate the lowest concentration or amount that can be determined with a specific degree of confidence.

The lower limit of detection can be expressed in either of two ways. The absolute limit is the smallest detectable weight of the substance expressed in micrograms, nanograms, and so on. The relative limit is the lowest detectable concentration expressed as a percentage, parts per million, micrograms per gram, micrograms per milliliter, and the like, of the sample. The choice of the absolute versus the relative method is usually made on the basis of convenience or pertinence to the problem.

As a result of widely varying pathways of development of many analytical techniques, there

Trace analysis methods and limits of detection

Method	Limits of detection Absolute, g	Limits of detection Concentration, ppm	Method	Limits of detection Absolute, g	Limits of detection Concentration, ppm
Chromatography			Optical-emission		
Thin-layer	10^{-5} to 10^{-3}		AC spark		10^1 to 10^3
Gas-liquid		10 to 10^6	DC arc		10^{-2} to 10^2
Liquid-liquid		10^{-3} to 10^0	Electrical plasma		
Electrochemical			DC jet		10^{-4} to 10^2
Coulometry	10^{-9} to 10^0		RF-induced		10^{-4} to 10^2
Ion selective electrode		10^{-2} to 10^2	Microwave-induced	10^{-9} to 10^{-6}	10^{-2} to 10^1
Polarography			Optical-fluorescence		
Conventional		10^0 to 10^3	Atomic		
Modern		10^{-3} to 10^3	Flame		10^{-3} to 10^2
Laser probe microanalysis		10^2 to 10^4	Nonflame	10^{-15} to 10^{-9}	
Nuclear			Molecular		10^{-3} to 10^{-1}
Neutron activation		10^{-3} to 10^{-1}	Optical-phosphorescence		10^{-3} to 10^2
Electron probe		10^2 to 10^3	Optical-Raman		10^0 to 10^5
Ion probe		10^{-1} to 10^1	Spectrometric-resonance		
Mass spectrometry			Nuclear magnetic		10^1 to 10^5
Isotope dilution		10^{-5} to 10^6	Electron spin	10^{-9} to 10^{-6}	
Spark source		10^{-3} to 10^1	Thermal analysis	10^{-5} to 10^{-4}	
Organic microanalysis	$>10^{-5}$		Wet chemistry		
Optical-absorption			Gravimetry	10^{-3} to 10^{-2}	
Atomic			Titrimetry		10^{-2} to 10^4
Flame		10^{-3} to 10^1	X-ray spectrometry		
Nonflame	10^{-15} to 10^{-9}		Auger		10^3 to 10^5
Molecular			ESCA		10^3 to 10^5
UV-visible		10^{-3} to 10^2	Fluorescence		10^{-1} to 10^2
Infrared		10^3 to 10^6	Mössbauer		10^0 to 10^3
Microwave		10^0 to 10^3	Photoelectron		10^0 to 10^3

is a lack of consistency in the definition or specification of detection limits in the analytical literature. Consequently, it is difficult to make critical comparisons between methods for a particular element. However, since a particular element may be determined by a number of different techniques, depending upon the matrix in which it is being sought, a summary of the experimental values that have been published is pertinent. These values are given in the table. SEE ACTIVATION ANALYSIS.

Bibliography. P. Benes and V. Majer, *Trace Chemistry of Aqueous Solutions: General Chemistry and Radiochemistry*, 1980; K. Beyermann, *Organic Trace Analysis*, 1984; G. Christian and J. B. Challis (eds.), *Trace Analysis: Spectoscopic Methods for Molecules*, 1986; W. J. Kirsten, *Organic Elemental Analysis: Ultramicro, Micro and Trace Methods*, 1983; J. F. Lawrence (ed.), *Trace Analysis*, vols. 1–4, 1981–1985.

SPECTROPHOTOMETRIC ANALYSIS

JAMES N. LITTLE

A method of chemical analysis based on the absorption or attenuation by matter of electromagnetic radiation of a specified wavelength or frequency. The region of the electromagnetic spectrum most useful for chemical analysis is that between 200 nanometers and 300 micrometers. Since the sample being analyzed absorbs the radiation, spectrophotometric analysis is sometimes referred to as absorptimetric analysis.

The instruments used in this work are referred to as spectrophotometers. A simple spectrophotometer consists of a source of radiation, such as a light bulb; a monochromator containing a

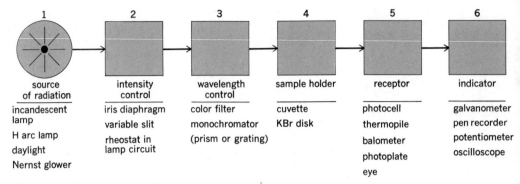

Fig. 1. Block diagram of generalized spectrophotometer.

prism or grating which disperses the light so that only a limited wavelength, or frequency, range is allowed to irradiate the sample; the sample itself; and a detector, such as a photocell, which measures the amount of light transmitted by the sample. See **Fig. 1**.

Bouguer-Lambert-Beer law. By using a spectrophotometer, the intensity of the light transmitted through an absorbing substance may be compared with the light intensity when no such substance is in the light beam. Two fundamental laws govern the intensity of the light transmitted by an absorbing material. The first law, called the Bouguer-Lambert law, is given as Eq. (1). Here I_0 is the intensity of the light beam with no sample present, I is the intensity of the

$$\log (I_0/I) = Kb \tag{1}$$

light beam after passing through the sample, K is a constant depending on the sample and wavelength of the light, and b is the thickness of the absorbing solution. The second law, called Beer's law, is shown by Eq. (2). Here I and I_0 are as above, K' is a constant depending on the sample

$$\log (I_0/I) = K'c \tag{2}$$

and wavelength of the light, and c is the concentration of absorbing material in the sample. Usually these two laws are combined in the form of Eq. (3), where I_0, I, b, and c are as described

$$\log (I_0/I) = abc \tag{3}$$

above, and a is a constant called the absorptivity or extinction coefficient.

Two other terms are commonly used in spectrophotometric analysis. These are transmittance T and absorbance A as defined by Eqs. (4) and (5). The absorbance A is directly proportional

$$T = I/I_0 \tag{4} \qquad A = \log (1/T) = \log (I_0/I) = abc \tag{5}$$

to the length of the light path through the sample and to the concentration of the absorbing material. It is the term most used in quantitative spectrophotometric work.

The Bouguer-Lambert-Beer law is strictly obeyed only when monochromatic radiation, that is, radiation of a single wavelength or frequency, is used. The monochromators of most commercial spectrophotometers produce radiation which is close enough to monochromatic so that the deviations from the law from this source are minor, except in the infrared region. There are, however, occasional deviations from this law for both chemical and instrumental reasons.

In most quantitative analytical work, a calibration or standard curve is prepared by measuring the absorption of known amounts of the absorbing material at the wavelength at which it strongly absorbs. Such a calibration curve is shown in **Fig. 2** for the absorbing material whose absorption spectrum is shown in **Fig. 3**. The absorbance of the sample is read directly from the measuring circuit of the spectrophotometer. Calibration curves are usually linear, as in Fig. 2. Occasionally, instrumental or chemical factors lead to nonlinear curves.

Absorption spectra. When the transmittance of the absorbance of a sample is measured and plotted as a function of wavelength, an absorption spectrum (Fig. 3) is obtained. This spec-

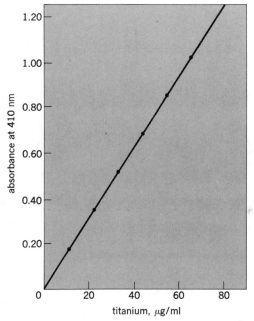

Fig. 2. Calibration curve for determination of titanium by its color formed with hydrogen peroxide.

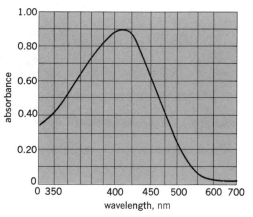

Fig. 3. Absorption spectrum of the peroxytitanate complex in region 340–700 nm.

trum indicates that the sample transmits the least light at 410 nm and transmits the most around 700 nm. Because of instrumental sensitivity limits, absorption spectra are obtainable only on samples which are relatively transparent to the radiation that is being used.

Reflectance spectra. In opaque samples, such as solids or highly absorbing solutions, the radiation reflected from the surface of the sample may be measured and compared with the radiation reflected from a nonabsorbing or white sample. If this reflectance intensity is plotted as a function of wavelength, it gives a reflectance spectrum. Reflectance spectra are used most often in matching colors of dyed fabrics or painted surfaces; they are used occasionally in qualitative analysis but seldom in quantitative analysis.

Chromogen. A molecule which absorbs radiation in a particular spectral region, usually in the visible or ultraviolet, is called a chromogen.

Chromophore. Groups of atoms within a molecule which are responsible for the absorption of light in the visible or ultraviolet regions are called chromophores. These chromophores are usually resonating structures which absorb at approximately the same wavelength, or frequency, regardless of the molecule to which they are attached. Examples of such groups are the phenyl group, C_6H_5, which absorbs at 270 nm and the azo group $N{=}N$, which absorbs at 370 nm.

Auxochrome. Substituent groups which affect the wavelength of the spectral regions of strong absorption of chromophores are called auxochromes. Auxochromes cause two types of wavelength shifts. A shift to longer wavelength, or lower frequency, is called a bathochromic shift. Conversely, a shift to shorter wavelength is called a hypsochromic shift.

Infrared spectrophotometry. The interaction with matter of electromagnetic radiation of wavelength between 1 and 300 μm induces either rotational or vibrational energy level transitions, or both, within the molecules involved. This region from 1 to 300 μm is usually referred to as the infrared. The frequencies of infrared radiation absorbed by a molecule are determined by its rotational energy levels and by the force constants of the bonds in the molecule. Since these energy levels and force constants are usually unique for each molecule, so also the infrared spectrum of each molecule is usually unique. The qualitative analytical use of the infrared region is

Fig. 4. High-resolution infrared spectrophotometer. (*Beckman, Scientific Instruments Division*)

based on this fact. Because of their individuality, infrared spectra of organic compounds are considered equivalent to, or superior to, the preparation of chemical derivations for the identification of species in organic chemistry. For this reason the infrared portion of the spectrum is often called the fingerprint region (**Figs. 4** and **5**).

For radiation sources, infrared instruments usually use a hot filament called a Nernst glower, or a hot carborundum rod called a Globar. Various inorganic prisms are used in the monochromators for different regions, for example, rock salt (sodium chloride) from 2 to 15 μm, potassium bromide from 15 to 27 μm, or cesium bromide from 12 to 40 μm. Gratings are also used in monochromators, either alone or in conjunction with prisms. A wide variety of detectors are used; examples are thermocouples and thermistors. These detectors must be very sensitive, as the amount of energy they must detect is quite small. Infrared cells, or sample containers, are prepared from materials transparent in the region of interest, and usually are made of rock salt, potassium bromide, or some other inorganic salt.

Although almost every compound has a unique infrared spectrum, various groupings within a molecule have well-defined regions of absorption. For example, hydroxyl groups in alcohols absorb strongly at 2.8, 7.3, and about 8.5 μm; ester carbonyls absorb from 5.7 to 5.8 μm; and free amino groups absorb at 3.0 and from 6.1 to 6.4 μm. A typical infrared spectrum, that of acetophenone ($C_6H_5COCH_3$), is shown in Fig. 5. The band at 3.3 μm is due to the methyl group (CH_3),

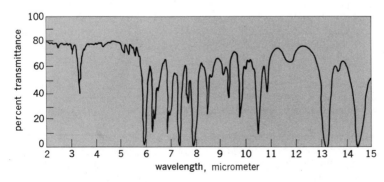

Fig. 5. Infrared spectrum of acetophenone.

that at 5.9 μm to the conjugated carbonyl group (CO), and those at 13.2 and 14.4 μm are due to monosubstituted benzene (C_6H_5).

Quantitative analysis is based on the Bouguer-Lambert-Beer law as applied to a specific absorption band, usually unique to the compound being determined, as shown in Figs. 2 and 3. Occasionally, adequate quantitative results can be obtained using data for a similar compound containing the same functional group as in the material measured. The accuracy and precision of most infrared analyses is ±3–5% of the amount of material present.

Infrared spectrophotometry can be applied to gaseous, liquid, or solid samples. For gaseous samples, cells (sample containers) from 1 cm (0.4 in.) up to 50 m (160 ft) long are used in order to get enough molecules in the light path to measure. In the long cells, the length is obtained by using mirrors to make the light traverse the cell numerous times before it is measured. Liquid samples are handled in cells whose thicknesses vary from 0.1 mm to 1 cm (0.04 to 0.4 in.). The most common solvents for liquids are carbon tetrachloride and carbon disulfide, since these solvents have few absorption bands in the infrared region. Water is a very poor solvent because it absorbs strongly in this spectral region. Solid samples are analyzed by (1) preparing a thin film which may be either a free film or a film cast on a salt plate; (2) preparing a paste or mull by grinding up the solid with a viscous material such as mineral oil (Nujol), which has few infrared bands; or (3) pressing a disk of an intimate mixture of the solid with potassium bromide, KBr. This latter, the so-called KBr disk method, appears to be the best for quantitative infrared work with solids.

The infrared region is used primarily for analyses of organic compounds because they are readily soluble in a desirable solvent and because they have unique and complex spectra. However, work has been done on the infrared spectra of inorganic compounds in the forms of KBr disks or Nujol mulls.

Attenuated total reflectance. A technique that is useful, especially for infrared measurements, is attenuated total reflectance, also called frustrated internal reflectance and internal reflectance spectroscopy. With this technique, the infrared spectrum of a surface can be obtained without any chemical treatment of the sample; its principle is based on the phenomenon of energy reflection at the interface of two media which have different refractive indexes and are in optical contact with each other.

The sample to be analyzed is positioned on a reflectance prism and placed in the light path of the infrared source. The beam of infrared energy is absorbed at those wavelengths where the sample normally absorbs; at other wavelengths the infrared energy is reflected. The resulting spectrum is similar to that of a transmission spectrum of a material. This technique is used extensively for analyzing coatings, pastes, paints, fibers, and fabrics.

Near-infrared spectrophotometry. This designates work carried out between 0.78 and 3 μm. The instruments in use in this region have quartz prisms in their monochromators and lead sulfide photoconductor cells as detectors. The absorption bands in this region are mainly overtones (harmonics) of bands in the infrared region. These bands are quite sharp and are of great value in quantitative analysis for various functional groups, particularly those containing hydrogen atoms; examples are terminal methylene groups (CH_2), hydroxyl groups, and amines. The cells used in this region are usually made in quartz, and hence are more durable than infrared cells. The near-infrared spectra give some of the same information as those in the infrared region, but occasionally more specifically, more inexpensively, or more rapidly. The accuracy and precision of ±1–3% of the amount present is also somewhat better than in the infrared spectrophotometry.

Visible spectrophotometry. The visible region of the spectrum covers the narrow range from about 380 to 780 nm. The spectrophotometers for this region use tungsten lamps as light sources, glass or quartz prisms or gratings in the monochromators, and photomultiplier cells as detectors. Within this narrow portion of the electromagnetic spectrum, a majority of the spectrophotometric analyses are made. The absorption of light in this region is caused by the excitation of the outer electrons of the molecule by the impinging light beam. Figure 3 shows a typical visible absorption spectrum. The substance, peroxytitanate ion, absorbs light in the region below 500 nm; that is, it absorbs violet, blue, and green light and transmits red, orange, and yellow. For analytical work, the wavelength of maximum absorption is usually used, in this case 410 nm. The calibration curve in Fig. 2 was made by plotting the absorption at 410 nm versus the concentration of the absorbing material in the solution. Unknown samples are then analyzed by measuring the absorbance of the solution after appropriate reagents have been added. The amount of material in question, in this case titanium, is obtained from the calibration curve.

Although few materials, in particular inorganic ions, have visible colors, there are spectrophotometric methods utilizing visible colors for most of them. The method used above for titanium is typical. Hydrogen peroxide, when added to a colorless titanium solution, forms the highly colored peroxytitanium complex. Similar reactions are those of thiocyanate ion with ferric iron and of ammonia with copper. Organic reagents have been prepared which form intense colors with different metal ions. Many of these reagents are so specific that they form colors with but one or two inorganic ions. Examples of these are o-phenanthroline which reacts with ferrous ion and 2.2′-biquinoline with copper.

Visible spectrophotometry is used extensively because there are methods available for determining a wide variety of materials, especially inorganic cations, with great sensitivity and selectivity. Visible spectrophotometry is usually carried out with liquid samples and is usually used for quantitative, rather than qualitative purposes. Visible reflectance spectra of solid samples are run where the matching of colors is important or in cases where the samples are opaque. Gaseous samples seldom are run in the visible region since they seldom absorb visible light intensely.

Ultraviolet spectrophotometry. The spectral region from 200 to 400 nm, called the near ultraviolet, is commonly used in chemical analysis. The absorption of ultraviolet radiation by a molecule is usually the result of exciting the outer, or valence, electrons of the molecule in question. The more easily the electrons are excited, the longer the wavelength of the absorption peak.

Ultraviolet spectrophotometers usually have a hydrogen lamp as a radiation source; a quartz prism, or a grating in the monochromator; and a photomultiplier tube as a detector. Quartz or silica cells of 1-mm to 10-cm (0.04- to 4-in.) length are commonly used for the samples.

Simple inorganic ions and their complexes as well as organic molecules can be detected and determined in this region. Useful solvents are water, saturated hydrocarbons, aliphatic alcohols, and ethers. Organic compounds which absorb ultraviolet radiation have at least one unsaturated linkage, such as C=C, C=O, N=N, or S=O, which acts as a chromophore. The wavelength of the absorption peak increases with the degree of unsaturation within the chromophore.

Inorganic groups which absorb in the ultraviolet region owe their activity to numerous valence electrons, as in the complex $FeCl_4^-$, or to electrons in a single atom, possibly hydrated, as in the rare-earth elements.

Quantitative work in the ultraviolet region is common, and most substances obey the Bouguer-Lambert-Beer law over a wide range. In the field of organic analysis, the ultraviolet region is most applicable to aromatic compounds. The spectra of some compounds, such as phenols, may be greatly enhanced by using basic solutions of the samples. Since most compounds which absorb in the region have intense bands, it is possible to analyze either dilute solutions or extremely small samples. For example, the spectrum in **Fig. 6** is that of 0.01% acetophenone in isooctane in a 1-

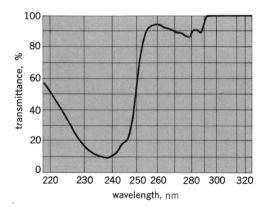

Fig. 6. Ultraviolet absorption spectrum of acetophenone in isooctane.

cm (0.4-in.) cell. By using longer cells, it is possible to detect 0.00001% acetophenone in isooctane. Many substances absorb much more intensely than acetophenone.

Similarly, ultraviolet spectrophotometry is especially useful for the determination of inorganic ions as simple complexes, such as $FeCl_4^-$, $PtCl_6^{2-}$, or I_3^-. Accuracy and precision are usually about $\pm 1-2\%$ of the amount of material being determined.

As indicated above, samples are usually liquids although gases may also be analyzed. Because of relatively greater scattering of short-wavelength radiation, transmission measurements on turbid samples are difficult to interpret, and opaque samples are seldom run by reflectance.

Filter photometry. In filter photometry, the monochromator of the spectrophotometer is replaced by a filter. This filter passes a band of light of a much wider range of wavelengths than those passed by even the poorest monochromator. For the example in Fig. 3, a filter which transmits blue light would be used in order to obtain maximum sensitivity. The filter chosen is usually of the color complementary to that of the solution, that is, the filter is chosen so as to transmit best the light which the sample absorbs most. In the visible region, colored glass or gelatin films containing dyes have been most widely used. Interference filters, based upon selective transmission of radiation through very thin metallic films between two glass plates, usually yield more nearly monochromatic bands and higher transmittance. They are used when their somewhat greater costs can be justified. Absorbing liquids and gases have been used as filters but are generally more cumbersome.

Filter photometers are generally much less expensive from spectrophotometers. Because they do not use monochromatic light, the calibration curves obtained often do not obey the Bouguer-Lambert-Beer law. However, by careful use of calibration curves, filter photometers can give sufficiently accurate and precise results for a wide variety of applications.

Although filter photometers are most often used in the visible region, some filters for the infrared and ultraviolet regions are available. *See* MOLECULAR STRUCTURE AND SPECTRA.

Bibliography. American Institute of Physics, *The Spectrophotometer*, 1975; A. M. Gillespie, *A Manual of Fluorometric and Spectrophometric Measurements*, 1985; Z. Marczenko, *Spectrophotometric Determination of Elements*, 1976; C. N. Rao, *Ultraviolet and Visible Spectroscopy: Chemical Applications*, 3d ed., 1975.

SPECTROCHEMICAL ANALYSIS
ANDREW T. ZANDER

Any of a number of techniques for the determination of the presence or the concentration of elemental or molecular constituents in a sample through the use of spectrometric measurements. Spectrometric measurements entail the monitoring of electromagnetic radiation as it is caused to emanate from, or interact with, the sample of interest.

Physical basis for measurements. The interaction of an analyte with electromagnetic radiation is based on changes in the level of some characteristic energy state of the analyte, for example, the oscillatory motion of a chemical bond, or the orbital location of a valence electron, or the rotational motion of the magnetic vector of an atomic nucleus. All of these types of characteristic states are quantized in energy by the principles of quantum mechanics. The change in energetic state of an analyte can be caused by the absorption or emission of energy of an amount exactly equal to the difference in energies of the two states.

Electromagnetic radiation is that manifestation of energy which is described by Planck's equation, $E = hc/\lambda = h\nu$ where E is the energy, h is Planck's constant, c is the speed of light, λ is the wavelength, and ν is the frequency of radiation. Energy is thus characterized by discrete wavelengths or frequencies, and the range of all wavelengths is called the electromagnetic spectrum.

Therefore, a spectrometric measurement is the result of the interaction of a particular wavelength and quantity of radiation with some characteristic energy state of an analyte so as to cause the energy state of the analyte to change, as depicted in **Fig. 1**. Considering hydrogen as an example, the location of the electron (*e*) of a hydrogen atom in the 1*s* orbital corresponds to a specific energy, here called E_1. The 1*p* orbital has a different, higher energy associated with it. If

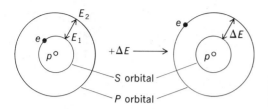

Fig. 1. Change in energy state of the electron in a hydrogen atom when light of wavelength λ_H is absorbed by the atom.

an amount of energy exactly equal to the difference between E_1 and E_2, that is, $\Delta E = E_2 - E_1$, is imparted to the hydrogen atom, then the electron may be raised to the higher orbital. Since $\Delta E = hc/\lambda_H$, the energy transition corresponds to electromagnetic radiation of a specific wavelength, here called λ_H. Therefore, if a beam of light of wavelength λ_H is caused to impinge on a hydrogen atom, that light will be absorbed from the beam as the electron changes orbitals, and the atom will have gained energy. An atom other than hydrogen will have different levels of energy corresponding to the same wavelength of radiation would be used. This is the basis of qualitative analysis. More atoms of any kind will require more light to cause the transitions. The amount of light used is directly proportional to the number of atoms present. This is the basis of quantitative analysis. *See* Energy level.

Spectrometric interactions. There are different types of spectral interaction, or radiative transitions, which can occur, depending upon the condition of the characteristic energy state involved prior to the interaction. The interactions are absorption, emission, and luminescence of radiation, depicted schematically in **Fig. 2**.

Transitions between the lowest energy level and another level are called resonance transitions. Nonresonance transitions occur between two levels, neither of which is the lowest level. Absorption occurs when a system gains energy through the retention of the energy associated with electromagnetic radiation incident on the system. Emission is the opposite of absorption. A system loses energy, and the energy loss is manifested as electromagnetic radiation. Fluorescence is a process in which energy gained radiatively is immediately lost by a system, also radiatively. Phosphorescence is like fluorescence, but upon gaining energy from incident electromagnetic

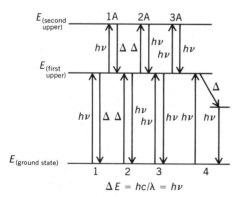

key:
1 = resonance absorption
1A = nonresonance absorption
2 = resonance emission
2A = nonresonance emission
3 = resonance fluorescence
3A = nonresonance fluorescence } luminescence
4 = phosphorescence
Δ = any transition mechanism other than radiative

Fig. 2. Types of spectral interaction.

radiation, the system loses part of it by nonradiative means (for example, through collisions if the system is made up of atoms). The loss of the rest of the energy gained in the absorption process is then emitted as electromagnetic radiation. Since the energy lost radiatively is less than that gained radiatively, the wavelength of the emitted electromagnetic radiation will be longer than that of the incident electromagnetic radiation.

In addition to the monitoring of these transitions, it is possible to monitor the manner in which the polarization of the radiation changes as it is absorbed (circular dichroism) or emitted (optical rotary dispersion). As a consequence, there are many different techniques of spectrometric measurement. They differ depending upon what type of radiation is monitored, what type of transition is involved, and whether some characteristic of the radiation is also observed.

Qualitative and quantitative. The analytical determination of the presence of a constituent (qualitative determination) results from sensing the energetic transition as it occurs. Each particular element or molecule will possess its own energetic characteristics, so that monitoring a specific transition will identify a constituent, exactly analogous to identifying people through the use of fingerprints. The determination of the concentration of a constituent (quantitative determination) is a result of the direct relationship between the amount of radiation which is emitted or absorbed and the amount of element or molecule present.

Electromagnetic spectrum regions utilized. The electromagnetic spectrum extends both higher and lower in energy and wavelength on either side of the visible spectrum, which is only a small portion of it, as shown in **Fig. 3**. As equipment was developed for doing experiments in any part of the electromagnetic spectrum, that portion became available for, and has been used in, spectrometric measurements. Consequently, there are spectrometric methods which range from gamma-ray spectrometry at the high-energy end of the spectrum, through ultraviolet/visible spectrophotometry, to radio-frequency resonance methods and sonic imaging. These methods, and all others, are separated just by the differences in resolving and detection equipment required, not by any inherent differences in the conduct of a spectrometric measurement. SEE GAMMA RAYS; SPECTROPHOTOMETRIC ANALYSIS.

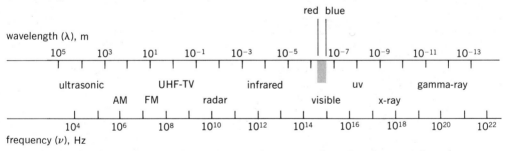

Fig. 3. Electromagnetic spectrum, showing approximate frequency and wavelength ranges of various regions.

Basic spectrometric measurement. In the basic spectrometric experiment which is performed for spectrochemical analysis, the sample is put into a state in which it will interact in the manner desired with the radiation of choice (**Fig. 4**). Examples are techniques in which a molecular compound is placed in a magnetic field so that the magnetic vector of the hydrogen nuclei in the molecule will line up with the magnetic field and thus be available to absorb radio-frequency radiation as they precess about the lines of force; or a solution of a sample is sprayed into a flame so that the sample is decomposed, leaving only its atoms, making the atoms available for absorption, emission, or luminescence of light; or the sample may not need preconditioning and only require being held in place. For absorption and luminescence measurements, a source of the radiation of choice, for example, a tungsten filament lamp, a gamma-ray emitter, or a laser, is focused onto the sample so that the radiation may be absorbed. For emission measurements

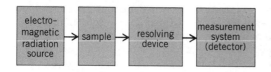

Fig. 4. Schematic diagram of the basic spectrometric experiment.

the sample is supplied with external energy which is not radiative, for example, the heat from a flame, or an applied voltage, or momentum from mechanical motion. The sample then emits the gained energy as photons (packets, quanta) of radiation.

The next stage of the experiment separates all the wavelengths of radiation present so that they may be measured independently of all others. Finally, there is a device for detecting the amount of radiation present (emission or luminescence) or absent (absorption). The identity of the wavelength or wavelengths gives qualitative information. The amount of radiation gives quantitative information.

Except for rather dramatic differences in the character of the equipment needed to perform the spectrometric measurements (for example, an x-ray source is not at all similar in construction or operation to a radio-frequency transmitter; nor is a gamma-ray detector similar to an ultrasonic receiver), all spectrometric measurements are performed in the sequence described. Spectrochemical analysis follows from relating the spectrometric measurement to the quality or quantity of sample constituent.

Bibliography. G. W. Castellan, *Physical Chemistry*, 3d ed., 1983; H. H. Willard, L. L. Merritt, and J. A. Dean, *Instrumental Methods of Analysis*, 5th ed., 1974; J. D. Winefordner, *Trace Analysis: Spectroscopic Methods for Elements*, 1976.

EMISSION SPECTROCHEMICAL ANALYSIS
Andrew T. Zander

A technique used in qualitative or quantitative chemical analysis and conducted by monitoring and measuring the spectrum of light caused to be emitted by the material to be analyzed. In general, there are many ways in which to conduct an emission spectrometric measurement; the differences among approaches result mainly from the choice of location within the electromagnetic spectrum at which to observe emitted radiation. However, emission spectrochemical analysis traditionally refers to those analytical determinations based on radiation in the visible through vacuum ultraviolet region of the electromagnetic spectrum (wavelengths about 800 to 100 nanometers). The technique is used principally to detect (qualitative analysis) and determine (quantitative analysis) metals and some nonmetals. Under optimum conditions, as little as 10^{-10} g of an element per gram of sample can be determined. Routine concentration ranges in which emission spectrometry is used are from approximately 10^{-8} g per gram of sample to 10^{-2} g per gram of sample (1% by weight).

The steps in emission spectrochemical analysis are: vaporization and atomization of sample; excitation of atomic vapor; resolution of emitted radiation; and observation and measurement of resolved radiation.

Vaporization and atomization. A number of approaches to emission spectrochemical analysis have been developed, all of which follow the four steps listed, but which differ fundamentally in the first two steps. These steps are involved with the transfer of energy to the sample so that the sample will emit characteristic radiation. The application, control, and transfer of different forms of energy pose significantly different equipment requirements. Therefore, different versions of emission spectrometry have developed. The principal versions of vaporization, atomization, and excitation are by: ac spark, dc arc, electrical plasmas (dc plasma jet, radio-frequency-induced plasma, microwave-induced plasma), and chemical flame.

It is necessary first to produce a vapor (vaporization) in which all the compounds of the sample are broken down into their constituent atoms (atomization). It is these individual atoms which are detected and quantitated. If the sample is originally a solid, it can be made part of an anode-cathode pair. When an electric discharge is struck across these electrodes, for example producing an arc spark or a dc arc, the sample is vaporized into the electrode gap. There are numerous ways to configure an electrode such that the sample is an integral part of it; the desirable ones make it possible to reproduce the procedure simply and easily so that the technique is routinely applicable.

A more exotic way to produce a cloud of atoms from a solid sample is to fire a high-power pulse of laser radiation at the sample. The material vaporized when the small laser crater is blasted out is in just the state required for an atomic emission experiment. This approach has not yet been used extensively because high-pulsed power lasers are extremely expensive, and sample vapor cannot be reproduced in quantity. Consequently, it is not a cost-effective approach, and it is restricted to qualitative or semiquantitative analyses at best.

A solid sample may also be disolved to bring it into an accessible solution state. The liquid solution sample can then be made part of an electrode. However, there are simpler ways in which to use liquid samples.

Dissolved solids or liquid samples can be broken down into microscopic-sized droplets, in a process called nebulization, and a stream of these micrometer-sized droplets can be injected directly into a discharge of hot gas or a flame. The hot gas then evaporates the solvent in the droplet, vaporizes the resulting solid particle, and atomizes the compounds of which the particle is made.

The hot gas can be produced in many ways. The complex mixture of atmospheric constituents and the electrode material in a dc arc discharge form a gas which reaches about 7500 K. Nebulized sample has been sprayed into dc arcs, but this approach tends to destabilize the discharge, resulting in poor signal precision.

A variety of techniques, in which sample nebulization systems and components for generation of a stable hot gas work together, have been developed. The earliest and most successful of these techniques is that of nebulizing the sample, mixing it homogeneously with the fuel and oxidant in a burner manifold, so that as the flame burns, the sample is vaporized and atomized. Many approaches for nebulizing the sample, mixing the droplets and gases, and configuring the burner manifold and burner head have proved successful. However, chemical flames (at least the types which are easily managed in the laboratory) reach temperatures of only 3000–4000 K. As a consequence, many compounds, for example refractory oxides and carbides, are not atomized because they are capable of existing at these temperatures. *See* F*lame photometry.*

The general term electrical plasmas is used to refer to a group of discharges developed for the generation of temperatures close to 10,000 K. No known chemical compounds can exist at this temperature. These electrical plasmas, which will be discussed below, represent nearly optimal devices to use for vaporization and atomization, for they have been developed in parallel with sample injection systems so that signal stability is retained.

Excitation of atomic vapor. In the majority of spectrochemical instruments, the same process is used to produce the atomic vapor and to excite the atoms to cause emission of characteristic radiation. Mechanisms other than collisional mechanisms contribute to transfer of energy to analyte atoms. These might be: resonance with an applied electromagnetic field, excited-state interlevel transfer, radiative transfer, or other suprathermal processes. The extent of contribution from steps such as these, particularly when the experiment is conducted at atmospheric pressure, is not clear, and also not likely to be large.

In any case, the energy which is used resides in a hot gas. The hot gas is produced in a number of different ways. A chemical flame produces a hot gas from the combustion of some sort of fuel, for example acetylene or nitrous oxide (laughing gas), with the release of many thousands of calories of heat. Chemical flames normally have temperatures between 1000 and 4000 K.

An electric discharge which is caused by applied potentials or applied currents across an electrode gap (ac spark or dc arc) produces a hot gas of atmospheric species, electrode material, and electrons. Temperatures up to about 7500 K are produced.

Newer kinds of electric discharges, termed electrical plasmas, produce hot gases of much higher temperature. A plasma is a cloud of vapor in which more than 1% of the atoms are ionized, that is, have their outer electrons stripped away from them. In general, the greater the extent of ionization, the greater the amount of energy resident in the hot gas.

Modern electrical plasmas are produced in two principal ways. Direct-current (dc) plasma

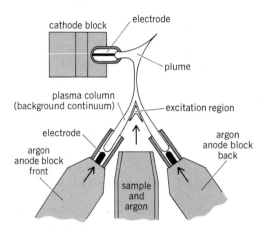

Fig. 1. Direct-current plasma jet device.

jets are electrode-based plasmas in which a conventional dc arc is ignited in an electrode gap by applying a current. However, unlike the dc arc mentioned above, a dc plasma jet also has a highly controlled flow of cooling gas surrounding the ac. This cooling flow decreases the electrical conductivity of the outer regions of the arc, causing the applied current to be carried in a path of reduced volume. This is called the thermal pinch effect. As a result, the current density of the arc rises dramatically and the temperature of the gas through which the arc is carried increases. Temperatures of up to 50,000 K have been recorded with dc plasma jets, but routine laboratory devices regularly generate plasmas of between 4000 and 10,000 K (**Fig. 1**). In a dc plasma jet device, each electrode block consists of an electrode, either pyrolytic graphite or thoriated tungsten, and an electrode sleeve through which the cooling gas causing the thermal pinch flows. Consequently, this device actually produces three plasma jets. This is done to ensure stability of operation and produce a large excitation region.

Inductively coupled plasmas are non-electrode-based plasmas in which a carrier gas, usually argon or helium, is caused to reach a plasma state through the interaction of electrons, carrier gas atoms, and an applied radio-frequency or microwave-frequency field. Electrons will resonate with the magnetic vector impressed through an induction coil surrounding a flow of carrier gas, or with the electrical vector impressed into a resonant cavity through which the carrier gas flows. The electrons eventually gain extremely high kinetic energies through this resonance. In spite of the fact that the number of electrons that gain this high energy is small, the result is that they impart this energy through collisions to the much more numerous and massive carrier gas atoms. The carrier gas atoms eventually obtain a sufficient amount of energy to reach excited states. At this point the plasma has ignited, and the carrier gas can release its energy by emitting radiation, losing electrons (ionizing), or colliding with other atoms (thermal transfer). Temperatures routinely attained in induced plasmas are also between 4000 and 10,000 K, although much higher temperatures have been observed (**Fig. 2**).

Once the hot gas is produced, the vaporized and atomized sample is injected into it, or alternatively the hot gas functions as the vaporizer and atomizer. In any case, the plasma is used to impart energy to the analyte atoms so that they may reach any number of excited states. While more than one mechanism is followed as the analyte releases this energy, the route which is principally desired is that of radiational deactivation. That is, the analyte loses the energy by ejecting photons (emitting light).

Because the analyte atoms are allowed to attain a large number of excited states, a wide range of radiations (different wavelengths of light) are emitted. The energy gained and lost relates directly, through Planck's equation, $E = hc/\lambda$, to wavelengths between the infrared and vacuum ultraviolet regions.

Resolution. Identification of the emitting element and quantitative measurement of the radiation emitted requires that the light emitted be resolved into its component wavelengths; that

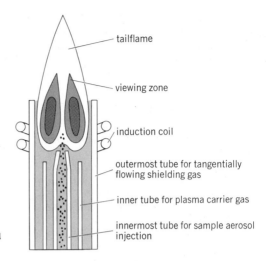

Fig. 2. Inductively coupled radio-frequency plasma.

is, a spectrum of the light must be produced. This is done with a spectrograph or spectrometer using photographic or photoelectric detection, respectively. The instrument dispersion must be on the order of 0.1 nm/mm to 1.0 nm/mm, so that individual wavelengths are adequately separated from each other at the detector. Prism spectrometers are no longer used very frequently, since they have inadequate dispersion over the entire wavelength range. Diffraction-grating instruments have uniform wavelength dispersion and can be produced so as to give quite high resolution without wasting much input radiation. In general, emission spectrochemical analysis requires higher spectral resolution than other types of spectrometric experiments because of the larger number of spectral lines produced.

Observation and measurement. Visual observation is rarely used in routine spectrochemical analysis. The human eye does not possess high enough resolving power to sort out the complex spectra usually produced.

Photographic observation has long been the method of choice for spectrochemical detection. It is used mainly for qualitative analysis and quantitative analysis at the minor- and high-trace-constituent levels. Glass plates or film may be employed. Spectrograms usually are viewed with a projection comparator, which projects sample and standard plates on a split field, ground-glass screen, permitting any spectrum on one plate to be brought adjacent to and in register with any spectrum on the other. Many comparators also incorporate a microphotometer. Any emulsion is suitable for the 230–430-nm range; at higher wavelengths, special sensitization is needed. Emulsion sensitivity varies directly with grain size; thus, high sensitivity is obtained at the cost of resolving power, and vice versa. Sensitivity and contrast vary with wavelength and age for all emulsions. Aging conditions can be retarded by storage at 0°F (32°C).

Different production batches of a given emulsion are seldom identical; if quantitative work is to be done, a 6 to 12 months' supply of plates or film should be bought from a single batch and stored at 0°F (32°C).

Measurement of line intensity ratios requires calibration of the emulsion, that is, construction of a curve showing blackening of the emulsion as a function of the intensity of incident light of the wavelength used. Lines whose intensities are to be compared photographically should be as close as possible in wavelength.

Densitometers or microphotometers measure spectral-line blackening by scanning the illuminated spectrum with a fine receiving slit or by projecting a fine illuminated line onto and through the spectrogram and onto a photocell.

The major advantage of a photographic observation is that it is both the least expensive and the simplest method for obtaining the entire spectrum simultaneously (multielement analysis) and storing it nearly indefinitely. It is not, however, the most sensitive method, and thus is not useful for the important realm of trace-element and ultratrace-element analysis.

Photoelectric observation affords extremely sensitive and rapid detection. But one photomultiplier tube must be used for each and every wavelength to be monitored. The photocurrent output is directly proportional to the rate at which photons strike the photoanode. This current is retrieved by various electronic means, and is amplified to give an electric signal which is then proportional to the amount of analyte present. Photoelectric detection is best suited to trace and ultratrace determinations because of its high sensitivity. But it presents significant problems for simultaneous multielement analyses due to its one tube per wavelength limitation. However, arrays of photomultiplier tubes have been constructed for the simultaneous observation of up to 60 wavelengths.

Another approach to multielement detection is to use only one photomultipler tube and to adjust the spectrometer precisely so that successive wavelengths pass by the tube. This is called spectrometer scanning. Its principal drawback is that each wavelength is viewed sequentially, and so it takes longer to monitor more wavelengths. Its principal advantage is that there is no limit to which wavelengths can be viewed, all of them being accessible.

Another approach to multielement, simultaneous detection is the use of imaging electronic devices, such as TV tubes, photodiode arrays, and charge-coupled and charge-injection devices. These solid-state devices can all be considered to be electronic versions of photographic plates. Consequently, all the simultaneity and simplicity of photographic detection and the high sensitivity and speed of photoelectric detection are obtained at once. The drawbacks of these devices are that they have low resolution and are quite expensive. Thus they require even higher-resolution spectrometers. SEE SPECTROCHEMICAL ANALYSIS.

Bibliography. G. W. Ewing, *Instrumental Methods of Chemical Analysis*, 5th ed., 1984; J. C. VanLoon (ed.), *Analytical Atomic Absorption Spectroscopy: Selected Methods*, 1980; J. D. Winefordner (ed.), *Trace Analysis: Spectroscopic Methods for Elements*, 1976.

FLAME PHOTOMETRY
ANDREW T. ZANDER

A branch of spectrochemical analysis in which samples in solution are excited to luminescence by introduction into a chemical flame. Flame photometry is particularly useful in determining small amounts (for example, microgram of analyte per gram of sample) of lithium, sodium, potassium, and other alkali and alkaline-earth elements in solution.

Flame photometry can be regarded as the precursor to the modern techniques of plasma emission spectrometry and atomic absorption spectrophotometry. The techniques and applications of emission flame photometry and absorption flame spectrophotometry are identical to modern instrumental emission and absorption atomic spectrochemical techniques. The sole qualification is the use of a chemical flame as the atom reservoir. SEE ATOMIC STRUCTURE AND SPECTRA; EMISSION SPECTROCHEMICAL ANALYSIS; SPECTROCHEMICAL ANALYSIS.

COMBUSTION
BERNARD LEWIS AND RODERICK S. SPINDT
R. S. Spindt is author of the section Spectroscopy.

The burning of any substance, whether it be gaseous, liquid, or solid. In combustion, a fuel is oxidized, evolving heat and often light. The oxidizer need not be oxygen per se. The oxygen may be a part of a chemical compound, such as nitric acid (HNO_3) or ammonium perchlorate (NH_4ClO_4), and become available to burn the fuel during a complex series of chemical steps. The oxidizer may even be a non-oxygen-containing material. Fluorine is such a substance. It combines with the fuel hydrogen, liberating light and heat. In the strictest sense, a single chemical substance can undergo combustion by decomposition, with emission of heat and light. Acetylene, ozone, and hydrogen peroxide are examples. The products of their decomposition are carbon and hydrogen for acetylene, oxygen for ozone, and water and oxygen for hydrogen peroxide.

Solids and liquids. The combustion of solids such as coal and wood occurs in stages. First, volatile matter is driven out of the solid by thermal decomposition of the fuel and burns in the air. At usual combustion temperatures, the burning of the hot, solid residue is controlled by the rate at which oxygen of the air diffuses to its surface. If the residue is cooled by radiation of heat, combustion ceases.

The first product of combustion at the surface of char, or coke, is carbon monoxide. This gas burns to carbon dioxide in the air surrounding the solid, unless it is chilled by some surface. Carbon monoxide is a poison and it is particularly dangerous because it is odorless. Its release from poorly designed, or malfunctioning, open heaters constitutes a serious hazard to human health.

Liquid fuels do not burn as liquids but as vapors above the liquid surface. The heat evolved evaporates more liquid, and the vapor combines with the oxygen of the air.

Gases. At ordinary temperatures, molecular collisions do not usually cause combustion. At elevated temperatures the collisions of the thermally agitated molecules are more frequent. More important as a cause of chemical reaction is the greater energy involved in the collisions. Moreover, it has been reasonably well established that there is very little combustion attributable to direct reaction between the molecules. Instead, a high-energy collision dissociates a molecule into atoms, or free radicals. These molecular fragments react with greater ease, and the combustion process proceeds generally by a chain reaction involving these fragments. An illustration will make this clear. The combustion of hydrogen and oxygen to form water does not occur in a single step, reaction (1). In this seemingly simple case, some fourteen reactions have been identified. A hy-

$$2H + O_2 \rightarrow 2H_2O \tag{1}$$

drogen atom is first formed by collision; it then reacts with oxygen molecules, reaction (2), forming an OH radical. The latter in turn reacts with a hydrogen molecule, reaction (3), forming water and

$$H + O_2 \rightarrow OH + O \qquad (2) \qquad\qquad OH + H_2 \rightarrow H_2O + H \tag{3}$$

regenerating the H atom which repeats the process. This sequence of reactions constitutes a chain reaction. Sometimes the O atom reacts with a hydrogen molecule to form an OH radical and another H atom, reaction (4). Thus a single H atom can form a new H atom in addition to

$$O + H_2 \rightarrow OH + H \tag{4}$$

regenerating itself. This process constitutes a branched-chain reaction. Atoms and radicals recombine with each other to form a neutral molecule, either in the gas space or at a surface after being adsorbed. Thus, chain reactions may be suppressed by proximity of surfaces; and the number and length of the chains may be controlled by regulating the temperature, the composition of the mixtures, and other conditions.

Under certain conditions, where the rate of chain branching equals or exceeds the rate at which chains are terminated, the combustion process speeds up to explosive proportions; because of the rapidity of molecular events, a large number of chains are formed in a short time so that essentially all of the gas undergoes reaction at the same time; that is, an explosion results. The branched-chain type of explosion is similar in principle to atomic explosions of the fission type, where more than one neutron is generated by the reaction between a neutron and a uranium nucleus. Another cause of explosion in gaseous combustion arises when the rate at which heat is liberated in the reaction is greater than the rate at which the heat dissipates to the surroundings. The temperature increases, accelerates the reaction rate, liberates more heat, and so on, until the entire gas mixture reacts in a very short time. This type of explosion is known as a thermal explosion. There are cases intermediate between branched-chain and thermal explosions which depend upon the type and proportion of gases mixed, the temperature, and the density.

In slow combustion, intermediate products can be isolated. Aldehydes, acids, and peroxides are formed in the slow combustion of hydrocarbons, and hydrogen peroxide in the slow combustion of hydrogen and oxygen. At the relatively low temperature of combustion of paraffin hydrocarbons (propane, butane, ethers) a bluish glow is seen. This light from activated formaldehyde formed in the process is called a cool flame.

In the gaseous combustion and explosive reactions described above, the processes proceed simultaneously throughout the vessel. The gas mixture in a vessel may also be consumed by a combustion wave which, when initiated locally by a spark or a small flame, travels as a narrow intense reaction zone through the explosive mixture. The gasoline engine operates on this principle. Such combustion waves travel with moderate velocity, ranging from 1 ft/s (0.3 m/s) in hydrocarbons and air to 20–30 ft/s (6–9 m/s) in hydrogen and air. The introduction of turbulence or agitation accelerates the combustion wave. The accelerating wave sends out compression or shock waves which are reflected back and forth in the vessel. Under certain conditions these waves coalesce and change from a slow combustion wave to a high-velocity detonation wave. In hydrogen and oxygen mixtures, the speed is almost 2 mi/s (3.2 km/s). The pressure created by detonation can be very high and dangerous.

Combustion mixtures can be made to react at lower temperatures by employing a catalyst. The molecules are adsorbed on the catalyst, where they may be dissociated into atoms or radicals, and thus brought to reaction condition. An example is the catalytic combination of hydrogen and oxygen at ordinary temperatures on the surface of platinum. The platinum glows as a result of the heat liberated in the surface combustion.

Spectroscopy. The spectroscopy of combustion is an experimental technique for obtaining data from flames without interfering with the combustion process. The spectrum of light emitted or absorbed in a flame is a physical property of the materials present in the flame.

Flame spectra are used in interpreting combustion mechanisms and in determining flame temperatures. Various spectrographic techniques are used depending on the type of measurement desired. The method of line reversal, in which the radiation intensity from thermally excited metal atoms is compared to a blackbody lamp filament of controllable brightness, is one technique that gives a measure of temperature; it can be handled by simple equipment with only filters to isolate radiation. An example is the addition of small amounts of sodium salts to gaseous or liquid fuels—a technique known as sodium D-line reversal.

Band spectra from reactive combustion intermediates are usually studied with a spectroscope of either the prism or grating type; for extremely fine resolution, an interferometer is sometimes used. The dispersed radiation is detected either photographically or by photomultiplier tubes. Photography is of value in recording spectra over a range of wavelengths simultaneously. If only specific lines are of interest, two or more photomultiplier tubes can be used to record relative line brightnesses. A continuous scan of a spectrum can be achieved by moving one phototube and slit along the focus of the spectral radiation and displaying the output on an oscilloscope.

The band spectra from flames has been associated with the quantized energy changes in molecules due to rotation and vibration. Each spectral line in a band spectrum represents a discrete energy level. Under equilibrium temperature conditions, there is ideally a Maxwell-Boltzmann distribution of molecules in different energy states. According to the kinetic theory of gases, this is a dynamic equilibrium with molecules having their energy distributed in a specific way over these energy states. The radiation of light is a measure of the number of molecules in the process of changing energy levels. From these measurements, on band spectra, flame temperatures and reaction intermediates can be determined.

Bibliography. J. A. Barnard and J. N. Bradley, *Flame and Combustion*, 2d ed., 1985; N. A. Chigier (ed.), *Progress in Energy and Combustion Science*, vol. 1–10, 1976–1986; A. G. Gaydon and H. G. Wolfhard, *Flames: Their Structure, Radiation, and Temperature*, 4th ed., 1979; M. L. Parsons et al., *Handbook of Flame Spectroscopy*, 1975; M. L. Parsons and P. M. McElfresh, *Flame Spectroscopy: Atlas of Spectral Lines*, 1971; R. A. Strehlow, *Combustion Fundamentals*, 1984; F. J. Weinberg, *Optics of Flames*, 1963; F. A. Williams, *Combustion Theory*, 3d ed., 1985.

FLUOROMETRIC ANALYSIS

ROBERT F. GODDU, JAMES N. LITTLE, AND MICHAEL J. KURYLO

M. J. Kurylo wrote the section Flash-Photolysis Resonance Fluorescence.

A method of chemical analysis in which a sample, exposed to radiation of one wavelength, absorbs this radiation and reemits radiation of the same or longer wavelength. If this reemission occurs in about 10^{-9} s, it is called fluorescence. Reemission after about 10^{-6} s or more is called phosphorescence. Fluorescence analysis utilizes the reemitted light to determine the material which is the source of the fluorescence. SEE PHOSPHORESCENCE.

Fluorometer. The radiation source is usually a mercury arc, and one line of its spectrum is isolated by the primary filter. The light then passes through the sample, and the fluorescence is measured at an angle, usually a right angle, to the light beam. **Figure 1** is a diagram of the components involved in the technique. The secondary filter before the detector eliminates scattered radiation of the wavelength of the source. The arrangement of a fluorometer is very similar to that of a nephelometer, and it has the same advantage of sensitivity—being able to detect very small amounts of reemitted radiaton. Fluorometric analysis can also be employed in the x-ray region by using a molybdenum or tungsten source, appropriate metal filters, and, usually, a solid sample.

For quantitative work at very low fluorescent intensities, the intensity is almost directly proportional to the concentration of the fluorescing material. At high fluorescent intensities there are usually appreciable deviations from any linear function of fluorescent intensity and concentration. For this reason, fluorescence analysis is usually carried out in a very dilute solution, and a calibration curve of fluorescent intensity versus concentration of emitter is carefully prepared.

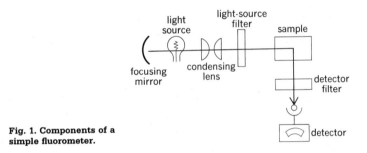

Fig. 1. Components of a simple fluorometer.

Variables. Among the many variables which must be closely watched in quantitative fluorescence analysis are scattering of the incident light by colloidal particles in solution, absorption of the fluorescent light by colored materials in solution, either quenching or intensification of the fluorescence of the compound in question by other ions or compounds present in the solution, fluorescence of other compounds present in the solution, and often the extreme dependence of fluorescence intensity on the temperature of the solution.

Applications. In spite of the drawbacks mentioned above, fluorescence is used in analytical work because of its sensitivity and selectivity in many systems. For example, vitamin B_2 (riboflavin) is usually determined by its fluorescence in solutions as dilute as 0.001 microgram/ml. Other compounds or ions commonly determined by fluorescence in solution are β-phenylnaphthylamine, an organic antioxidant; thiamin (vitamin B_1), after reaction with ferricyanide; aluminum, after complexing by 8-hydroxyquinoline; and zirconium, after reaction with morin (an organic compound). The fluorescence technique has been extended to the determination of amino acids, proteins, and nucleic acids. Solids are also often analyzed by fluorescence. In particular, uranium compounds fluoresce with a yellow color, and uranium in ores is often detected after fusion of the ore with sodium fluoride. Many sensitive qualitative tests for the presence of various inorganic ions have been developed, based on the selective formation of a fluorescing compound.

Atomic fluorescence method. A later atomic fluorescence method is based upon the intensity of the fluorescent emission when atoms in a flame are excited by absorption of radiation. With this technique metal ions can be determined directly and with remarkable sensitivity, for example, Zn 0.001 ppm, Cd 0.002 ppm, and Hg 0.1 ppm.

Fluorescence spectroscopy. Using spectrofluorometers, fluorescence spectroscopy has been developed to increase the selectivity of fluorometry. In this technique the emitted fluorescent light is passed through a monochromator so that the fluorescence emission spectrum can be

Fig. 2. Turner 210 "Spectro." (*G. K. Turner Associates*)

recorded. It is possible to measure several fluorescing compounds in the same solution, provided that they have sufficiently different fluorescence emission spectra. In obtaining this selectivity, some sensitivity is lost since some fluorescent light is lost in the monochromator, and only a small portion of the total fluorescent energy emitted is measured at any one wavelength.

This loss in sensitivity can be minimized by the use of a more powerful xenon arc lamp as the radiation source. A further increase in selectivity is effected by using different wavelengths of incident radiation to excite the fluorescence. Advanced spectrofluorometers, which give absolute excitation and emission spectra by correcting for any nonlinearity in the radiation source and detector, are commercially available (**Fig. 2**).

Flash-photolysis resonance fluorescence. The use of fluorescence analysis has extended beyond standard analytical applications. The sensitivity and selectivity of the technique, coupled with the easily characterized relationship between intensity and concentration, make it well suited for following the chemical reactivity of selected species. More specifically, recognition of these analysis traits led to the development of a novel experimental technique for measuring absolute rate constants for gas-phase reactions involving free radicals (principally atoms and diatomic radicals). The method combines pulsed ($\sim 10^{-5}$ s duration) photolytic production of the radical with real-time resonance fluorescence monitoring of its subsequent temporal behavior. This technique, flash-photolysis resonance fluorescence, revolutionized gas-phase chemical kinetics research in the 1970s.

The typical experimental apparatus (**Fig. 3**) is by no means unique, and there are numerous variations. The basic components include a vacuum reaction chamber, a pulsed photolysis source (flash lamp), an analytical light source (reasonance lamp), and a detection system. The latter three components are situated on mutually perpendicular axes of the cell. In a typical experiment a gas mixture consisting of a photolytic radical source, molecular reactant, and inert diluent are flash-photolyzed in the cell under preselected conditions of light intensity and wavelength distribution. In this way the desired uniform concentration of transient reactive species is produced across the central region of the cell. The analytical light source probes an intersection of this central region continuously, thereby generating a fluorescence emission from the radical species. Conditions are generally chosen such that the reaction is first-order in radical concentration. In order to achieve these conditions as well as to minimize secondary reactions between the radical and either reaction or photolytic products, radical concentrations less than 10^{11} cm^{-3} must frequently be used. The linearity of fluorescence intensity with concentration under these conditions permits real-time tracking of the reactive behavior of the transient species.

The versatility of this technique for probing atomic reactions is a result of the development of intense vacuum ultraviolet atomic line sources. The intensity and spectral purity of these electrodeless discharge lamps make them ideally suited for measuring very low atom concentrations. Such resonance lamps, when operated under optically thin conditions, are an extremely sensitive analytical tool (particularly for the very strong atomic resonance transitions in the vacuum ultraviolet spectral region). The use of rotational-vibrational bands within the electronic absorption

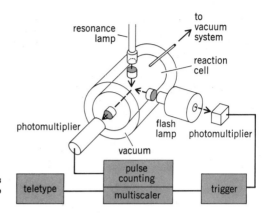

Fig. 3. Flash-photolysis resonance fluorescence apparatus.

spectra of a free radical has shown similar utility for resonance fluorescence detection of diatomic radical species. Due to difficulties in matching lamp spectral outputs with the absorption band, the technique is less sensitive for such radicals than for atoms. Consequently, laser sources tuned to specific rotational-vibrational transition of the radical have proved more effective due to the increased intensity and accompanying reduction in scattered light.

Other modifications of the flash-photolysis resonance fluorescence technique have been concerned with variations in the radical production system. The replacement of the spark discharge flash lamp (designed originally for vacuum ultraviolet operation) with a near-ultraviolet or visible laser has increased the versatility of reactant production. By selecting an appropriate absorption band of the precursor, reactive transients can be prepared in specific energy states.

The application of resonance fluorescence analysis for real-time kinetics investigations is limited only by the fluorescing ability of the free radical and the availability of a suitable resonance light source. This versatility led to the adaptation of this measurement technique to other kinetics experiments utilizing entirely different production and analysis methodologies. Discharge flow reactors are a prime example of such expanded use. *See* FLUORESCENCE.

Bibliography. W. Braun and M. Lenzi, Resonance fluorescence method for kinetics of atomic reactions, *Disc. Faraday Soc.*, 44:252–262, 1967; D. Davis and W. Braun, Intense vacuum ultraviolet atomic line sources, *Appl. Opt.*, 7:2071–2074, 1968; A. M. Gillespie, *A Manual of Fluorometric and Spectrophotometric Experiments*, 1985; M. A. Konstantinova-Schlezinger, *Fluorimetric Analysis*, 1965; J. R. Lakowicz, *Principles of Fluorescence Spectroscopy*, 1983; E. L. Wehry (ed.), *Modern Fluorescence Spectroscopy*, vols. 3–4, 1981.

ACTIVATION ANALYSIS
W. S. LYON

A technique in which a neutron, charged particle, or gamma photon is captured by a stable nuclide to produce a different, radioactive nuclide which is then measured. The technique is specific, highly sensitive, and applicable to almost every element in the periodic table. Because of these advantages, activation analysis has been applied to chemical identification problems in many fields of interest.

Neutron method. Neutron activation analysis (NAA) is the most widely used form of activation analysis. In neutron activation analysis the sample to be analyzed is placed in a nuclear reactor where it is exposed to a flux of thermal neutrons. Some of these neutrons are captured by isotopes of elements in the sample; this results in the formation of a nuclide with the same atomic number, but with one more mass unit of weight. A prompt gamma ray is immediately emitted by the new nuclide, hence the term (n,γ) reaction, which can be expressed as reaction (1), where z

$$^{z}_{m}A(n,\gamma) \rightarrow \,^{z}_{m+1}A + \gamma \tag{1}$$

refers to the atomic number and m the atomic weight. Usually the product nuclide ($_{m+1}^{z}A$) is radioactive, and by measuring its decay products one can identify and quantify the amount of target element in the sample. The basic activation equation is given by Eq. (2), and enables one

$$A = Nf\sigma \left(1 - e^{\frac{-0.693t}{t_{1/2}}}\right) \qquad (2)$$

A = activity of product nuclide (disintegrations per second)
N = atoms of target element
f = flux of neutrons (neutrons per cm^2-s)
σ = cross section of the target nuclide (cm^2)
$t_{1/2}$ = half-life of induced radioactive nuclide
t = time of irradiation

to calculate the number of atoms of an unknown target element by measuring the radioactivity of the product. These radioactive products usually decay by emission of a beta particle (negative electron) followed by a gamma ray (uncharged). Cross sections and half-lives are well known, and neutron fluxes can be measured by irradiation and measurement of known materials. When such techniques are used, the method is known as absolute activation analysis. A simple (and older) method is to simultaneously irradiate and compare with the unknown sample a standard containing known amounts of the elements in question. This is called the comparator technique.

Measurement techniques. Measurement of the induced radioactivities is the key to activation analysis. This is usually obtained from the gamma-ray spectra of the induced radionuclides. Gamma rays from radioactive isotopes have unique, discrete energies, and a device that converts such rays into electronic signals that can be amplified and displayed as a function of energy is a gamma-ray spectrometer. It consists of a detector [germanium doped with lithium, GeLi, or sodium iodide doped with thallium, NaI(Tl)] and associated electronics. Gamma rays interact in the detector to form photoelectrons, and these are then amplified and sorted according to energy. Peaks in the resulting gamma-ray spectra are called gamma-ray photo-peaks. By taking advantage of the different half-lives and different gamma-ray energies of the induced radionuclides, positive identification of many elements can be made. A computer, or provision for recording the spectrometric data in computer-compatible form, is almost a necessity. Calibration of counting conditions with a particular detector enables the activation analyst to relate the area under each gamma-ray photo-peak to an absolute disintegration rate; this supplies the A in Eq. (2). Such techniques are multielement, instrumental, and absolute. Where the sought element captures a neutron to produce a non-gamma-emitting nuclide, the analyst must make chemical separations and then beta-count the sample, or must go to another technique. SEE GAMMA-RAY DETECTORS.

Charged-particle method. Activation analysis can also be performed with charged particles (protons or He^{3+} ions, for example), but because fluxes of such particles are usually lower than reactor neutron fluxes and cross sections are much smaller, charged-particle methods are usually reserved for special samples. Charged particles penetrate only a short distance into samples, which is another disadvantage. A variant called proton-induced x-ray emission (PIXE) has been highly successful in analyzing air particulates on filters. Here the samples are all similar, and are low in total mass, and many of the elements of interest such as sulfur, calcium, iron, zinc, and lead are not easy to determine by neutron activation analysis. The protons excite prompt x-rays characteristic of the element, and these are measured. Prompt gamma rays have been used for measurement in some neutron activation analysis studies, but that method has had only limited success. Photon activation, using photons produced by electron bombardment of high-z targets such as tungsten, is another special variant of rather minor interest. Neutron sources other than reactors are sometimes used: ^{252}Cf is an element that emits neutrons as it decays, and can be used for neutron activation analysis; and small accelerators called 14-MeV neutron generators produce a low flux of high-energy neutrons which are used primarily for determination of oxygen and nitrogen.

Applications. Activation analysis has been applied to a variety of samples. It is particularly useful for small (1 mg or less) samples, and one irradiation can provide information on 30 or more elements. Samples such as small amounts of pollutants, fly ash, very pure experimental alloys, and biological tissue have been successfully studied by neutron activation analysis. Of particular interest has been its use in forensic studies; paint, glass, tape, and other specimens of physical evidence have been assayed for legal purposes. In addition, the method has been used for authentication of art objects and paintings where only a small sample is available. Activation analysis services are available in numerous university, private, commercial, and United States government laboratories, and while it is not an inexpensive method, for many special samples it is the best and cheapest method of acquiring necessary data. SEE TRACE ANALYSIS.

Bibliography. S. Amiel (ed.), *Nondestructive Activation Analysis: With Nuclear Reactors and Radioactive Neutron Sources*, 1981; P. Kruger, *Principles of Activation Analysis*, 1971; W. S. Lyon and H. H. Ross, Nucleonics, *Anal. Chem. (Annu. Rev.)*, 54(4):227R, 1982; 1972 International Conference on Modern Trends in Activation Analysis, *J. Radioanal. Chem.*, vol. 16, no. 1 and 2, 1973; 1976 International Conference on Modern Trends in Activation Analysis, *J. Radioanal. Chem.*, vols. 37–39, 1977; 1981 International Conference on Modern Trends in Activation Analysis, *J. Radioanal. Chem.*, vols. 69–71, 1982.

PROTON-INDUCED X-RAY EMISSION (PIXE)
LEE GRODZINS

A highly sensitive analytic technique for determining the composition of elements in small samples. Proton-induced x-ray emission (PIXE) is a nondestructive method capable of analyzing many elements simultaneously at concentrations of parts per million in samples as small as nanograms. PIXE has gained acceptance in many disciplines; for example, it is the preferred technique for surveying the environment for trace quantities of such toxic elements as lead and arsenic. There has also been a rapid development in the use of focused proton beams for PIXE studies in order to produce two-dimensional maps of the elements at spatial resolutions of micrometers.

Apparatus. The typical PIXE apparatus uses a small Van de Graaff machine to accelerate the protons which are then guided to the sample. Nominal proton energies are between 1 and 4 MeV; too low an energy gives too little signal while too high an energy produces too high a background. The energetic protons ionize some of the atoms in the sample, and the subsequent filling of empty inner orbits results in the characteristic x-rays. These monoenergetic x-rays emitted by the sample are then efficiently counted in a high-resolution silicon (lithium) detector which is sensitive to the x-rays of all elements heavier than about sodium.

Advantages. The advantages of PIXE over electron-induced x-ray techniques derive from the heaviness of the proton which permits it to move through matter with little deflection. The absence of scattering results in negligible continuous radiation (bremsstrahlung) so that the backgrounds under the characteristic x-ray lines come mainly from secondary effects. As a result, proton-induced x-ray techniques are two to three orders of magnitude more sensitive to trace elements than are techniques based on electron beams. Another important consequence is that the proton beam can be extracted from the vacuum so that samples, in particular of biological material, can be studied in controlled atmospheric environments.

Application. PIXE was developed in European laboratories in the early 1970s. At the University of Lund, Sweden, and at the Niels Bohr Institute in Copenhagen, the emphasis was on using broad beams to obtain great sensitivity to trace elements in bulk samples; at the Harwell National Laboratory in England the emphasis was on focused beams to obtain spatial resolution at some sacrifice in analytic sensitivity. Some 40 to 50 PIXE laboratories now carry out investigations in such diverse fields as materials science, earth and planetary sciences, biomedicine, environment, and archeology. A number of these laboratories have sophisticated facilities in which samples are scanned under computer control by micrometer-size beams. Previously the high cost of the proton accelerator confined PIXE to laboratories which used existing Van de Graaffs, constructed mainly in the 1950s and 1960s for nuclear studies. Several companies now market small accelerators which are suited to PIXE, and commercial laboratories offer analytic PIXE services to the technology community. SEE ACTIVATION ANALYSIS; TRACE ANALYSIS.

SECONDARY ION MASS SPECTROMETRY (SIMS)
C. A. Evans, Jr.

A technique for microchemical analysis that permits the visualization of a three-dimensional chemical image of the solid state using primary ions. It is also referred to as ion probing.

The ion probing of a material is accomplished by bombarding the sample with a 5–20-keV beam of primary ions. This process sputters off the surface layers of the sample, producing a variety of secondary species including neutral atoms and molecules, secondary electrons, photons, and positive and negative ions. Mass spectrometric analysis of the positive and negative sputtered secondary ions can provide microchemical characterization with lateral resolution better than that of the electron probe and can produce sampling depths similar to those of the ion scattering and Auger electron spectrometry techniques. In addition, mass spectrometric analysis provides full elemental coverage from hydrogen to uranium, including the capability of isotopic characterization. This combination of ion-bombardment excitation and mass spectrometric detection can provide detectabilities in the range of 10^{-15} to 10^{-19} g. SEE ELECTRON SPECTROSCOPY; MASS SPECTROMETRY; MASS SPECTROSCOPE.

Ion production. Ion bombardment of a solid can produce positive ions by two different processes—kinetic and chemical. When a primary ion of 5–20-keV energy strikes the surface layers of a material, it transfers kinetic energy to the sample atoms, initiating a collision cascade, which results in sample atoms being ejected from the surface as well as being excited to metastable and ionized states in the outer atomic layers. Any unbound electrons in the sample will have a much higher velocity than the solid-state ions have, and all these ions will be neutralized before they can escape into the vacuum. However, an atom can escape from the sample surface as a neutral particle while maintaining its own internal energy in a metastable state. This metastable atom can eject an Auger electron (can quantum-deexcite) in the vacuum above the sample surface and become an ion capable of detection by the mass spectrometer. This is, in essence, the kinetic ionization process, which predominates in the inert gas bombardment of metals and some semiconductors, and in such cases most of the ions which are to be analyzed are produced just outside the sample surface in the vacuum. SEE AUGER EFFECT.

The chemical ionization process depends on the presence of one or more chemically reactive species in the sample to reduce the number of conduction-band electrons available for neutralization of the ions produced in the solid. With the reduction of neutralization events by the presence of this chemical compound, more of the ions produced in the solid state escape the sample surface for mass analysis. Thus, when chemical ionization predominates, most ions are produced in the outer 5 nanometers of the sample.

Instrumentation. Ion probe instrumentation ranges from custom-designed, laboratory-constructed devices to general-purpose, commercially available instruments. All these instruments require a primary ion source, primary ion extraction and focusing systems, a sample chamber and sample mounting facilities, a mass spectrometer for mass-charge separation of the secondary ions, and an ion-detection system.

There are three different types of ion probes, or SIMS: (1) The secondary ion mass analyzer provides general surface analysis and depth-profiling capabilities. (2) The ion microprobe mass spectrometer uses primary ions of argon or oxygen from a duoplasmatron ion source that are focused to a 1–2 micrometer spot before the sample is bombarded. The sputtered secondary ions are extracted into a double-focusing mass spectrometer or spectrograph for mass-charge separation. Ion detection is accomplished by electrical or photographic means or both. The ion microprobe produces a magnified image of elemental or isotopic distributions, in a manner similar to that of the electron probe, using synchronous rastering of the primary ion beam and an oscilloscope. A diagram of an instrument of this type is shown in **Fig. 1**. (3) The direct-imaging mass analyzer developed by R. Castaing and G. Slodzian is unique among the instruments employed for secondary ion microchemical analysis. In the Castaing-Slodzian design the sample area (10–300 μm in diameter) to be analyzed is bombarded by the primary ions. The secondary ions are extracted and imaged by an electrostatic immersion lens. The image bears a point-to-point relation to the ion's place of origin on the sample surface and is magnified approximately 10 times. The secondary ion image then traverses a magnetic sector, an electrostatic mirror, and a second

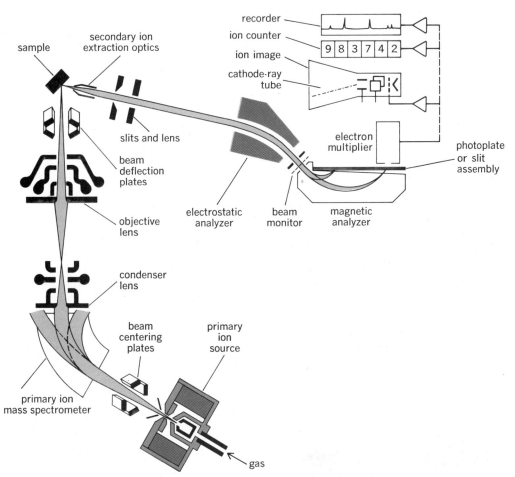

Fig. 1. Ion microprobe. (*After C. A. Evans, Jr., Secondary ion mass analysis, Anal. Chem., 44:67A, November 1972*)

magnetic sector. This sophisticated system provides mass and energy separation while accomplishing many of the same functions of the double-focusing mass spectrometers used with ion microprobes. The addition of another electrostatic lens and image converter allows the resultant mass-resolved image to be visually observed on a fluorescent screen or recorded on a photographic film for future reference. The same mass-resolved image can be directed onto a scintillator for electrical measurement by a photon-counting system. Secondary ion emission from a localized area, that is, microanalysis, can be quantitatively measured by placing a mechanical aperture in an ion or electron image plane and allowing only those electrons (resulting from the ion-to-electron converter) representing a selected area on the specimen to fall on the scintillator-counting system. The important concept to keep in mind is that the direct-imaging design acquires all data points in the secondary ion image simultaneously, whereas the ion microprobe acquires image data points sequentially.

Analytical applications. The ion probe has been applied to a wide variety of analytical problems. Like the electron microprobe, the ion probe is used to analyze samples on a microscale. This work has included the analysis of small mineral samples or patches in geological and lunar materials for isotopic age dating as well as trace, minor, and major element composition. Other applications include the analysis of airborne particulates and semiconductor devices. The ion probe provides much better elemental sensitivity and broader elemental coverage than does the electron probe in these applications. In addition, the ion probe provides isotopic detection, which is unavailable with the electron probe.

Another feature of the microanalytical capability of the ion probe is the ability to acquire secondary ion images of the elements present in the sample under study. Through the study of these images a great deal of information on the qualitative and semiquantitative elemental distributions can be obtained with 1–2-μm resolution. **Figure 2** is the $^{27}Al^+$ image from a Cr-Al microcircuit test pattern. The light areas result from the presence of Al, and the dark regions present the absence of Al. In this example, Cr is present in the dark regions. The resolution of these images is illustrated by the fact that the lines in the most closely spaced set of lines are 1.5 μm wide and 1.5 μm apart. Secondary ion images are extremely valuable in the study of concentration variations on the micrometer level.

As a result of the layer-by-layer removal of the sample surface by the primary ion beam, subsurface features are exposed and ionized. The continuous monitoring of the secondary ions provides information on the in-depth variation of the sample composition. The in-depth profiling can be used to examine each of the outer monolayers if low primary ion current densities are employed, or to examine variations over the outer 1–5 μm if high primary beam densities are employed.

The realization of optimum depth resolution depends on a variety of instrument operating conditions. The most important of these relate to the interaction of the primary ion beam and the sample. During the bombardment process, many of the primary ions interact with several atomic layers and "stir" the upper atomic layers as the ions dissipate their kinetic energy.

At the higher kinetic energies (> 10 keV) the primary ions can actually push some material deeper into the sample and distort the true depth profile. A similar effect has been noted when attempts have been made to profile the outermost 10–20 nm of a sample. If a reactive primary ion is being used, it can penetrate below the outermost layer and come to rest some distance into the sample (for example, 10–20 nm). Thus the chemical enhancement effect will not be established until that depth is reached by the erosion of the primary beam. As a consequence the ion intensities from the outer 10 nm layers cannot be related to those encountered further in. Recently it has been demonstrated that the use of a high residual partial pressure of oxygen can provide a

Fig. 2. $^{27}Al^+$ secondary ion image of an Al-Cr microcircuit test pattern. (*From C. A. Evans, Jr., Secondary ion mass analysis, Anal. Chem., 44:67A, November 1972*)

superficial oxygen layer which causes the chemical enhancement effect to occur and allows the continuous comparison of ion intensities.

The other important consideration for high-resolution depth profiling is the shape of the crater produced by the primary ion beam. Since the crater produced by a stationary primary ion beam is gaussian, at any given moment ions are being produced from a variety of depths into the sample rather than from one specific depth. Three methods are generally employed to obtain a flat-bottomed crater, or the effect of a flat-bottomed crater. The first is to appropriately focus the primary beam to obtain a flat-bottomed, vertical-sided crater. This technique is limited since it tends to give very low ion etching rates and is difficult to accomplish. The other two methods depend on the use of primary beam rastering and mechanical aperturing of the secondary ion beam (in the direct-imaging instruments) or electronic aperturing of the secondary ion detector (in the ion microprobe). These techniques produce a flat-bottomed crater with nonvertical sides and allow only those ions from the flat area to be detected.

A variety of materials have been characterized by the above depth-profiling methods. The systems studied include oxides, metals, and a wide number of electronic device–oriented and semiconductor samples.

Bibliography. C. A. Andersen and J. R. Hinthorne, *Anal Chem.*, 45:1421, 1973; A. Benninghoven et al., *Secondary Ion Mass Spectrometry: Basic Concepts, Instrumental Aspects, Applications*, 1986; A. Benninghoven et al. (eds.), *Secondary Ion Mass Spectrometry SIMS V: Proceedings*, 1982; M. Bernheim, G. Blaise, and G. Slodzian, *Int. J. Mass Spectrom. Ion Phys.*, 10:293, 1972–1973; C. A. Evans, Jr., *Anal. Chem.*, 44:67A, November 1972; J. M. Morabito and R. K. Lewis, *Anal. Chem.*, 45:869, 1973.

CONTRIBUTORS

CONTRIBUTORS

Andrews, Dr. Lester. Department of Chemistry, University of Virginia.
Aspnes, Dr. David. Bell Telephone Laboratories, Murray Hill, New Jersey.

Backstrom, Prof. Gunnar. Institute of Physics, University of Umea, Sweden.
Bashkin, Dr. Stanley. Department of Physics, University of Arizona.
Bayfield, Dr. James E. Department of Physics, University of Pittsburgh.
Bienenstock, Dr. Arthur. Stanford Linear Accelerator Center, Stanford University.
Billings, Dr. Bruce H. Special Assistant to the Ambassador for Science and Technology, Embassy of the United States of America, Taipei, Taiwan.
Birks, Dr. L. S. Head, X-Ray Optics Branch, Radiation Technology Division, Naval Research Laboratory, Washington, D.C.
Brockhouse, Dr. Bertram N. Institute for Materials Research, McMaster University.
Bursey, Prof. Maurice M. Department of Chemistry, University of North Carolina.

Caldwell, Prof. C. Denise. Department of Physics, Yale University.
Clark, Dr. Walter. Research Laboratories, Eastman Kodak Company, Rochester, New York.

Dehmelt, Dr. Hans. Department of Physics, University of Washington.
Dieke, Prof. G. H. Deceased; formerly, Chairperson, Department of Physics, Johns Hopkins University.

Evans, Dr. Charles A., Jr. President, Charles Evans and Associates, San Mateo, California.
Evans, Dr. John W. National Solar Observatory, Sunspot, New Mexico.

Fano, Prof. U. James Franck Institute, University of Chicago.
Feldman, Leonard C. Electronic Laboratories, Great Neck, New York.
Firk, Dr. Frank W. K. Department of Physics, Yale University.

Gerjuoy, Dr. Edward. Department of Physics, University of Pittsburgh.
Goddu, Dr. Robert F. Manager, Fibers and Film Research Division, Hercules, Inc., Wilmington, Delaware.
Gove, Dr. Harry E. Department of Physics and Astronomy, University of Rochester.
Grodzins, Dr. Lee. Physics Department, Massachusetts Institute of Technology.

Hansch, Dr. Theo W. Department of Physics, Stanford University.

Happer, Dr. William. Department of Physics, Columbia University.

Harrison, George R. Deceased; formerly, Dean Emeritus, School of Science, Massachusetts Institute of Technology.

Harvey, Dr. John A. Oak Ridge National Laboratory, Oak Ridge, Tennessee.

Herber, Prof. Rolfe H. Department of Chemistry, Rutgers University.

Herzberger, Max J. Deceased; formerly, Consulting Professor, Department of Physics, Louisiana State University.

Horen, Dr. Daniel J. Nuclear Division, Oak Ridge National Laboratory, Oak Ridge, Tennessee.

Hurst, Dr. G. S. Oak Ridge National Laboratory, Oak Ridge, Tennessee.

Jenkins, Prof. Francis A. Deceased; formerly, Department of Physics, University of California, Berkeley.

Jennings, Dr. Donald A. Time and Frequency Division, National Bureau of Standards, Boulder, Colorado.

Klick, Dr. Clifford C. Superintendent, Solid State Division, U.S. Naval Research Laboratory.

Koch, Dr. Peter M. Department of Physics, State University of New York, Stony Brook.

Koller, Dr. Noémie. Department of Physics, Rutgers University.

Kurylo, Dr. Michael J. National Measurements Laboratory, National Bureau of Standards.

Kusch, Prof. Polykarp. Department of Physics, University of Texas, Dallas.

Lewis, Dr. Bernard. Combustion and Explosives Research, Inc., Pittsburgh, Pennsylvania.

Little, Dr. James N. Research Division, Waters Associates, Milford, Massachusetts.

Lord, Prof. Richard C. Department of Chemistry, Massachusetts Institute of Technology.

Lyon, Dr. W. S. Analytical Division, Oak Ridge National Laboratory, Oak Ridge, Tennessee.

McMath, Dr. Robert R. Deceased; formerly, Director, McMath-Hulbert Observatory, University of Michigan.

Madison, Dr. Vincent. Department of Medicinal Chemistry, School of Pharmacy, University of Illinois.

Mantler, Michael. Technical University of Vienna, Austria.

Meggers, Dr. William F. Deceased; formerly, National Bureau of Standards.

Miller, Dr. Glenn H. Weapons Effects Division, Sandia Laboratories, Albuquerque, New Mexico.

Moscowitz, Dr. Albert. Department of Chemistry, University of Minnesota.

Mulliken, Dr. Robert S. Professor Emeritus, Departments of Physics and Chemistry, University of Chicago.

Nier, Prof. Alfred O. School of Physics and Astronomy, University of Minnesota.

Olness, Dr. John W. Brookhaven National Laboratory, Associated Universities, Inc., Upton, New York.

Otten, Prof. Ernst Wilhelm. Institut für Physik, Johannes Gutenberg-Universitat Mainz, West Germany.

Park, Prof. David. Department of Physics, Williams College.

Parrish, Dr. William. IBM Almaden Research Laboratory, San Jose, California.

Patel, Dr. C. K. N. Bell Laboratories, Murray Hill, New Jersey.

Pecora, Prof. Robert. Department of Chemistry, Stanford University.

Pipkin, Prof. Francis M. Department of Physics, Harvard University.

Roberts, Prof. Louis D. Department of Physics, University of North Carolina.

Rowley, Dr. W. R. C. Department of Trade and Industry, Division of Mechanical and Optical Metrology, National Physical Laboratory, Teddington, Middlesex, England.

Schulman, Dr. James H. U.S. Naval Research Laboratory.

Sellin, Dr. Ivan A. Department of Physics, University of Tennessee.

Siegbahn, Dr. Hans. Institute of Physics, University of Uppsala, Sweden.

Siegbahn, Prof. Kai. Institute of Physics, University of Uppsala, Sweden.

Skogerboe, Dr. Rodney K. Department of Chemistry, Colorado State University.

Slichter, Dr. Charles P. Department of Physics, University of Illinois, Urbana.

Slomp, Dr. George. Physical and Analytical Chemistry Division, Upjohn Company, Kalamazoo, Michigan.

Spindt, Dr. Roderick S. Gulf Research and Development Company, Pittsburgh, Pennsylvania.

Veillon, Dr. Claude. Biophysics Research Laboratory, Peter Bent Brigham Hospital, Harvard Medical School.

Walsh, Dr. Walter M. Bell Laboratories, Murray Hill, New Jersey.

Watson, Dr. W. W. Professor Emeritus of Physics, Yale University.

Weissman, Prof. Samuel Isaac. Department of Chemistry, Washington University.

Winick, Dr. Herman. Stanford Linear Accelerator Center, Stanford Synchrotron Radiation Laboratory, Stanford, California.

Wyant, Prof. James C. Optical Sciences Center, University of Arizona.

Zander, Dr. Andrew T. *Perkin-Elmer Spectroscopy Division, Ridgefield, Connecticut.*

INDEX

INDEX

Asterisks indicate page references to article titles.

Absorption spectrum 13
  crystals *see* Crystal absorption spectra
Activation analysis 265-267*
  applications 266-267
  charged-particle method 266
  measurement techniques 266
  neutron method 265-266
Afterglow *see* Phosphorescence
Alpha particles: scattering experiments 19-20
  spectra 179
Amici prism system 94
Angerman, N.S. 212
Angus, J. 132
Anomalous Zeeman effect 39-40
Antiferromagnetic resonance 208
Arc discharge 80-81*
  arc production 80
  regions of an arc 80-81
Atom *see* Atomic structure and spectra
Atomic absorption spectrometer 114
Atomic emission spectroscopy 115-116

Atomic fluorescence spectrometry 115
Atomic gas: luminescence in 61-62
Atomic spectroscopy 114-116*
  atomic absorption spectrometer 114
  atomic emission spectroscopy 115-116
  atomic fluorescence spectrometry 115
  resonance-ionization spectroscopy 131-134*
Atomic structure and spectra 17-32*
  beam-foil spectroscopy 120-122*
  Bohr atom 20-22
  coupling schemes for multielectron atoms 24-25
  Doppler spread 29
  electromagnetic nature of atoms 17-18
  energy level 13-15*
  hydrogen spectrum 25-27
  infrared spectroscopy 137-145*
  Lamb shift and quantum electrodynamics 27-28
  linewidth 16-17*
  mass spectroscope 230-234*

Atomic structure and spectra (*cont.*):
  multielectron atoms 22-25
  nuclear magnetism and hyperfine structure 27
  Pauli's exclusion principle for multielectron atoms 22-23
  planetary atomic models 18-19
  radiationless transitions 28-29
  recoil ion spectroscopy 29
  relativistic Dirac theory and superheavy elements 29-30
  scattering experiments 19-20
  spectroscopy *see* Atomic spectroscopy
  spin-orbit coupling for multielectron atoms 23-24
  uncertainty principle 30-31
Auger, P. 28
Auger effect 28-29, 33-34*
  x-ray spectra 161
Auger electron spectroscopy 34

Balmer, J. J. 14
Balmer series 26-27

Band spectrum 13, 55-56*
  diatomic electronic spectra 52-54
  electronic band structures 54-55
  vibrational and rotational states 50-52
Barkla, C. 19
Beam-foil spectroscopy 120-122*
  beam-foil interaction 122
  degree of isolation 121-122
  highly excited states 29
  Lamb shift 122
  mean lives of electronic levels 121
  optical spectra 121
  relativistic effects 122
Bell, A.G. 116
Beta rays 180-182*
  detection 181
  interaction with matter 180-181
  spectrometers 181-182
  spectrum 178-179
Blewett, J.P. 81
Bohr, N. 13, 18
Bohr atom 20-22
  elliptical orbits 22
  quantization 20-21
Borde, C. 126
Bouguer-Lambert-Beer law 248
Boyle, R. 17
Buys-Ballot, C.H.D. 128

Carbon-13 nuclear magnetic resonance 215-218
  applications 216-217
  carbon counting 217
  dynamic effects 218
  proton counting 217-218
  pulsing Fourier transformation spectrometers 216
  sensitivity 215-216
  tracer studies 218
Chromophore, optically active 76-77

Circular dichroism: Cotton effect 75-77*
  measurement of optical activity 72
CMR see Carbon-13 nuclear magnetic resonance
Coblentz, W.W. 137
Cockroft, A.L. 132
Combustion 260-262*
  gases 261-262
  solids and liquids 261
  spectroscopy 262
Compton, A.H. 145
Concave diffraction grating 97
Condon, E.U. 53
Continuous spectrum 13
Cotton effect 75-77*
Crystal absorption spectra 56-59*
  extrinsic absorption 59
  free-carrier absorption 58
  general properties 57
  intrinsic absorption 58-59
  lattice absorption 57-58
Curran, S.C. 132
Cyclotron resonance experiments 226-227*
  microwave spectroscopy 203

Dalton, J. 17
Davisson, C. 31
Dehmelt, H.G. 221
Diffraction grating 95-98*
  concave grating 97
  echelette grating 97
  echelle grating 97
  grating spectroscopes 96-97
  mountings 97-98
  production of 95-96
  properties of 96
Dipole moment: molecules 45-46
Dirac, P.A.M. 22, 34
Dispersing prism 94-95
Division of amplitude interferometer 98-99
Division of wavefront interferometer 98-99

Doppler, C. 128
Doppler effect 128-131*
  gamma rays 185
  linewidth broadening 16, 29
Doppler-free spectroscopy 125, 129
Doppler-free two-photon spectroscopy 127, 130-131

Echelette grating 97
Echelle grating 97
Electron microprobe: x-ray fluorescence analysis 165-166
Electron-nuclear double resonance 207
Electron paramagnetic resonance spectroscopy 222-226*
  liquids 223
  microwave spectroscopy 203
  motion effects 224-225
  multiple resonance methods 225
  rates of electron transfer 223-224
  solids 223
  transient methods 225
Electron spectroscopy 169-175*
  applications 171-172
  Auger electron spectroscopy 34
  chemical shifts 173-175
  modes of excitation 169-171
Emission spectrochemical analysis 256-260*
  excitation of atomic vapor 257-258
  observation and measurement 259-260
  resolution 258-259
  vaporization and atomization 256-257
Emission spectrum 12-13
Enantiomer see Optical activity

INDEX

ENDOR *see* Electron-nuclear double resonance
Energy level 13-15*
  linewidth 16
  molecules 47-49
  Mössbauer effect 186
  Ritz's combination principle 15-16*
EXAFS *see* Extended x-ray absorption fine structure
Exclusion principle: multielectron atoms 22-23
Extended x-ray absorption fine structure 89-90, 169*

Fabry-Perot interferometry 103-104, 152-153
Ferrimagnetic resonance 208-209
Ferromagnetic resonance 208
Field shift 41
Filter photometry 253
Fine structure 34-35*
  extended x-ray absorption fine structure 169*
Fizeau interferometer 101-102
Flame photometry 260*
Flash-photolysis resonance fluorescence 264-265
Fluorescence 60, 69*
  fluorometric analysis 262-265*
  Mössbauer effect 186-190*
Fluorescence spectroscopy 124, 263-264
  atomic fluorescence spectrometry 115
  x-rays *see* X-ray fluorescence analysis; X-ray spectrometry
Fluorometric analysis: applications 263
  atomic fluorescence method 263
  flash-photolysis resonance fluorescence 264-265
  fluorescence spectroscopy 263-264
  fluorometer 263

Fluorometric analysis (*cont.*): variables 263
Fourier transform spectroscopy 140
Franck-Condon principle: electronic band spectra 53-54
  polyatomic electronic spectra 55

Gamma-ray detectors 107-109*
  detector-spectrometer devices 107-108
  Ge(Li) semiconductor detectors 108-109
  miscellaneous detectors 109
  miscellaneous spectrometers 109
  NaI(Tl) detector 109
Gamma rays 182-186*
  applications to nuclear research 184-185
  Doppler shift 185
  interaction with matter 186
  Mössbauer effect 186-190*
  nature of 182
  origin 183
  practical applications 185
  spectra 179
  use as nuclear labels 183-184
  wave-particle duality 183
Germer, L. 31
Goudsmit, S.A. 22
Grant, D. 215
Grating spectroscope 96-97

Hansch, T.W. 126
Heisenberg, W. 30
Helium: fine structure 34-35*
Hole-burning spectroscopy 128
Holographic interferometry 104-105
Hydrogen: energy levels 14
  fine structure 34-35*
  spectrum 25-27

Hydrogen atom *see* Bohr atom
Hyperfine structure 35-36*
  atoms and molecules 35
  due to nuclear spin 35-36
  isotope effect 35
  liquid and solid systems 36
  measurement of nuclear moments 43-44
  nuclear magnetism 27
Hyper-Raman effect 149-150

IKES *see* Ion kinetic energy spectroscopy
Infrared spectrophotometry 249-251
Infrared spectroscopy 137-145*
  applications 142-145
  instrumentation and techniques 138
  interferometric methods 140-141
  typical spectra 138-140
  use of tunable lasers 141-142
Intensity fluctuation spectroscopy 151-152
Interferometric spectroscopy 140-141
Interferometry 98-106*
  basic classes of interferometers 98-99
  Fabry-Perot interferometer 103-104
  Fizeau interferometer 101-102
  holographic 104-105
  Mach-Zehnder interferometer 102
  Michelson interferometer 99-100
  Michelson stellar interferometer 103
  phase shifts in 106
  shearing interferometers 102-103
  speckle 105

284 SPECTROSCOPY SOURCE BOOK

Interferometry (*cont.*):
 Twyman-Green interferometer 100-101
Intermolecular Stark effect 37
Inverse Stark effect 37
Inverse Zeeman effect 40
Ion kinetic energy spectroscopy 241-242
Isoelectronic sequence 33*
Isotope shift 35, 41-42*

Johnson, L. 216
Kerl, R.J. 118
King, W.H. 42
Krishnan, K.S. 145
Kronig, R.de L. 89
Kruger, H. 221

Lamb, W. 28, 35
Lamb, W.E., Jr. 122, 125
Lamb shift 35
 beam-foil spectroscopy 122
 quantum electrodynamics 27-28
Landsberg, G. 145
Laser spectroscopy 123-128*
 absorption spectroscopy 123-124
 Doppler-free spectroscopy 125, 129
 Doppler-free two-photon spectroscopy 127, 130-131
 fluorescence spectroscopy 124
 frequency mixing 124
 high-resolution 124-128
 hole-burning spectroscopy 128
 intercavity absorption 124
 multiphoton absorption 124
 photoacoustic spectroscopy 119-120
 polarization spectroscopy 126-127

Laser spectroscopy (*cont.*):
 Raman spectroscopy 124, 145-146
 resolution limits of high-resolution methods 127-128
 resonance-ionization spectroscopy 131-134*
 saturation spectroscopy 125-126, 130
 time-resolved 128
 tunable sources 123
Light scattering, quasielastic 151-154
Line spectrum 13, 32-33*
Linear Stark effect 36
Linewidth 16-17*
Lloyd's mirror 99
Loeb, L.B. 80
Luminescence 59-68*
 activators and poisons 60-61
 atomic gases 61-62
 configuration coordinate curve model 62-66
 configuration coordinate curves for dielectrics 65-66
 fluorescence *see* Fluorescence
 involving electron motion 67-68
 luminescent substances 60
 phosphorescence *see* Phosphorescence
 quantum-mechanical corrections for configuration coordinate curves 64-65
 sensitized 66-67
 type of radiation emitted 60
Lyman series 26-27

Mach-Zehnder interferometer 102
Magnetic resonance 204-209*
 antiferromagnetic resonance 208

Magnetic resonance (*cont.*):
 electron-nuclear double resonance 207
 electron paramagnetic resonance spectroscopy 222-226*
 ferrimagnetic resonance 208-209
 ferromagnetic resonance 208
 nuclear 205-207
 observation 205
 origin 204-205
 paramagnetic resonance 207-208
Mandelstam, L. 145
Mass-analyzed ion kinetic energy spectroscopy 241-243
Mass shift 41
Mass spectrometry 235-244*
 analysis of inorganic solids 244
 analysis of mixtures 243-244
 assignment of empirical formula 238-240
 computer applications 243
 metastable ions 241-242
 methods of ion production 235-236
 secondary ion mass spectrometry 268-271*
Mass spectroscope 230-234*
 applications 232
 miscellaneous types 233
 operation 230-232
 tandem accelerator mass spectrometers 233-234
Matrix isolation 110-112*
 applications 110-112
 experimental apparatus 110
Michelson interferometer 99-100
Michelson stellar interferometer 103
Microwave spectroscopy 202-204*
 apparatus 202
 applications 203-204

Microwave spectroscopy (*cont.*):
enhancement of population differences 202-203
interaction of microwaves with matter 202
measurement of nuclear moments 44
MIKES *see* Mass-analyzed ion kinetic energy spectroscopy
Molecular beam spectroscopy 135-137
Molecular beams 134-137*
beam-foil spectroscopy 120-122*
Doppler effect 129
Doppler-free spectroscopy 125
laser excitation 137
measurement of nuclear moments 44
molecular beam spectroscopy 135-137
production and detection 134-135
Molecular structure and spectra 45-55*
correlation of structure with optical activity 72-73
Cotton effect 76-77
dipole moments 45-46
electronic band spectra 52-54
electronic band structures 54-55
energy level 13-15*
infrared spectroscopy 137-145*
linewidth 16-17*
mass spectroscope 230-234*
microwave spectroscopy 203
molecular beams 134-137*
molecular electronic states 54-55
molecular energy levels 47-49
molecular polarizability 46
molecular sizes 45

Molecular structure and spectra (*cont.*):
polyatomic electronic spectra 55
pure rotation spectra 50
Raman spectroscopy 124, 145-146, 150
spectra 49-54
vibration-rotation bands 50-52
Molecule *see* Molecular structure and spectra
Mössbauer, R.L. 186
Mössbauer effect 186-190*
application 188-190
energy modulation 186-187
experimental realization 187-188
theory 186

NAA *see* Neutron activation analysis
Near-infrared spectrophotometry 251
Neutron activation analysis 265-266
Neutron spectrometry 190-196*
applications 196
monoenergetic neutron spectrometers 195
neutron cross sections 192-194
neutron reactions and resonance parameters 192
slow neutrons *see* Slow neutron spectroscopy
techniques 194-195
time-of-flight spectrometers 194-195
unbound and bound states of nuclides 191
Neutron spectroscopy *see* Neutron spectrometry
NMR *see* Nuclear magnetic resonance
Normal Zeeman effect 38-39
NQR *see* Nuclear quadrupole resonance

Nuclear gamma resonance fluorescence *see* Mössbauer effect
Nuclear magnetic resonance 205-207, 209-220*
applications 206-207
carbon-13 NMR 215-218
imaging 220
instrument development 215
lanthanide shift reagents 212-215
nuclei with quadrupole moments 206
spectroscopy 209-212
two-dimension 218-220
Nuclear moments 42-45*
effects of 43
measurement 43-44
Nuclear Overhauser effect 212
Nuclear quadrupole resonance 221-222*
application 222
techniques 222
Nuclear spectra 178-180*
alpha particles *see* Alpha particles
beta rays *see* Beta rays
gamma rays *see* Gamma rays
from nuclear reactions 179-180
spontaneous fission 179
Nuclear Zeeman effect 40-41

One-atom detection 132-133
Optical activity 71-74*
chromophores 76-77
circular dichroism 72
correlation with molecular structure 72-74
Cotton effect 75-77*
measurement 71-72
optical rotation 71-72
rotatory dispersion 74-75*
Optical interferometer *see* Interferometry

Optical mixing spectroscopy *see* Intensity fluctuation spectroscopy
Optical prism 94-95*
  dispersing 94-95
  reflecting 94
Optical rotation: measurement of optical activity 71-72
Optical spectroscopy: measurement of nuclear moments 44
Optoacoustic spectroscopy *see* Photoacoustic spectroscopy

Paramagnetic resonance 207-208
  electron *see* Electron paramagnetic resonance spectroscopy
Patel, K.N. 118
Pauli, W. 22
Pauli exclusion principle *see* Exclusion principle
Phase-shifting interferometry 106
Phosphorescence 60, 69-70*
Photoacoustic spectroscopy 116-120*
  absorption by gases 117-118
  condensed-phase spectroscopy 118-120
  methods of measuring absorption 116-117
Photoemission spectroscopy 90
Photometry *see* Flame photometry; Spectrophotometric analysis
Photon correlation spectroscopy *see* Intensity fluctuation spectroscopy
PIXE *see* Proton-induced x-ray emission
Planck, M. 13
Plasma: microwave spectroscopy 203

Polarization spectroscopy 126-127
Polyatomic electronic spectra 55
Prism *see* Optical prism
Proton-induced x-ray emission 267*
Pulsing Fourier transformation spectrometer 216

Quadratic Stark effect 36-37
Quadratic Zeeman effect 40
Quantum electrodynamics: Lamb shift 27-28
Quasielastic light scattering 151-154*
  applications 154
  Fabry-Perot interferometry 152-153
  intensity fluctuation spectroscopy 151-152
  rotational diffusion coefficients 153
  static light scattering 151
  translational diffusion coefficients 153

Radio-frequency spectroscopy 204*
Raman, C.V. 145
Raman effect 145-151*
  discovery 145
  hyper-Raman effect 149-150
  Raman spectroscopy 124, 145-146
  resonance Raman effect 148-149
  special forms 148-150
  stimulated Raman effect 150
  theory 146-148
Raman spectroscopy 124, 145-146
  applications 150-151
Rayleigh prism system 94
Recoil ion spectroscopy 29
Reflecting prism 94

Reflection grating 95
Resonance-ionization spectroscopy 131-134*
  applications 133-134
  laser schemes 132
  one-atom detection 132-133
  theory 131-132
Resonance Raman effect 148-149
Ritz, W. 13, 15
Ritz's combination principle 15-16*
Roberts, J. 216
Rotatory dispersion 74-75
  Cotton effect 75-77*
Rutherford, E. 18
Rydberg, J. 13

Saturation spectroscopy 125-126, 130
Schwinger, J. 81
Secondary ion mass spectrometry 268-271*
  analytical applications 269-271
  instrumentation 268-269
  ion production 268
Sellin, I.A. 29
Sensitized luminescence 66-67
  concentration quenching 67
  resonant transfer 66-67
Shearing interferometer 102-103
SIMS *see* Secondary ion mass spectrometry
Slow neutron spectroscopy 196-198*
  applications 197
  continuous spectra interpretation 197
  experiments 196
  interpretation of diffusion phenomena 197
  quasiparticle interpretation 197
Speckle interferometry 105

INDEX

Spectrochemical analysis 253-256*
  basic spectrometric measurement 255-256
  electromagnetic spectral regions 255
  emission *see* Emission spectrochemical analysis
  flame photometry 260*
  physical basis for measurements 253-254
  qualitative and quantitative 255
  spectrometric interactions 254-255
Spectrofluorometer 263-264
Spectrography 107*
Spectrohelioscope 106-107*
Spectrophotometer 247-248
Spectrophotometric analysis 247-253*
  absorption spectra 248-249
  attenuated total reflectance 251
  Bouguer-Lambert-Beer law 248
  filter photometry 253
  infrared spectrophotometry 249-251
  near-infrared spectrophotometry 251
  ultraviolet spectrophotometry 252-253
  visible spectrophotometry 251-252
Spectrum 12*
  band *see* Band spectrum
  fine structure 34-35*
  hyperfine structure 35-36*
  line *see* line spectrum
  Stark effect 36-37*
  synchrotron radiation 87-88
  wavelength standards 93-94*
  Zeeman effect 37-41*
  see also Atomic structure and spectra; Molecular structure and spectra
Stark, J. 36

Stark effect 36-37*
  intermolecular 37
  inverse 37
  linear 36
  quadratic 36-37
Static light scattering 151
Stimulated Raman effect 150
Superheavy elements: relativistic Dirac theory 29-30
Synchrotron radiation 81-91*
  bending magnet sources 85
  experimental facilities 82-84
  extended x-ray absorption fine structure 89-90
  photoemission spectroscopy 90
  properties 84-89
  research applications 89-91
  spectrum 87-88
  surface structure studies 90
  time-resolved fluorescence spectroscopy 91
  undulator magnet sources 85-87
  wiggler magnet sources 85
  x-ray diffraction topography 90
  x-ray fluorescence trace analysis 91

Tandem accelerator mass spectrometer 233-234
Thomson, J.J. 17, 80
Time-of-flight spectrometers 198-199*
  neutron spectrometers 194-195
Time-resolved fluorescence spectroscopy 91
Time-resolved laser spectroscopy 128
Trace analysis 246-247*
  goals 246
  methods 246
  proton-induced x-ray emission 267*
  sensitivity and detection limits 246-247

Trace analysis (*cont.*):
  techniques 246
  x-ray fluorescence analysis 91, 167-168
Transmission grating 95
Two-photon spectroscopy 127, 130-131
Twyman-Green interferometer 100-101

Uhlenbeck, G.E. 22
Ultraviolet spectrophotometry 252-253
Uncertainty principle 30-31

Visible spectrophotometry 251-252
Wavelength measurement 92-93*
  dispersion methods 92
  Fourier transform method 92
  use of interferometers 92
Wavelength standards 93-94*
Wieman, C. 126

X-ray diffraction topography 90
X-ray fluorescence 156-158
X-ray fluorescence analysis 160-168*
  applications 168
  basis of method 160
  crystal spectrometer 163-164
  energy dispersive systems 164-165
  microanalysis 165-166
  quantitative analysis 167-168
  radiation sources 162-163
  specimen preparation 166-167
  supplemental methods 168
  trace analysis 91, 167-168
  x-ray spectra 161-162

X-ray spectrometry 154-160*
  bulk samples 159
  improvements and limitations 159-160
  spectrum analysis 156-158
  thin samples 158-159
  scattering experiments 19

Young's double pinhole interferometer 99
Zeeman, P. 37
Zeeman effect 37-41*
  anomalous 39-40
  in crystals 40
  in molecules 40

Zeeman effect (cont.):
  inverse 40
  normal 38-39
  nuclear 40-41
  quadratic 40